T0262717

Encyclopedia of Biotechnology

Encyclopedia of Biotechnology

Edited by **Suzy Hill**

New York

Published by Callisto Reference,
106 Park Avenue, Suite 200,
New York, NY 10016, USA
www.callistoreference.com

Encyclopedia of Biotechnology
Edited by Suzy Hill

International Standard Book Number: 978-1-63239-216-9 (Hardback)

Printed in the United States of America.

Contents

Preface

Every book is a source of knowledge and this one is no exception. The idea that led to the conceptualization of this book was the fact that the world is advancing rapidly; which makes it crucial to document the progress in every field. I am aware that a lot of data is already available, yet, there is a lot more to learn. Hence, I accepted the responsibility of editing this book and contributing my knowledge to the community.

A variety of information regarding the field of biotechnology is presented in this book. Biotechnology is the scientific field of researching the most resourceful techniques to get helpful products for human society by using practical micro-organisms, tissues of plants or animals. It is very significant to make the precise distinction between biotechnology as a separate science of getting precious products from cells or tissues of viable organisms, and any other applications of bioprocesses that are based on using the whole living plants or animals. The book is a compilation of current advances of selected studies that are ongoing in certain biotechnological applications. This book has several well researched chapters related to the achievements made in the field of biotechnology.

While editing this book, I had multiple visions for it. Then I finally narrowed down to make every chapter a sole standing text explaining a particular topic, so that they can be used independently. However, the umbrella subject sinews them into a common theme. This makes the book a unique platform of knowledge.

I would like to give the major credit of this book to the experts from every corner of the world, who took the time to share their expertise with us. Also, I owe the completion of this book to the never-ending support of my family, who supported me throughout the project.

Editor

Part 1

Biotechnology of Agricultural Wastes - Recycling, Saving Energy and Food Quality Preservation

Biotechnology of Agricultural Wastes Recycling Through Controlled Cultivation of Mushrooms

Marian Petre and Alexandru Teodorescu
Department of Natural Sciences,
Faculty of Sciences, University of Pitesti,
Romania

1. Introduction

The agricultural wastes recycling with applications in agro-food industry is one of the biological challenging and technically demanding research in the biotechnology domain known to humankind so far. Annually, the accumulation of huge amounts of vineyard and winery wastes causes serious environmental damages nearby winemaking factories. Many of these ligno-cellulose wastes cause serious environmental pollution effects, if they are allowed to accumulate in the vineyards or much worse to be burned on the soil. At the same time, the cereal by-products coming from the cereal processing and bakery industry are produced in significant quantities all over the world (Moser, 1994; Verstraete & Top, 1992).

To solve the environmental troubles raised by the accumulation of these organic wastes, the most efficient way is to recycle them through biological means (Smith, 1998). As a result of other recent studies, the cultivation of edible and medicinal mushrooms was applied using both the solid state cultivation and controlled submerged fermentation of different natural by-products of agro-food industry that provided a fast growth as well as high biomass productivity of the investigated strains (Petre& Teodorescu, 2009; Stamets, 2000).

These plant wastes can be used as the main ingredients to prepare the organic composts for edible and medicinal mushrooms growing in order to get organic food and biological active compounds from the nutritive fungal biomass resulted after solid state cultivation or submerged fermentation of such natural materials (Petre & Petre, 2008; Petre et al., 2010).

Taking into consideration this biological advantage there were tested some variants of biotechnology for agricultural wastes recycling through the controlled cultivation of edible and medicinal mushrooms *Ganoderma lucidum* (Curt.:Fr.) P. Karst (folk name: Reishi or Ling-zhi), *Lentinus edodes* (Berkeley) Pegler (folk name: Shiitake) and *Pleurotus ostreatus* (Jacquin ex Fries) Kummer (folk name: Oyster Mushroom) on organic composts made of cereal grain by-products as well as winery and vineyard wastes (Petre & Teodorescu, 2010).

2. The solid state cultivation of mushrooms on winery and vineyard wastes

The main aim of this work was focused on screening the optimal biotechnology of edible and medicinal mushrooms growing through the solid-state cultivation by recycling different kind of agricultural by-products and wastes coming from vineyard farms and winemaking industry (Petre *et al.*, 2011).

Taking into consideration that most of the edible and medicinal mushrooms species requires a specific micro-environment including complex nutrients, the influence of all physical and chemical factors upon fungal pellets production and mushroom fruit bodies formation has been studied by testing new biotechnological procedures (Oei, 2003).

To establish the laboratory biotechnology of recycling the winery and vineyard wastes by using them as a growing source for edible mushrooms, two mushroom species of Basidiomycetes group, namely *L. edodes* (Berkeley) Pegler and *P. ostreatus* (Jacquin ex Fries) Kummer were used as pure mushroom cultures isolated from the natural environment and being preserved in the local collection of the University of Pitesti. The stock cultures were maintained on malt-extract agar (MEA) slants (20% malt extract, 2% yeast extract, 20% agar-agar). Slants were incubated at 25°C for 120-168 h and stored at 4°C.

The pure mushroom cultures were expanded by growing in 250-ml flasks containing 100 ml of liquid malt-extract medium at 23°C on rotary shaker incubators at 110 rev min $^{-1}$ for 72-120 h. After expanding, the pure mushroom cultures were inoculated into 100 ml of 3-5% (v/v) malt-yeast extract liquid medium, previously poured in 250 ml rotary shake flasks and then were maintained at 23-25°C (Petre & Teodorescu, 2010).

The experiments of inoculum preparation were set up under the following conditions: constant temperature, 25°C; agitation speed, 90-120 rev min^{-1}; initial pH, 5.5–6.5. All the seed mushroom cultures were incubated for 120–168 h.

After that, the seed cultures of these mushroom species were inoculated in liquid culture media (20% malt extract, 10% wheat bran, 3% yeast extract, 1% peptone) at pH 6.5 previously distributed into rotary shake flasks of 1,000 ml. During the incubation time, all the spawn cultures were maintained in special culture rooms, designed for optimal incubation at 25°C. Three variants of culture compost were prepared from marc of grapes and vineyard cuttings in the following ratios: 1:1, 1:2, 1:4 (w/w).

The winery and vineyard wastes were mechanically pre-treated by using an electric grinding device to breakdown the lignin and cellulose structures in order to make them more susceptible to the enzyme actions. All the culture compost variants made of winery and vineyard wastes were transferred into 1,000 ml glass jars and disinfected by steam sterilization at 120°C for 60 min. When the jars filled with composts were chilled they were inoculated with the liquid spawn already prepared (Petre *et al.*, 2010).

Each culture compost variant for mushroom growing was inoculated using such liquid spawn having the age of 72–220 h and the volume size ranging between 3–9% (v/w). During the period of time of 18–20 d after this inoculation, the mushroom cultures had developed a significant mycelia biomass on the culture substrates (Carlile & Watkinson, 1996).

According to the registered results of the performed experiments the optimal laboratory-scale biotechnology for edible mushroom cultivation on composts made of marc of grapes and vineyard cuttings was established (Fig. 1).

The effects induced by the composts composition, nitrogen and mineral sources as well as the inoculum amount upon the mycelia growing during the incubation period were investigated. There were made three variants of composts which were tested by comparing them with the control sample made of poplar sawdust (Petre & Teodorescu, 2010).

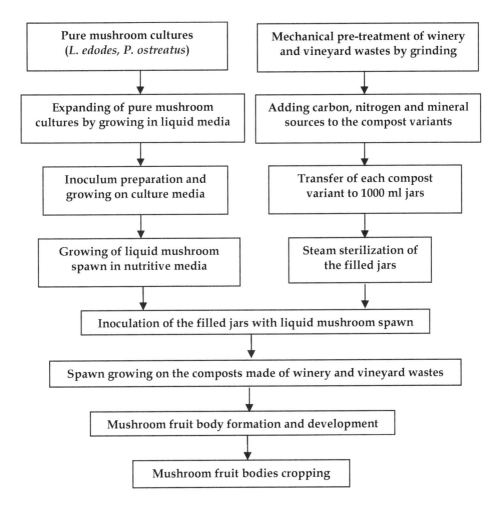

Fig. 1. Scheme of laboratory-scale biotechnology for edible mushroom production by recycling winery and vineyard wastes

The first variant of compost composition was prepared from vineyard cuttings, the second one from a mixture between marc of grapes and vineyard cuttings in equal proportions and the third one was made only from marc of grapes as full compost variant. The experiments were carried out for 288 h at 25°C with the initial pH 6.5 and the incubation period lasted for 168-288 h (Petre *et al.*, 2007).

2.1 Results and discussion

As it can be noticed in figure 2, the registered results show that from all tested compost variants the most suitable substrate for mycelia growing was that one prepared from marc of grapes, because it showed the highest influence upon the mycelia growing and fresh mushroom production (32–35 g%). All registered data represent the means of triple determinations.

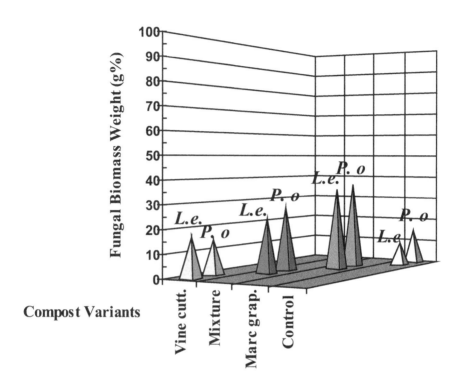

Fig. 2. Comparative effects of composts composition upon mycelia growing of *P. ostreatus* (*P.o.*) and *L. edodes* (*L.e.*)

This compost variant was followed by the mixture prepared from marc of grapes and vineyard cuttings in equal amounts (20-23 g %) and, finally, by the variant made of only vineyard cuttings (12-15 g %). From the tested nitrogen sources, barley bran was the most efficient upon the mycelia growing and fruit mushroom producing at 35-40 g % fresh fungal biomass weight, being closely followed by rice bran at 25–30 g %. Wheat bran is also a well known nitrogen source for fungal biomass synthesis but its efficiency in these experiments was relatively lower than the ones induced by the barley and rice bran added as natural organic nitrogen sources (Stamets, 2000). All registered data are the means of triple determinations. The effects of nitrogen sources were registered as they are presented in figure 3. Among the tested mineral sources, the natural calcium carbonate ($CaCO_3$) from marine shells yielded the best mycelia growing as well as fungal biomass production at 28-32 g% and, for this reason, it was registered as the most appropriate mineral source, being followed by the natural gypsum ($CaSO_4 \cdot 2 H_2O$) at 20-23 g %, as it is shown in figure 4.

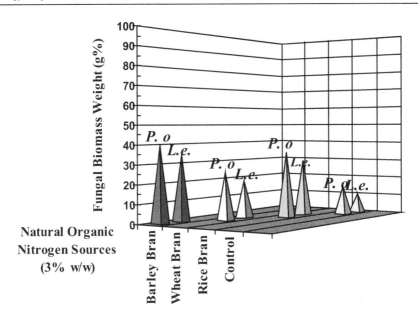

Fig. 3. Comparative effects of organic nitrogen sources upon mycelia growing of *P. ostreatus* (*P.o.*) and *L. edodes* (*L.e.*)

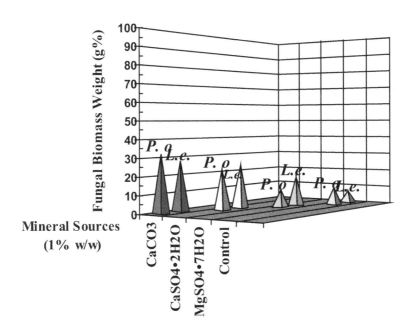

Fig. 4. Comparative effects of mineral sources upon mycelia growing of *P. ostreatus* (*P. o.*) and *L. edodes* (*L.e.*)

The mineral sources like hepta-hydrate magnesium sulfate ($MgSO_4 \cdot 7 H_2O$) showed a quite moderate influence upon the fungal biomass growing as other researchers have already reported so far. All data are the means of triple determinations (Stamets, 2000; Chahal, 1994).

The whole period of mushroom growing from the inoculation to the fruit body formation lasted between 25-30 d in case of *P. ostreatus* cultivating and 50-60 d for *L. edodes*, depending on each fungal species used in experiments (Chahal, 1994). However, during the whole period of fruit body formation, the culture parameters were set up and maintained at the following levels, depending on each mushroom species: air temperature, 15–17ºC; the air flow volume, 5–6m³/h; air flow speed, 0.2–0.3 m/s; the relative moisture content, 80–85%, light intensity, 500–1,000 luces for 8–10 h/d. The final fruit body production of these mushroom species used in experiments was registered between 1.5 kg for *L .edodes* and 2.8 kg for *P. ostreatus*, relative to 10 kg of composts made of vineyard and winery wastes, comparing with 0.7-1.2 kg on 10 kg of poplar sawdust used as control samples.

3. The controlled submerged cultivation of mushrooms on winery wastes

The submerged cultivation of mushroom mycelium is a promising biotechnological procedure which can be used for synthesis of pharmaceutical substances with anticancer, antiviral and immuno-stimulatory effects from the nutritive mushroom biomass (Wasser & Weis, 1994). As result of other recent studies, the continuous cultivation of edible and medicinal mushrooms was applied by using the submerged fermentation of different natural by-products of agro-food industry (Bae, et al., 2000; Jones, 1995; Moo-Joung, 1993). The biotechnology of controlled cultivation of medicinal mushrooms was established and tested in different variants of culture media that were made of different sorts of bran and broken seeds resulted from the industrial food processing of wheat, barley and rye seeds. This biotechnology can influence the faster growth as well as higher biomass productivity of *G. lucidum* and *L. edodes* mushroom species (Petre *et al.*, 2010).

The main stages of biotechnology to get high nutritive fungal biomass by controlled submerged fermentation were the followings:

1. Preparation of culture media and pouring them into the cultivation vessel of the bioreactor.
2. Steam sterilization of bioreactor vessel at 121°C and 1.1 atm. for 20 min.
3. Inoculation of sterilized culture media with mycelium from pure cultures of selected strains inside the bioreactor vessel for submerged cultivation, using the sterile air hood with laminar flow.
4. Running the submerged cultivation cycles under controlled conditions: temperature 23 ± 2°C, speed 70 rpm and continuous aeration at 1.1 atm.
5. Collecting, cleaning and filtering the fungal pellets obtained by the submerged fermentation of substrates made of by-products resulted from cereal grains processing.

Two mushroom species belonging to Basidiomycetes Class, namely *G. lucidum* (Curt.:Fr.) P. Karst and *L. edodes* (Berkeley) Pegler were used as pure cultures in experiments. The stock cultures were maintained on malt-extract agar (MEA) slants. Slants were incubated at 25°C for 5-7 d and then stored at 4°C. The fungal cultures were grown in 250-ml flasks containing 100 ml of MEYE (malt extract 20%, yeast extract 2%) medium at 23°C on rotary shaker incubators at 110 rev min[-1] for 5-7 d. The fungal cultures were prepared by aseptically inoculating 100 ml in three variants of culture media by using 3-5% (v/v) of the seed culture and then cultivated at 23-25°C in 250 ml rotary shake flasks. The biotechnological experiments were conducted under the following conditions: temperature, 25°C; agitation speed, 120-180 rev min⁻¹; initial pH, 4.5–5.5. After 10–12 d of incubation the fungal cultures were ready to be inoculated aseptically into the glass vessel of 20 l laboratory-scale bioreactor, that was designed to be used for controlled submerged cultivation of edible and medicinal mushrooms on substrata made of wastes resulted from the industrial processing of cereal grains (Fig. 5).

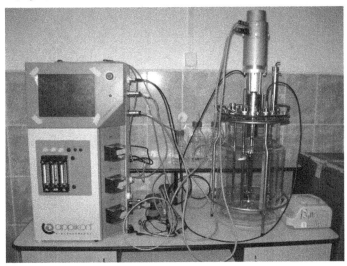

Fig. 5. General view of the Laboratory scale bioreactor (15 L)

After a period of submerged fermentation lasting up to 120 h, small mushroom pellets developed inside the nutritive broth (Fig. 6, 7).

Fig. 6. Mycelial biomass of *G. lucidum* collected after submerged fermentation

Fig. 7. Mycelial biomass in the shape of fungal pellets of *L. edodes*, collected after submerged fermentation

The fermentation process was carried out by inoculating the growing medium volume (10,000 ml) with mycelia inside the culture vessel of the laboratory-scale bioreactor. The whole process of growing lasts for a single cycle between 5-7 days in case of *L. edodes* and between 3 to 5 days for *G. lucidum*. The strains of these fungal species were characterized by morphological and cultural stability, proven by their ability to maintain the phenotypic and taxonomic identities. The experiments were carried out in three repetitions. Observations on morphological and physiological characters of these two tested species of fungi were made after each culture cycle, highlighting the following aspects:

- sphere-shaped structure of fungal pellets, sometimes elongated, irregular, with various sizes (from 2 to 5 mm in diameter), reddish-brown colour – *G. lucidum* culture (Fig. 8).

Fig. 8. Stereomicroscopic view of *G. lucidum* pellets after controlled submerged fermentation

- elliptically-shaped structures of fungal pellets, with irregular diameters of 4 up to 7 mm showing mycelia congestion, which developed specific hyphae of *L. edodes* (Fig. 9).

Fig. 9. Stereomicroscopic view of *L. edodes* pellets after controlled submerged fermentation

Samples for analysis were collected at the end of the fermentation process, when pellets formed specific shapes and characteristic sizes. The fungal biomass was washed repeatedly with double distilled water in a sieve with 2 mm diameter eye, to remove the remained bran in each culture medium.

3.1 Results and discussion

Biochemical analyses of fungal biomass samples obtained by submerged cultivation of mushrooms were carried out separately for the solid fraction and liquid medium remained after the separation of fungal biomass by filtering. The percentage distribution of solid substrate and liquid fraction in the samples of fungal biomass are shown in table 1.

Mushroom species	Total volume of separated liquid per sample (ml)	Total biomass weight per sample (g)	Water content after separation (%)
L. edodes	83	5.81	83.35
L. edodes	105	7.83	82.50
L. edodes	95	7.75	82.15
L. edodes	80	5.70	79.55
G. lucidum	75	7.95	83.70
G. lucidum	115	6.70	82.95
G. lucidum	97	5.45	80.75
G. lucidum	110	6.30	77.70

Table 1. Percentage distribution of solid substrate and liquid fraction in the preliminary samples of fungal biomass

In each experimental variant the amount of fresh biomass mycelia was determined. The percentage amount of dry biomass was determined by dehydration at 70°C, up to constant weight. Total protein content was determined by biuret method, whose principle is similar to the Lowry method, this method being recommended for the protein content ranging from 0.5 to 20 mg/100 mg sample. In addition, this method required only one sample incubation period (20 min) and by using them was eliminated the interference with various chemical agents (ammonium salts, for example).

The principle method is based on reaction that takes place between copper salts and compounds with two or more peptides in the composition in alkali, which results in a red-purple complex, whose absorbance is read in a spectrophotometer in the visible domain (λ - 550 nm). The registered results are presented as the amounts of fresh and dry biomass as well as protein contents for each fungal species and variants of culture media (Tables 2, 3).

Culture variants	Fresh biomass (g)	Dry biomass (%)	Total protein (g % d.w.)
I	20.30	5.23	0.55
II	23.95	6.10	0.53
III	22.27	4.79	0.73
IV	20.10	4.21	0.49
Control	4.7	0.5	0.2

Table 2. Fresh and dry biomass and protein content of L. edodes after submerged fermentation

Culture variants	Fresh biomass (g)	Dry biomass (%)	Total protein (g % d.w.)
I	25.94	9.03	0.67
II	22.45	10.70	0.55
III	23.47	9.95	0.73
IV	21.97	9.15	0.51
Control	5.9	0.7	0.3

Table 3. Fresh and dry biomass and protein content of G. lucidum after submerged fermentation

According to the registered data, using wheat bran strains the growth of G. lucidum biomass was favoured, while the barley bran led to the increased growth of L. edodes mycelium and G. lucidum as well. In contrast, dry matter content was significantly higher when using barley bran for both species used. Protein accumulation was more intense in case of using barley bran compared with those of wheat and rye, at both species of mushrooms.

The sugar content of dried mushroom pellets collected at the end of experiments was determined by using Dubois method (Wasser & Weis, 1994). The mushroom extracts were prepared by immersion of dried pellets inside a solution of NaOH pH 9, in the ratio 1:5.

All dispersed solutions containing the dried pellets were maintained 24 h at a precise temperature of 25°C, in full darkness, with continuous homogenization to avoid the oxidation reactions. After removal of solid residues by filtration, the samples were analyzed by the previous mention method. The nitrogen content of mushroom pellets was analyzed by Kjeldahl method (Table 4).

Mushroom species	Culture variant	Sugar content (mg/ml)	Kjeldahl nitrogen (%)	Total protein (g % d.w.)
L. edodes	I	5.15	6.30	0.55
L. edodes	II	4.93	5.35	0.53
L. edodes	III	4.50	5.70	0.73
L. edodes	IV	4.35	5.75	0.49
	Control	0.55	0.30	0.2
G. lucidum	I	4.95	5.95	0.67
G. lucidum	II	5.05	6.15	0.55
G. lucidum	III	5.55	6.53	0.73
G. lucidum	IV	4.70	5.05	0.51
	Control	0.45	0.35	0.3

Table 4. The sugar, total nitrogen and total protein contents of dried mushroom pellets

Comparing all registered data resulted from triple determinations, it can be noticed that the biochemical correlation between dry weight of mushroom pellets and their sugar and nitrogen contents is kept at a balanced ratio for each tested mushrooms (Stamets, 2000).

Among all mushroom samples that were tested in biotechnological experiments G. lucidum G-3 showed the best values of their composition in sugars, total nitrogen and total protein contents. In this stage, 70-80% of the former fungal pellets were separated by collecting them from the culture vessel of the bioreactor and separating from the broth by slow vacuum filtration. On the base of these results, the optimal values of physical and chemical factors which influence the mushroom biomass synthesis were taken into consideration in order to established the following schematic flow of the biotechnology for mushroom biomass producing by submerged fermentation, as it is shown in figure 10.

The main advantages of the submerged fermentation of winery wastes under the metabolic activity of selected mushrooms, by comparison with the solid state cultivation are the followings:

a. the shortening of the biological cycle and cellular development in average from 8-10 weeks to at mostly one week per cellular culture cycle;
b. the ensuring of the optimal control of physical and chemical parameters which are essential for producing important amounts of mushroom pellets in a very short time;
c. 20–30% reduction of energy and work expenses as well as the volume of the volume of raw materials materials which are manipulated during each culture cycle;
d. 15-20% increasing of fungal biomass amount per medium volume unit for each mushrooms species;
e. the whole removing of any pollutant sources during the biotechnological flux;
f. the culture media for mushroom growing are integrally natural without using of artificial additives as it is used in classical cultivating procedures;
g. the mushroom pellets produced by applying this biotechnology for ecological treatment of agricultural wastes was 100% made by natural means and will be used for food supplements production with therapeutic properties which will contribute to the increasing of health level of human consumers having nutritional metabolic deficiencies.
h. the biochemical correlation between the dry weight of mushroom pellets and their sugar and nitrogen contents is kept at a balanced ratio for each tested mushroom species.

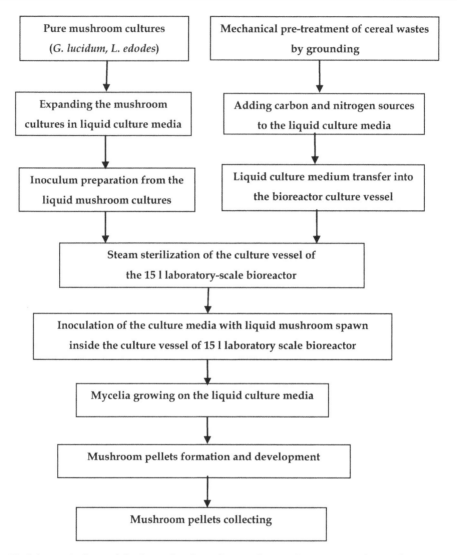

Fig. 10. Schematic flow of the biotechnology for mushroom biomass producing by submerged fermentation.

4. The controlled cultivation of mushrooms in modular robotic system

The agricultural works as well as industrial activities related to plant crops and their processing have generally been matched by a huge formation of wide range of lignocellulose wastes. All these vegetal wastes cause serious environmental troubles if they accumulate in the agro-ecosystems or much worse to be burned on the soil. For the human–operational farms, all processes are made by human personnel exclusively, starting from filling of cultivation beds with compost, up to fruit-bodies harvesting (Reed *et al.*, 2001).

In this respect, a strong tendency for increasing the number of researches in the field of mushroom's automated cultivation, harvesting and processing technologies as well as for continuously development of new robotic equipments can be noticed (Reed *et al.*, 2001).

The solid state cultivation of edible and medicinal mushrooms *Lentinula edodes* and *Pleurotus ostreatus* could be performed by using a modular robotic system that provides the following fully automatic operations: sterilization of composts, inoculation in aseptic chamber by controlled injection device containing liquid mycelia as inoculum, incubation as well as mushroom fruit bodies formation in special growing chambers with controlled atmosphere and the picking up of edible and medicinal mushroom fruit bodies (Petre *et al.*, 2009).

The biotechnology concerning the controlled cultivation of edible mushrooms in continuous flow depends on the strictly maintaining of biotic as well as physical and chemical factors that could influence the bioprocess evolution. The proceeding of edible mushroom cultivation consists in a continuous biotechological flow, having a chain of succesive stages that are working in the non-sterile zone and mostly in the sterile zone of the modular robotic system. In this way, there is provided the technological security both from the structural and functional points of view in order to produce organic foods in highest security and food quality. The functional biotechnological model of the modular robotic system was designed for controlled cultivation and integrated processing of edible mushrooms to get ecological food in highest safety conditions (Petre *et al.*, 2009).

The modular robotic system designed for edible mushroom cultivation provides the automatic sterilization of composts, the automatic inoculation inside the aseptic room by a special device of controlled injection of liquid mycelia, the incubation and fruit bodies formation in special chambers under controlled atmosphere as well as the automatic harvesting of mushroom fruit bodies (Petre *et al.*, 2011).

This system includes three major zones, respectively, the non-sterile zone, the sterile zone and the fruit-body processing zone (Fig. 11).

Thus, during the first stage of the biotechnological flow, in the non-sterile zone of the cultivation system, a natural and nutritive compost is prepared from sawdust or shavings of deciduous woody species in the ratio of 30-40 parts per weight (p.p.w.), marc of grapes chemically untreated, in 20-30 p.p.w., brans of organic cereal seeds (wheat, barley, oat, rye, rice), in 10-20 p.p.w., yeasts, in 3-5 p.p.w., and powder of marine shells, in 1-3 p.p.w., for pH adjustment, which then, it is hidrated with demineralized water, in 20-30 p.p.w. In the next stage, such prepared compost is decanting in polyethylene thermoserilizable bags, which have round orifices of 0,3-0,5 mm in diameter, uniform distribuited between them, at 10-15 cm distance, each one of them having a working volume of 10-20 kg (Petre *et al.*, 2011).

Beforehand, special devices for uniform distribution of mycelia as liquid inoculum are mounted inside of these bags. Then, these bags are fitted out with supporting devices on the transfer and transport systems and special devices for coupling to the automatic inoculation subdivision by controlled injection of liquid mycelia (Fig. 11).

Each one of these zones is linked with next one by an interfacing zone. In this way, the non-sterile zone is linked with the sterile zone through the first interfacing zone and this one is connected with the fruit body processing zone by the second interfacing area, as it is shown in figure 11.

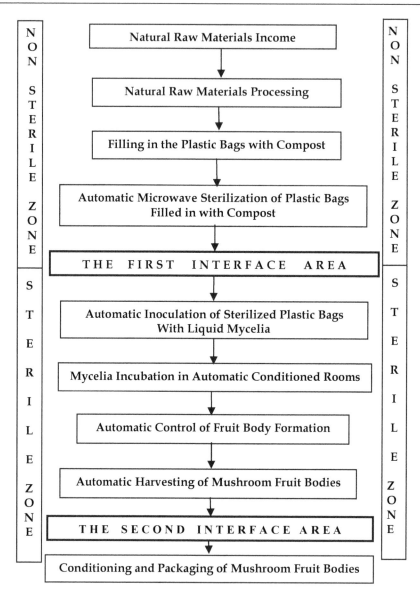

Fig. 11. Schematic flow of the modular robotic system for controlled cultivation of edible mushrooms

Inside the non-sterile zone, the bags filled with composts are placed on the supporting devices, mounted on the transfer pallets, which are inserted in the first part of the sterile zone, respectively, in the module of the automatic sterilization with microwave at 120-125°C, and the pallets with bags are automatically chilled in the zone of controlled cooling of sterilized composts up to the room temperature. These pallets with sterilized bags are automatically transferred into the aseptic room to make the inoculation with liquid mycelia by using a robotic device of controlled injection. Further on, the pallets with the inoculated bags either are evacuated from the sterile zone or they are automatically transferred to the incubation and fruit body formation rooms. In these rooms of incubation and fruit body formation, both the optimal temperature of mycelia growing and the relative air humidity are provided as well as a constant steril air flow introduced under pressure by using an automatic device and an adecquate lighting level (Petre *et al.*, 2011; Petre *et al.*, 2009).

In this way, the bags are maintained from 15 up to 30 days, during this time a mycelial net being formed from the hypha anastomosis having a compact structure and a white-yelowish color, that covers the whole surface of compost and from which the mushroom fruit bodies will emerge and develop soon as specific morphological structures of the origin species. These mushroom fruit bodies were grown and maturated in almost 3-10 days, depending on the cultivated mushroom species, at constant temperature of 18-21^{0}C, air relative humidity 90-95% and controlled aeration at 3-5 air volume exchanges per hour and the suitable lighting at 2.000-3.000 luxes per hour, for 12 h daily. For the fruit bodies picking-up, the pallets are automatically discharged by the same robotic system and transferred to the automatic harvesting zone, where another robotic system automatically collects all the mushroom fruit bodies by a special designed device to be conditioned and packaged aseptically (Fig. 11). The modular robotic system designed for edible mushroom cultivation provides the automatic sterilization of composts, the automatic inoculation inside the aseptic room by a special device of controlled injection of liquid mycelia, the incubation and fruit bodies formation in special chambers under controlled atmosphere as well as the automatic picking-up of mushroom fruit bodies (Reed *et al.*, 2001).

Both interfacing zones were designed to keep the sterile zone at the highest level of food safety against the microbial contamination. Using this robotic biotechnological model of mushroom cultivation, the economical efficiency can be significantly increased comparing to the actual conventional technologies, by shorting the total time of mushroom cultivation cycles in average with 5-10 days, depending on the mushroom strains that were grown and providing high quality mushroom fruit bodies produced in complete safety cultivation system (Petre *et al.*, 2009).

4.1 Results and discussion

To increase the specific processes of cellulose biodegradation of winery and vineyard wastes and finally induce their bioconversion into protein of fungal biomass, there were performed experiments to cultivate the mushroom species of *P. ostreatus* and *L. edodes* on the following variants of culture substrata (see Table 5).

Variants of culture substrata	Composition
S1	Winery wastes
S2	Mixture of winery wastes and rye bran 2.5%
S3	Mixture of winery wastes and rise bran 2%
S4	Mixture of vine cuttings and wheat bran 1%
S5	Mixture of vine cuttings and barley bran 1.5%
Control	Pure cellulose

Table 5. The composition of five compost variants used in mushroom culture

The fungal cultures were grown by inoculating 100 ml of culture medium with 3-5% (v/v) of the seed culture and then cultivated at 23-25°C in 250 ml rotary shake flasks. The experiments were conducted under the following conditions: temperature, 25°C; agitation speed, 120-180 rev min $^{-1}$; initial pH, 4.5–5.5. After 10–12 d of incubation the fungal cultures were inoculated aseptically into glass vessels containing sterilized liquid culture media in order to produce the spawn necessary for the inoculation of 10 kg plastic bags filled with compost made of winery and vineyard wastes (Petre *et al.,* 2011; Petre *et al.,* 2009).

These compost variants were mixed with other natural ingredients in order to improve the enzymatic activity of mushroom mycelia and convert the cellulose content of winery and vineyard wastes into protein biomass. Until this stage, all the technological operations were handmade. In the next production phases, all the operations were designed to be carried out automatically by using a robotic modular system, which makes feasible the safety culture of edible mushrooms in continuous flow using as composts the winery and vineyard wastes.

The modular robotic system designed for edible mushrooms cultivation provides the automatic sterilization of composts, the automatic inoculation inside the aseptic room by a special device of controlled injection of liquid mycelia, the incubation and fruit bodies formation in special chambers under controlled atmosphere and the automatic picking-up of mushroom fruit bodies. In this way, the whole bags filled with compost have to be sterilized at 90-100°C, by introducing them in a microwave sterilizer. In the next stage, all the sterilized bags must be inoculated with liquid mycelia, which have to be pumped through an aseptic injection device (Fig. 12).

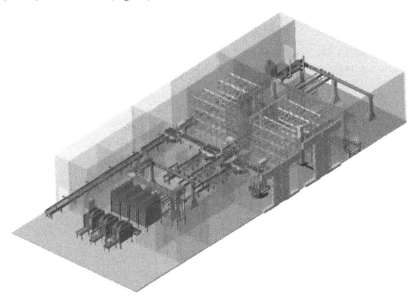

Fig. 12. General overview of the modular robotic system for controlled cultivating of mushrooms

Then, all the inoculated bags have to be transferred inside the growing chambers for incubation. After a time period of 10-15 d from the sterilized plastic bags filled with compost, the first buttons of the mushroom fruit bodies emerged.

For a period of 20-30 d there were harvested between 1.5 – 3.5 kg of mushroom fruit bodies per 10 kg compost bag. The specific rates of cellulose biodegradation were determined using the direct method of biomass weighing the results being expressed as percentage of dry weight (d.w.) before and after their cultivation. The registered data are presented in Table 6.

Variants of culture substrata	Before cultivation (g% d.w.)		After cultivation (g% d.w.)	
	L. edodes	P. ostreatus	L. edodes	P. ostreatus
S1	2.6-2.7	2.7-2.9	0.5	0.9
S2	2.3-2.5	2.5-2.8	0.4	0.7
S3	2.3-2.5	2.3-2.5	0.5	0.4
S4	2.5 -2.7	2.5 -2.7	0.7	0.8
S5	2.7-2.9	2.5-2.7	0.5	0.7
Control	3.0	3.0	1.4	1.5

Table 6. The rate of cellulose degradation of culture substrata during the growing cycles of *L. edodes* and *P. ostreatus*

The registered data revealed that by applying this biotechnology, the winery and vineyard wastes can be recycled as useful raw materials for mushroom compost preparation in order to get significant production of mushrooms.

In this respect, the final fruit body production during the cultivation of these two mushroom species was registered as being between 20–28 kg relative to 100 kg of composts made of winery wastes.

Significant bioconversion increasing of the winery and vineyard wastes by using the modular robotic system of continuous controlled cultivation of edible mushrooms can be achieved by:

a. using pure strains of the mushroom species *P. ostreatus* and *L. edodes* whose biomass has got nutritive and functional properties proved by the research results of some achieved projects or others that are running now;
b. excluding any potential contamination sources for the edible mushrooms by using total sterilization or filtration equipments in each production module, by controlling all raw and auxiliary materials, water and air;
c. keeping the high precision and accuracy of the inoculation operations, incubation and fruit body formation of edible mushrooms which induce constant biomass composition of either fungal mycelia or mushroom fruit bodies;
d. avoiding all errors in the sterile zone of production flow as well as the potential risk of edible mushroom contamination by the human operators.

5. Conclusions

According to the previous mentioned results, the following conclusions can be drawn:

1. Most suitable organic compost for mycelia growing was prepared from marc of grapes, showing the highest influence upon the mycelia growing and fresh mushroom production of 32–35 g%.

2. From the tested nitrogen sources, barley bran was the most efficient upon the mycelia growing and fruit mushroom producing at 35-40 g%, being closely followed by rice bran at 25-30 g% both in case of *P. ostreatus* and *L. edodes*, all data being reported as fresh biomass.

3. Among the tested mineral sources, the natural calcium carbonate ($CaCO_3$) yielded the best mycelia growing as well as fungal biomass production at 28-32 g%; for this reason it was registered as the most appropriate mineral source being followed by natural gypsum ($CaSO_4 \cdot 2 H_2O$) at 20-23 g%.

4. The originality and novelty of this biotechnology of winery and vineyard wastes recycling was confirmed by the Patents no 121717/2008 and 121718/2008 issued by the Romanian Office of Patents and Trade Marks

5. The mushroom pellets produced by applying the controlled cultivation of mushrooms as biotechnology for ecological treatment of winery wastes was 100% made by natural means and will be used for food supplements production with therapeutic properties which will contribute to the increasing of health level of human consumers with nutritional metabolic deficiencies.

6. The biochemical correlation between the dry weight of mushroom pellets and their sugar and nitrogen contents was kept at a balanced ratio for each tested mushroom species.

Among all mushroom samples that were tested in biotechnological experiments *G. lucidum* G-3 had shown the best values of its composition in sugars, total nitrogen and total protein content.

7. The originality and novelty of these biotechnological procedures to recycle the cereal wastes in order to get high nutritive biomass of mushroom pellets were confirmed through the Patents no 121677/2008, 121678/2008 and 121679/2008 issued by the Romanian Office of Patents and Trade Marks

8. By applying the biotechnology of controlled cultivation of edible mushrooms in modular robotic system, the final fruit body productions of both mushroom species *P. ostreatus* as well as *L. edodes* were registered as being between 20–28 kg relative to 100 kg of composts made of winery wastes.

9. The continuous controlled cultivation of edible mushrooms by using the modular robotic system can be achieved by:

    a. using pure strains of the mushroom species *P. ostreatus* and *L. edodes* whose biomass has got nutritive and functional properties proved by the research results of some achieved projects or others that are running now;

    b. excluding any potential contamination sources for the edible mushrooms by using total sterilization or filtration equipments in each production module, by controlling all raw and auxiliary materials, water and air;

    c. keeping the high precision and accuracy of the inoculation operations, incubation and fruit body formation of edible mushrooms which induce constant biomass composition of either fungal mycelia or mushroom fruit bodies;

    d. avoiding all errors in the sterile zone of production flow as well as the potential risk of edible mushroom contamination by the human operators.

10. The originality and novelty of this biotechnology of controlled cultivation of edible mushrooms in modular robotic system were confirmed by the Patent no 123132/20010, issued by the Romanian Office of Patents and Trade Marks.

6. Acknowledgment

All these works were supported by the Romanian Ministry of Education and Research through the Projects no. 51-002/2007 and 52143/2008 in the frame-work of the 4th Programme of Research and Development – „Partnership in priority domains"

7. References

Bae, J.T.; Sinha, J.; Park, J.P.; Song, C.H. & Yun, J.W. (2000). Optimization of submerged culture conditions for exo-biopolymer production by *Paecilomyces japonica*. *Journal of Microbiology and Biotechnology*, Vol. 10, pp. 482-487, ISSN: 1017-7825

Carlile, M.J. & Watkinson, S.C. (1996). Fungi and biotechnology. In: *The Fungi*, M.J. Carlile, S.C. Watkinson (Eds.), 253-264, Academic Press, ISBN: 0-12-159960-4, London, England

Chahal, D.S. (1994). Biological disposal of lignocellulosic wastes and alleviation of their toxic effluents. In: *Biological Degradation and Bioremediation of Toxic Chemicals*, G.R. Chaudry (Ed.), 347-356, Chapman & Hall, ISBN: 978-0-412-62290-8, London, England

Jones, K. (1995). *Shiitake – The Healing Mushroom*. Healing Arts Press, Rochester, ISBN: 0-89281-499-3, Vermont, USA

Moo-Young, M. (1993). Fermentation of cellulose materials to mycoprotein foods, *Biotechnology Advances*, Vol. 11, No. 3, pp. 469-482, ISSN: 0734-9750

Moser, A. (1994). Sustainable biotechnology development: from high-tech to eco-tech. *Acta Biotechnologica*, Vol. 12, No. 2, pp. 10-15, ISSN: 0138-4988

Oei, P. (2003). Mushroom Cultivation. 3rd Edition, Backhuys Publishers, ISBN: 90-5782-137-0, Leiden, The Netherlands

Petre, M.; Teodorescu, A.; Bejan, C.; Giosanu, D. & Andronescu, A.(2011). Enhanced Cultivation of Edible and Medicinal Mushrooms on Organic Wastes from Wine Making Industry. In: Proceedings of the International Conference „*Environmental Engineering and Sustainable Development*", pp. 234-239, ISBN: 978-606-613-002-8, Alba Iulia, Romania, May 26-28, 2011

Petre, M. & Teodorescu, A. (2010). *Handbook of submerged cultivation of eatable and medicinal mushrooms*. CD Press, ISBN: 978-606-528-087-8, Bucharest, Romania

Petre, M.; Teodorescu, A.; Tuluca, E.; Bejan, C. & Andronescu, A. (2010). Biotechnology of Mushroom Pellets Producing by Controlled Submerged Fermentation. *Romanian Biotechnological Letters*, Vol 12, No. 2, pp. 50-56, ISSN: 1224-5984

Petre, M. & Teodorescu, A. (2009): Ecological Biotechnology for Agro-Food Wastes Valorization. In: *Biotechnology of Environmental Protection*, M. Petre (Ed.), vol. 2, 2nd Edition, 143-150, CD Press, ISBN: 978-606-528-040-3; 978-606-528-042-7, Bucharest, Romania

Petre, M.; Teodorescu, A.; Nicolescu, A.; Dobre, M. & Giosanu , D. (2009). Food biotechnology for edible mushrooms producing by using a modular robotic system. In: Proceedings of 2nd International Symposium „*New Researches in Biotechnology*",

SimpBTH 2009, Biotechnology Series F–Suppl., pp. 261-269, ISSN:1224-7774, Bucharest, Romania, November 18-19, 2009

Petre, M. & Petre, V. (2008). Environmental Biotechnology to Produce Edible Mushrooms by Recycling the Winery and Vineyard Wastes. *Journal of Environmental Protection and Ecology*, Vol. 9, No.1, pp. 88-95, ISSN: 1311-5065

Petre, M.; Bejan, C.; Visoiu, E.; Tita, I. & Olteanu, A. (2007). Mycotechnology for optimal recycling of winery and vine wastes. *International Journal of Medicinal Mushrooms*, Vol. 9, No. 3, pp. 241-243, ISSN: 1521-9437

Reed, J.N.; Miles, S.J; Butler, J.; Baldwin, M. & Noble, R. (2001). Automation and Emerging Technologies for Automatic Mushroom Harvester Development. *Journal of Agricultural Engineering Research*, Vol 33, pp. 55-60, ISSN: 1095-9246

Smith, J. (1998). Biotechnology. 3rd Edition. Cambridge University Press, ISBN: 0-521-44911-1, London, England

Stamets, P. (2000). Growing Gourmet and Medicinal Mushrooms. Ten Speed Press, ISBN: 1-58008-175-4, Berkeley, Toronto, Canada

Verstraete, W. & Top, E. (1992). Holistic Environmental Biotechnology. Cambridge University Press, ISBN: 0-521-42078-4, London, England

Wasser, S.P. & Weis, A.L. (1994). Therapeutic effects of substances occurring in higher Basidiomycetes mushrooms: a modern perspective. *Critical Reviews in Immunology*, Vol. 19, pp. 65-96, ISSN: 1040-8401

Fermentation Processes Using Lactic Acid Bacteria Producing Bacteriocins for Preservation and Improving Functional Properties of Food Products

Grazina Juodeikiene[1], Elena Bartkiene[2], Pranas Viskelis[3],
Dalia Urbonaviciene[3], Dalia Eidukonyte[1] and Ceslovas Bobinas[3]
[1]Kaunas University of Technology,
[2]Veterinary Academy, Lithuanian University of Health Sciences,
[3]Institute of Horticulture, Lithuanian Research Centre for Agriculture and Forestry,
Lithuania

1. Introduction

During the recent years health-conscious consumers are looking for natural foods without chemical preservatives that will fit in their healthy lifestyle. The increasing consumption of precooked food, prone to temperature abuse, and the import of raw foods from developing countries are among the main causes of this situation. Biopreservation refers to extended shelf life and enhanced safety of foods using microorganisms and/or their metabolites (Ross et al., 2002). LAB is generally employed because they significantly contribute to the flavor, texture and, in many cases, to the nutritional value of the food products (McKay and Baldwin, 1990). LAB are used as natural or selected starters in food fermentations and exert the antimicrobial effect as a result of different metabolic processes (lactose metabolism, proteolytic enzymes, citrate uptake, bacteriophage resistance, bacteriocin production, polysaccharide biosynthesis, metal-ion resistance and antibiotic resistance) (Zotta, T., 2009; Corsetti, 2004). Lactic acid bacteria (LAB) play a key role in food fermentations where they not only contribute to the development of the desired sensory properties in the final product but also to their microbiological safety. LAB has a GRAS status (generally recognized as safe) and it has been estimated that 25% of the European diet and 60% of the diet in many developing countries consists of fermented foods (Stiles, 1996). Fermentation is one of the most ancient and most important food processing technologies. Fermentation is a relatively efficient, low energy preservation process, which increases the shelf life and decreases the need for refrigeration or other forms of food preservation technology. Currently, fermented foods are increasing in popularity (60% of the diet in industrialized countries) and, to assure the homogeneity, quality, and safety of products, they are produced by the intentional application in raw foods in different microbial systems (starter/protective cultures) (Holzapfel et al., 1995).

Examples of vegetable lactic acid fermentations are: sauerkraut, cucumber pickles, and olives in the Western world; Egyptian pickled vegetables in the Middle East; Indian pickled vegetables and Korean kim-chi, Thai pak-sian-don, Chinese hum-choy, Malaysian pickled vegetables and Malaysian tempoyak. Lactic acid fermented cereals and tubers (cassava) include: Mexican pozol, Ghanaian kenkey, Nigerian gari; boiled rice/raw shrimp/raw fish mixtures: Philippine balao-balao, burong dalag; lactic fermented/leavened breads: sourdough breads in the Western world; Indian idli, dhokla, khaman, Sri-lankan hoppers; Ethiopian enjera, Sudanese kisra and Philippine puto; Lactic acid fermented cheeses in the Western world and Chinese sufu/tofu-ru. Lactic acid fermented yogurt/wheat mixtures: Egyptian kishk, Greek trahanas, Turkish tarhanas.

Moreover, because of the improved organoleptic qualities of traditional fermented food, extensive research on its microbial biodiversity has been carried out with the goal of reproducing these qualities, which are attributed to native microbiota, in a controlled environment.

Recent years the interest increased in bacteriocin-like inhibitory substances (BLIS) producing LAB because of their potential use as natural antimicrobial agents to enhance the safety of food products. Bacteriocins from LAB are described as "natural" inhibitors, in regard to LAB having a GRAS status. Bacteriocin-like inhibitory substances (BLIS) from LAB are antimicrobial compounds that possess bacteriocin requisites but that have not yet been characterized for their amino acid sequence (Jack et al., 1995). Bacteriocins from the generally recognized as safe LAB, have received significant attention as a novel approach to the control of pathogens in foods (Klaenhammer, 1993; Settani et al., 2005). Nisin is the first antimicrobial polypeptide found in LAB (Rogers, 1928); at the time of discovery, the producer strains were identified as *Streptococcus lactis* [later classified as *Lactococcus lactis*] (Schleifer et al., 1985). Today nisin is a permitted preservative in at least 48 countries, in which it is used in a variety of products, including cheese, canned food and cured meat (Delves-Broughton, 1990). Another commercially produced bacteriocin is pediocin PA-1 produced by *Pediococcus acidilactici* and marketed as ATTA ™ 2431 (Kerry Bioscience, Carrigaline, Co, Cork, Ireland). The source of natural or controlled microbiota and/or antimicrobial compounds could be traditional fermented foods. The health benefits attributed to peptides in these traditional products have, so far, not been established, however. Several factors can affect the bacteriocin activity including interaction with other bacteriocins, constituents from the cells as well as the growth medium, purity and concentration of exogenous added enzymes (Moreno et al., 2000). Enzymes present during food making originate from different sources: there are those that already exist in the plant raw material, those associated with the metabolic activity of yeasts or LAB, and those intentionally added to the formulations. For example, amylases are added for the intensification of the saccharification stage (Stauffer, 1994), while microbial lipases induce changes in lipid and short-chain fatty acid compositions (Martınez-Anaya, 1996). Up to date, numerous recent review articles focused on the isolation of novel LAB bacteriocin-like inhibitory substances produced by LAB, evaluation their inhibitory activities, classification, biochemical and genetic characterization, studying the sensitivity of the BLIS antimicrobial activity to different factors, e.g. enzymes in fermentation media and the mode of action of LAB bacteriocins, as well as on some of their food application (mainly animal origin) have been published. Unfortunately literature lacks of concentrated articles dealing with the use of bacteriocins or bacteriocin-producing strains for the biopreservation and improving functional properties of vegetable and fruit products.

Fermentation Processes Using Lactic Acid Bacteria Producing Bacteriocins for Preservation and Improving
Functional Properties of Food Products

25

2. Characterization of bacteriocins produced by lactic acid bacteria

The antimicrobial ribosomal synthesized peptides produced by bacteria, including members of the LAB, are called bacteriocins. Such peptides are produced by many, if not all, bacterial species and kill closely related microorganisms (Jack et al., 1995). Due to their nature, they are inactivated by proteases in the gastrointestinal tract. Most of the LAB bacteriocins identified so far are thermo stable cationic molecules that have up to 60 amino acid residues and hydrophobic patches. Electrostatic interactions with negatively charged phosphate groups on target cell membranes are thought to contribute to the initial binding, forming pores and killing the cells after causing lethal damage and autolysin activation to digest the cellular wall (Gálvez et al., 1990). Example of damage caused by bacteriocin on *L. monocytogenes* CECT 4032 cells is presented in Figure 1.

The LAB bacteriocins have many attractive characteristics that make them suitable candidates for use as food preservatives, such as:

- Protein nature, inactivation by proteolytic enzymes of gastrointestinal tract
- Non-toxic to laboratory animals tested and generally non-immunogenic
- Inactive against eukaryotic cells
- Generally thermo resistant (can maintain antimicrobial activity after pasteurization and sterilization)
- Broad bactericidal activity affecting most of the Gram-positive bacteria and some, damaged, Gram-negative bacteria including various pathogens such as *L. monocytogenes, Bacillus cereus, S. aureus,* and *Salmonella*
- Genetic determinants generally located in plasmid, which facilitates genetic manipulation to increase the variety of natural peptide analogues with desirable characteristics

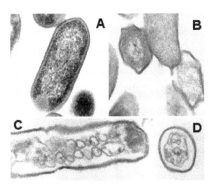

Fig. 1. Example of damage caused by bacteriocin on *L. monocytogenes* CECT 4032 cells. (A) cells without enterocin AS-48; (B) cells treated with 0.1 μg/ml of AS-48 for 2 h; (C and D) cells treated with 3 μg/ml of enterocin AS-48 for 10 min (adapted from Mendoza et al., 1999).

For these reasons, the use of bacteriocins has, in recent years, attracted considerable interest for use as biopreservatives in food, which has led to the discovery of an ever-increasing potential of these peptides. Undoubtedly, the most extensively studied bacteriocin is nisin, which has gained widespread applications in the food industry. This FDA approved bacteriocin is produced by the GRAS microorganism *Lactococcus lactis* and is used as a food additive in at least 48 countries, particularly in processed cheese, dairy products and canned foods. Nisin is effective against food-borne pathogens such as *L. monocytogenes* and many other Gram-positive spoilage microorganisms (Thomas et al., 2000; Thomas and Delves-Broughton, 2001). Nisin is listed as E-234, and may also be cited as nisin preservative or natural preservative. In addition to the work on nisin, several authors have outlined issues involved in the approval of new bacteriocins for food use (Harlander, 1993).

2.1 The biosynthetic pathway of bacteriocins

The biosynthetic pathways of bacteriocins with the focus on class II bacteriocins (mainly produced by LAB from fermented plant products) will be discussed in this section. All bacteriocins are synthesized as a biologically inactive prepeptide carrying an N-terminal leader peptide attached to the C-terminal propeptide (Hoover and Chen, 2005).

The mode of action of lactic acid bacteria bacteriocins belonging to class I, II and III are presented in Figure 2.

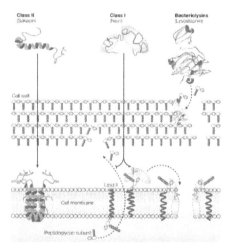

Fig. 2. Mode of action of bacteriocins by lactic acid bacteria.

Some bacteriocins (or lantibiotics) of class I (e.g. nisin) have a dual mode of action: (1) they prevent correct cell wall synthesis that leads to cell death by binding to lipid II – the main transporter of peptidoglycan subunits from the cytoplasm to the cell wall and (2) they employ lipid II as a docking molecule to initiate a process of membrane insertion and pore formation leading to a rapid cell death (Gillor et al., 2008; Cotter et al., 2005; Wiedemann et al., 2001). The majority of class II bacteriocins kill by inducing membrane permeabilization and the subsequent leakage of molecules from target bacteria (Gillor et al., 2008). Bacteriolysins (bacteriolytic proteins belonging to class III) function directly on the cell wall of Gram-positive targets leading to death and lysis of the target cell (Cotter et al., 2005).

Class II bacteriocins are synthesized as a prepeptide containing a conserved N-terminal leader and a characteristic double-glycine proteolytic processing site, and in contrast to lantibiotics, they do not undergo extensive posttranslational modification (Hoover and

Fermentation Processes Using Lactic Acid Bacteria Producing Bacteriocins for Preservation and Improving
Functional Properties of Food Products

27

Chen, 2005). The examples of the class II bacteriocins are Sakacin P, G, A and Pediocin PA-1/AcH.

A general scheme of the biosynthetic pathway of class II bacteriocins is shown in Figure 3.

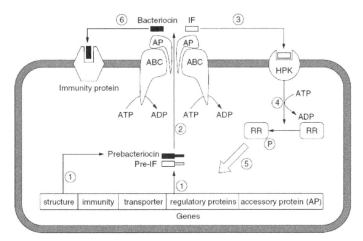

Fig. 3. The biosynthesis of class II bacteriocins. 1. Formation of prebacteriocin and prepeptide of induction factor (IF). 2. The processing of the prebacteriocin and pre-IF, and translocation by the ABC-transporter, resulting in the release of mature bacteriocin and IF. 3. Histidine protein kinase (HPK) senses the presence of IF and autophosphorylates. 4. The transfer of the phosphoryl group (P) to the response regulator (RR). 5. The activation of the regulated genes transcription by the RR. 6. Producer immunity.

The production of most class IIa bacteriocins is regulated by a three-component system which includes a histidine protein kinase, a response regulator, and an induction factor. Some class IIa bacteriocins are autoregulated by a two-component signal transduction system, which is a well-known phenomenon in lantibiotics (Sahl et al., 1998). A threshold concentration of the bacteriocin, which functions as a signal molecule accumulating during growth, triggers the transcription of the genes coding for bacteriocin production, suggesting a self-inducing cell density (quorum-sensing)-regulated system (Sahl et al., 1998).

Class IIc bacteriocins are an exceptional case due to their production with a typical N-terminal signal sequence of the sec-type, processing and excretion through the general secretory pathway (Leer et al., 1995; Worobo et al., 1995). Once the prepeptide is formed, it is processed to remove the leader peptide concomitant with export from the cell through a dedicated ABC-transporter and its accessory protein (Nes et al., 1996).

2.2 Activity spectra and biochemical properties of LAB and their produced bacteriocins

As mentioned before, LAB bacteriocins tend to be active against a wide range of mostly closely related Gram-positive bacteria (Jack et al., 1995). Gänzle (1998) corroborated this while reviewing the inhibitory spectrum of bavaricin A, BLIS C57, and plantaricin ST31, produced by sourdough LAB, indicating no inhibition of Gram-negative bacteria, whereas a variety of Gram-positive bacteria were sensitive. The insensitivity of Gram-negative bacteria to bacteriocins from LAB strains might be explained by their outer membrane providing them with a permeability barrier (Messens et al., 2002). Furthermore, the producer strains are found to be immune towards their own bacteriocin (Gänzle, 1998). Studies on the resistance of *List.monocytogenes* strains towards bavaricin A confirmed that only 3 of the 245 strains examined were resistant to bavaricin A (Larsen and Nørrung, 2003), meanwhile Rasch and Knøchel (1998) reported about the correlation between bavaricin A sensitivity and pediocin PA-1 sensitivity of the strains.

The seldom inhibition of commonly encountered enteropathogenic bacteria (*Enterobacter, Klebsiella,* or *Salmonella* was announced as the weakness of the bacteriocins produced by Gram-positive bacteria. Gram-positive bacteriocins are restricted to kill other Gram-positives and the killing range varies significantly (Gillor et al., 2008). Martínez-Cuesta et al. (2006) reported about the relatively narrow range of lactococcins A, B and M able to inhibit only *Lactococcus,* meanwhile some type A lantibiotics (e.g. nisin A, mutacin B-Ny266) had a wide range while killing *Actinomyces, Bacillus, Clostridium, Corynebacterium, Enterococcus, Gardnerella, Lactococcus, Listeria, Micrococcus, Mycobacterium, Propionibacterium, Streptococcus,* and *Staphylococcus* as shown by Mota-Meira et al. (2005). These particular bacteriocins were found to be active against some medically important Gram-negative strains of *Campylobacter, Haemophilus, Helicobacter,* and *Neisseria* (Morency et al., 2001).

Furthermore, IIa class bacteriocins have relatively narrow killing spectrum as compared to I class bacteriocins and inhibit only closely related Gram-positive bacteria (Heng et al. 2007). However, pediocin was reported to have a fairly broad activity spectrum while inhibiting *Streptococcus aureus* and vegetative cells of *Clostridium* spp., *Listeria* and *Bacillus* spp. (Nes and Holo, 2000; Eijsink et al., 2002). Besides, maximum antimicrobial activity against *Escherichia coli, Staphylococcus aureus* and *Bacillus cereus,* though it was more effective against *E. coli* than others, showed bacteriocin produced by strain CA 44 of *Lactobacillus* genus isolated from carrot fermentation medium. This bacteriocin was stable at up to 100°C but its activity declined compared to that at 68°C and was completely lost at 121°C. The maximum antimicrobial activity was retained within the pH range of 4–5, but it was adversely affected by the addition of papain. Bacteriocin was also effective against *B. cereus* in different fruit products (pulp, juice and wine) indicating its potential application as a biopreservative in fruit products (Joshi et al., 2006).

Quite a few studies on a bacterocin activity possessed by *L. sakei* strains have been performed. Schillinger and Lücke (1989) reported about 221 surveyed lactobacilli strains, among those 19 *L. sakei* strains, 3 *L. plantarum* strains and 1 *L. curvatus* strain were found to inhibit other lactobacilli. Bacteriocins were not identical according to the evaluation of supernatants antimicrobial spectra. Sakacin A produced by *L. sakei* Lb706 was reported to be active against *List. monocytogenes* strains 8732 and 17a, moreover 4 other strains of *L. sakei* and 1 strain of *L. plantarum* also shown antilisterial activity. Mørtvedt and Nes (1990) identified bacteriocin Lactosin S produced by *L. sakei* and described it as moderately heat stable, sensitive to protease and having antimicrobial activity against *Lactobacillus, Pediococcus* and *Leuconostoc* genera members. Furthermore, the instability of bacteriocin production and immunity was revealed by the plasmid biology investigation in *L. sakei* L45. Antagonistic effect of *L. sakei* CTC494 and sakacin K to different extents against *List.innocua* CTC1014 was demonstrated by Hugas and co-workers (1998). While Axelsson with co-

Fermentation Processes Using Lactic Acid Bacteria Producing Bacteriocins for Preservation and Improving
Functional Properties of Food Products

29

workers (1988) developed a system for heterologous expression using a bacteriocin-negative *L. sakei* Lb790 strain, where into Lb790 introduced two plasmids allowed the production of various bacteriocins (sakacin P, pediocin PA-1, and piscicolin 61) at levels equal to or exceeding levels in correspondence with the wild type cultures. Cuozzo et al. (2000) investigated II b class bacteriocin lactocin 705 from *L. casei* CRL 705 and reported about the required of both two peptides presence for the inhibitory activity.

The cloning, expression, and nucleotide sequence of the genes involved in the synthesis of pediocin PA-l was reported by Marugg et al. (1992). The genes were cloned and expressed in *E. coli* and 5.6-kbp fragment from the plasmid was found to be necessary for the bacteriocin production. Hoover and co-workers (1989) noted about the surveyed 37 pediococci cultures for the antagonistic effect against eight *List. monocytogenes* strains and indicated that a bacteriocin effect of these LABs against *List. monocytogenes* may not be limited to a few industrial starter cultures. 15 strains containing the *Lactobacillus, Pediococcus, Lactococcus,* and *Leuconostoc* genera were examined for the inhibition against eight strains of *List. monocytogenes* (Harris et al., 1989) and only cell-free supernatants from *Lactobacillus* species UAL11, *P. acidilactici* PAC 1.0, and *Leuconostoc* species UAL14 were reported to inhibit all eight strains of *List. monocytogenes*. The addition of proteolytic enzymes caused the prevention of the inhibition. Ennahar et al. (1996) found out that *Lactobacillus plantarum* WHE 92 produce a bacteriocin identical to pediocin AcH from *P. acidilactici* H, though pediocin AcH was produced more effectively in *L. plantarum* WHE 92 in the pH range of 5.0 to 6.0 as compared to *P. acidilactici* H. Moreover, *L. plantarum* WHE 92 seems to have more effective means of antagonism against *List. monocytogenes*, since dairy products are normally higher than pH 5.0 (Hoover and Chen, 2005). Miller and co-workers (1998) applied PCR random mutagenesis for the construction of pediocin AcH amino acid substitution mutants. One mutant peptide was found to have a 2.8-fold higher activity against *L. plantarum* NCDO955, while other mutations were inactive and shown a reduced antagonism. Johnsen et al. (2000) increased the stability of pediocin PA-1 with the replaced methionine residue with alanine, isoleucine, or leucine in order to protect from oxidation, since this peptide was found to lose its activity while stored as refrigeration or ambient temperatures.

Lacticin 3147 is a two peptide bacteriocin with a broad-spectrum activity and genetically resides on a self-transmissable plasmid that can be moved to other lactococci strains (Ross et al., 2000). McAuliffe and co-workers (1999) reported about a 3-log10 reduction in CFU/g of *List. monocytogenes* with the used *Lc. lactis* subs. *lactis* culture producing lacticin 3147. Bacteriocin S50 is another antimicrobial compound produced by lactococci, though it has a relatively narrow activity spectrum (Hoover and Chen, 2005). Lacticin FS92, containing 32 amino acids, is a heat-stable bacteriocin and appears to be active against *List. monocytogenes* as noticed by Mao et al. (2001). *List. monocytogenes* resistant mutants were found to remain sensitive to lacticin FS92, but not to pediocin PA-1, curvaticin FS47 and lacticin FS56. The susceptibility of *List. monocytogenes, List. innocua,* and *List. seeligeri* strains was found to strain-dependent at each pH examined in response to lactocin 705, enterocin CRL35, and nisin (Castellano et al., 2001).

3. Optimization of media and growth conditions for increased bacteriocin production

The incorporation of bacteriocins as a biopreservative ingredient into model food systems has been studied extensively and has been shown to be effective in the control of pathogenic and spoilage microorganisms (Neysen and De Vuyst, 2005). LAB can also be considered as protective cultures because they improve the microbiological quality as well as the safety of the food (Messens and De Vuyst, 2002) and can be a way to prevent product spoilage (Verluyten et al., 2004a). Lactic acid bacteria (LAB) are a group of microorganisms nutritionally exigent. They need a wide range of nutrients to grow and synthesize metabolic products, some nutritional requirements usually being strain specific. De Man-Rogosa-Sharpe (MRS) and yeast autolysate-peptone-tryptone-tween 80-glucose (LAPTg) are standard culture media commonly used to support the growth of lactobacilli in the starting point of their cultivation. These media contain carbon and energy sources (carbohydrates, e.g. glucose), complex nitrogen sources (yeast extract, meat extract, tryptone and peptone) and supplements derived from oleic acid (Tween 80). MRS also includes inorganic and organic salts that have shown a stimulating effect or are essential for the growth of most of the species of this genus. Different components of culture media strongly affect the growth and bacteriocin production of several microorganisms that are mainly considered for food applications and must be included in fermentation processes, which are used on a production scale. Extruded wheat material is one of the candidates to be included as fermentation media for cultivation of bacteriocins producing LAB (*Lactobacillus sakei* MI806, *Pediococcus pentosaceus* MI810 and *Pediococcus acidilactici* MI807), previously isolated from spontaneous Lithuanian sourdoughs, in fermented products preparation formula for wheat bread production to have a higher positive effect compared with the control medium on LAB growth and their antimicrobial activities (Juodeikiene et al., 2011).

Frequently, the conditions that lead to high bacteriocin production are similar to those prevailing during food fermentation processes (Leroy et al., 2002; Delgado et al., 2005; Neysen and De Vuyst, 2005). Bacteriocin production is usually proportional to growth and shows primary metabolite kinetics (Moretro et al., 2000) but often the correlation is weak (Delgado et al., 2005) and this is particularly evident for bacteriocins produced during the stationary phase (Jim´enez-D´ıaz et al., 1993). Food preservation using *in situ* bacteriocin production requires a better understanding of the relationship between growth and bacteriocin production. Different bacteriocin exhibits different inhibition profile on food spoilage and pathogenic microorganisms. Therefore, they could be natural replacements for synthetic food preservatives. In order to increase the productivity of the bacteriocins, a better understanding on the factors affecting their production is essential.

Bacteriocin titres change with environmental factors (Leal-S´anchez et al., 2002; Delgado et al., 2005), such as pH, temperature, and NaCl and ethanol concentrations. These environmental factors may influence growth negatively and thereby the secretion of the induction factor (Leal-S´anchez et al., 2002). Further, it has been suggested that some environmental factors reduce the binding of the induction factor to its receptor (Delgado et al., 2005). Bacteriocin production is strongly dependent on pH, nutrient sources and incubation temperature. Activity levels do not always correlate with cell mass or growth rate of the producer (Kim et al., 1997; Bogovic-Matijasic & Rogelj, 1998). Increased levels of bacteriocin production are often obtained at conditions lower than required for optimal growth (Bogovic-Matijasic & Rogelj, 1998; Todorov et al., 2000; Todorov & Dicks, 2004). Understanding the influence of food-related environmental factors on the induction of bacteriocins is essential for the effective commercial application of bacteriocin-producing LAB in the preservation of foods.

Leal et al. (1998) optimized the production of bacteriocins by *Lactobacillus plantarum* LPCO10 to allow the use of bacteriocins as natural food additives in canned vegetables and other

Fermentation Processes Using Lactic Acid Bacteria Producing Bacteriocins for Preservation and Improving
Functional Properties of Food Products

31

food systems. Results obtained indicated that the best conditions for bacteriocin production were shown with temperatures ranging from 22°C to 27°C, salt concentration from 2.3 to 2.5%, and *L. plantarum* LPCO10 inoculum size ranging from $10^{7.3}$ to $10^{7.4}$ CFU/ml, fixing the initial glucose concentration at 2%, with no aeration of the culture. Under these optimal conditions, about 3.2×10^4 times more bacteriocin per liter of culture medium was obtained than that used to initially purify plantaricin S from *L. plantarum* LPCO10 to homogeneity.

Delgado A. et al. (2007) by modeling studied the effects of some environmental factors on bacteriocin production by *Lactobacillus plantarum* 17.2b. Bacteriocin production by *L. plantarum* 17.2b was very sensitive to environmental conditions and uncoupled from growth. Maximum production required suboptimal growth temperatures, pH values above growth's optimum and no NaCl.

Many studies have focused on optimization of media and growth conditions of LAB for increased bacteriocin production. They have generally focused on the effects of pH, temperature, composition of the culture medium, and general microbial growth conditions (*in vitro* as well as in natural fermentations) on maximal bacteriocin production (FAO-WHO, 2002; ANVISA, 2010; Cruz et al., 2009; Silveira et al., 2009; Minei et al., 2008; Galvez et al., 2008). By supplementing the medium with growth limiting factors, such as carbohydrates, nitrogen, vitamins and potassium phosphate, or by adjusting the medium pH, levels of bacteriocin production is often increased.

Several mechanisms have been proposed for the bacteriocins activity: alteration of enzymatic activity, inactivation of anionic carriers through the formation of selective and non-selective pores and inhibition of spore germination (Parada et al. 2007; Martinez and De Martinis, 2006). Powell et al. (2006), Todorov and Dicks (2005a), (2005b), (2006a), (2006b), Todorov et al. (2000), (2007a), (2007b), (2004) and Todorov (2008) reported higher bacteriocin production levels for *L. plantarum* ST194BZ, *L. plantarum* ST13BR, *L. plantarum* ST414BZ, *L. plantarum* ST664BZ, *L. plantarum* ST23LD, *L. plantarum* ST341LD, *L. plantarum* 423, *L. plantarum* AMAK, *L. plantarum* ST26MS, *L. plantarum* ST28MS, *L. plantarum* ST8KF, *L. plantarum* ST31 in optimized growth media.

In general, the bactericidal/bacteriostatic action of bacteriocins encompasses the increased permeability of the cytoplasmic membrane of the target cells for a broad range of monovalent cations (e.g. Na^+, K^+, Li^+, Cs^+, Rb^+ and choline) leading to the destruction of proton motive force by dissipation of the transmembrane pH gradient and eventually to the cell death (Simova et al., 2009; Oppegård et al., 2007). The bactericidal or bacteriostatic activity possessed by bacteriocins is influenced by the following factors: bacteriocin dose and purification degree, physiological status of the indicator cells (e.g. growth phase) and experimental conditions (e.g. temperature, pH, presence of agents disrupting cell wall integrity and other antimicrobial compounds) (Deraz et al., 2007; Cintas et al., 2001). An increased antibacterial activity of non-lanthionine-containing bacteriocins at low pH can be explained by the following factors: (1) more molecules are available to interact with the sensitive cells due to a lesser probability of the aggregation of hydrophilic peptides; (2) more molecules are available for the bactericidal action, since fewer molecules remain bound to the wall; (3) an enhanced capacity of hydrophilic bacteriocins to pass through hydrophilic regions of the cell wall of the sensitive bacteria; (4) an inhibited interaction at higher pH values between the non-lanthionine-containing bacteriocins and putative membrane receptors (Parada et al., 2007).

These studies highlight the possibility of increase antimicrobial activity of fermented products with the aim to improve food safety and quality characteristics. Besides the optimization of bacteriocin production and enhancement of its activity are economically important to reduce the production cost.

4. Application of bacteriocins producing LAB for improving some safety characteristics of plant products

4.1 Possibilities to prolong microbiological spoilage of bread using novel BLIS producing LAB

Knowledge of fermenting microorganisms plays a defining role in the process of fermentation standardization and it is essential to have an exhaustive view of microbial interactions. The development of starter cultures for food fermentations follows a multidisciplinary approach and requires a thorough ecological study of these ecosystems. LAB are fundamental for the fermented product properties such as lactic fermentation, proteolysis, synthesis of volatile compounds, anti-mould and antiropiness effect. Since LAB are found to be the dominant microorganisms in sourdoughs, the rheology, flavour and nutritional properties of sourdough-based baked products greatly rely on their activity (Corsetti at al., 2003; Hammes and Gänzle, 1998; Gobbetti et al., 2005). Sourdough is used as an essential ingredient for acidification, leavening and production of flavour compounds and biopreservation of bread (Sadeghi, 2008; De Vuyst, 2007; Katina et al., 2005; Hansen, 2004; Clarke et al., 2004). In bakery practice, sourdough is usually sustained by repeated inoculation, whereby a reproducible and controlled composition and activity of the sourdough microflora is paramount to achieve a constant stability of sourdough as well as a constant quality of the end-product. Besides, many researchers have reported about the high resistance of sourdough breads to the microbiological spoilage by moulds and rope-forming bacilli (Valerio et al., 2009; Hassan and Bullerman, 2008; Sadeghi, 2008; Ryan et al., 2008; Mentes et al., 2007). Mould causes mouldiness and bacteria belonging to the genus *Bacillus* (Şimşek et al., 2006) are capable of causing massive economic losses due to the considerable resistance allowing them to survive food processing (Errington, 2003; Driks, 2002). The bacterial spoilage of bread, known as ropiness, occurs as an unpleasant fruity odour (Mentes et al., 2007), and is still of major economic concern in the baking industry.

4.1.1 Rope production in bread

Ropiness is mainly caused by *Bacillus subtilis* and *Bacillus licheniformis* (Collins et al., 1991) which reportedly originate from the raw materials, the bakery atmosphere and equipment surfaces (Bailey and von Holy, 1993). These strains are also known to be food-borne pathogens when present at levels of 105 CFU/g in bread crumb (Kramer and Gilbert, 1989). *Bacillus* is a genus of rod-shaped, endospore-forming aerobic or facultatively anaerobic, Gram-positive bacteria (in some species cultures may turn Gram-negative with age) and a member of the division *Firmicutes*. Many species of the genus exhibit a wide range of physiologic abilities that allow them to live in every natural environment. Under stressful environmental conditions, the cells produce oval endospores that can stay dormant for extended periods (Ravel and Fraser, 2005; Kunst et al., 1997). Bacteria belonging to the genus *Bacillus* are capable of causing economic losses to the baking industry due to the food spoilage condition known as rope (Valerio et al., 2008; Thompson et al., 1993). The predominant species involved in bread spoilage are *Bacillus subtilis* and *B. licheniformis*, though *B. pumilus*, *B. megaterium* and *B. cereus* are implicated as well (Şimşek et al., 2006; Rosenkvist and Hansen, 1995; Collins et al., 1991). These strains are also known to be food-borne pathogens when present at levels of 105 CFU/g in bread crumb (Kramer and Gilbert, 1989). Ropiness is noticed as an unpleasant odour, followed by a soft and sticky bread crumb caused by the enzymatic degradation and the production of extracellular slimy polysaccharides (Sadeghi, 2008; Valerio et al., 2008; Pepe et al., 2003).

Figure 4 illustrates an example of —ropy bread. *B. subtilis* spores have been isolated from ropy bread, meanwhile contamination of *Bacillus* have been reported to originate from raw materials, bakery environments and also from additives, including yeast, bread improvers, and gluten (Thompson et al., 1993; Rosenkvist and Hansen, 1995; Collins et al., 1991; Sorokulova et al., 2003; Bailey and von Holy, 1993).

Fig. 4. Rope production in bread.

B.subtilis spores being heat resistant can survive the baking process, since the maximum temperature in the loaf centre remains 97°C to 101°C for a few minutes (Östman, 2002; Rosenkvist and Hansen, 1995). During subsequent exposure to the warm (25°C to 30°C) and humid (water activity, ≥0.95) environmental conditions *Bacillus* spores germinate causing bread spoilage (Volavsek et al., 1992). The spore germination and growth of *Bacillus* vegetative cells during storage strongly depend on the water activity, pH and temperature (Condón et al., 1996; Quintavalla and Paroli, 1993).

4.1.2 Spore formation in *Bacillus subtilis*

Bacterial spores are very specialized, differentiated cell types and can survive the adverse conditions (e.g. starvation, high temperatures, ionizing radiation, mechanical abrasion, chemical solvents, detergents, hydrolytic enzymes, desiccation, pH extremes and antibiotics). Spores can cause massive problems in the food industry due to the considerable resistance allowing them to survive food processing and conservation methods (Errington, 2003; Driks, 2002).

The mature heat-resistant spore formation takes approx. 8 hours from the initial time of starvation. Numerous alterations in gene expression and a variety of physiological and morphological changes characterize the process of sporulation (Grossman and Losick, 1988). *B. subtilis* has been used as a model for the sporulation (the process of spore formation) studies (Errington, 2003; Eichenberger et al., 2004; Piggot and Hilbert, 2004; Phillips and

Strauch, 2002). Spore formation is a unique and complex process and can be divided into stages 0, II, III, IV, V, and VI (Grossman and Losick, 1988) involving asymmetric cell division, engulfment of the smaller cell and sacrifice of the original bacterial cell for the production of a single spore (Figure 5).

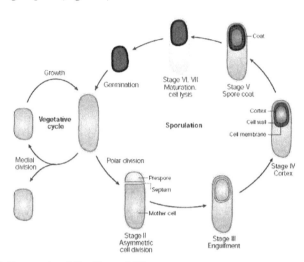

Fig. 5. The sporulation cycle of *Bacillus subtilis*.

Stage 0 is characterized as the cell commitment to sporulation that leads to the building of a septum (stage II). As a cell begins the process of forming an endospore, it divides asymmetrically, resulting in the creation of two compartments (the larger mother cell and the smaller prespore). Afterwards, the degradation of thepeptidoglycan in the septum occurs and mother cell engulfs the prespore, leading to the formation of a cell within a cell (stage III). The synthesis of the endospore-specific compounds, formation of the cortex and deposition of the coat (stages IV and V) proceeds due to the activities of the mother cell and prespore. Stages IV and V are followed by the final dehydration and maturation of the prespore (stages VI and VII). Finally, the mother cell is destroyed in a programmed cell death, and the endospore is released into the environment. The endospore remains dormant until it senses the return of more favourable conditions (Errington, 2003; Grossman and Losick, 1988; Phillips and Strauch, 2002).

4.1.3 Strategy for the control of *Bacillus* spp. by BLIS producing LAB in bread production

The initial *Bacillus* spore counts could be reduced by the recommended control procedures such as raw material quality, good sanitation of bakery equipment, stringent temperature control during baking, production cooling and storage environments (Bailey and von Holy, 1993; Viljoen and von Holy, 1993). The use of chemical preservatives (propionic and acetic acids) was reported to be one of the ways for the inhibition of *Bacillus* germination and growth in bread, although the current trend is to reduce the levels of these substances (Pattison et al., 2004; Marín et al., 2002). The increase in acidity by using traditional sourdough fermentation is an effective way to limit the germination and growth of rope forming bacteria (Sadeghi, 2008). Röcken (1996) reported about the enhanced thermal inactivation of *B. subtilis* spores by using an increased sourdough contents. Katina and co-workers 2005 examined the ability of LAB to inhibit the growth of rope forming strains in wheat bread and announced about the growth inhibition of *B. subtilis* and *B. licheniformis* by *Lactobacillus plantarum* VTT E-78076 and *Pediococcus pentosaceus* VTT E-90390. The added heat-treated cultures of *L. plantarum* E5 and *Leuc. mesenteroides* A27 were reported to prevent the growth of approximately 10^4 rope-producing *B. subtilis* G1 spores per cm^2 on bread slices for more than 15 days (Pepe et al., 2003). The inhibition of the rope spoilage of wheat bread was observed with the added 20–30 g of sourdough/100 g of wheat dough (Katina, K. et al., 2002). Kingamkono and co-workers (1994), Svanberg and co-workers (1992) shown that food fermentation by LAB to pH 4.0 or lower inhibited the growth of *Bacillus* as well as other pathogenic microorganisms. Suomalainen and Mäyrä-Makinen (1999) reported about the inhibitory effect of *Lactobacillus rhamnosus* LC705 against *Bacillus* spp. in bakery products. Bogovič-Matijašić and co-workers (1998) found out the antimicrobial activity of *Lactobacillus acidophilus* LF221 producing two bacteriocins against different pathogens including B. cereus. Røssland and co-workers (2003) demonstrated that a rapid decrease in pH during log phase of fermentation was related with the B. cereus growth inhibition. Meanwhile B. cereus sporulated and existed as endospores with the pH reduced at a slower rate in early log phase.

Several studies have been dedicated for the analysis of the antimicrobial activity of LAB and their produced bacteriocins, and have reported that the antifungal activity of LAB is lost after treatment with proteolytic enzymes. Batish et al. (1989) suggested that the antifungal substance produced by a LAB isolate was of proteinaceous nature since activity disappeared with proteinase treatment. Roy et al. (1996) isolated a *Lactococcus lactis* subsp. *lactis* with antagonistic activity against several filamentous fungi, that were lost after enzymatic treatment with chymotrypsin, trypsin and pronase Gourama (1997a) found that the inhibitory effect of a *Lactobacillus casei* strain against two *Penicillium* species was slightly reduced by treatment with trypsin and pepsin. Gourama and Bullerman (1995, 1997b) showed that a commercially available silage inoculant with a combination of *Lactobacillus species* (*L. plantarum, L. delbrueckii* subsp. bulgaricus and *L. acidophilus*) had antifungal and antiaflatoxin activity against *A. flavus*. The inhibitory activity was sensitive to treatments with trypsin and alpha-chymotrypsin, and it was concluded that the activity was due to a small peptide.

The changes in LAB antimicrobial effect upon the interactions with the enzymes is very important in the baking industry where commercial enzyme preparations often are used for fermentation processes intensification and recently became one of the research topics.

In one of the studies (Digaitiene et al., 2005), 270 bacterial strains were isolated from spontaneous sourdoughs and of these, five LAB (*Lactobacillus sakei* MI806, *Pediococcus pentosaceus* MI808, MI809 and MI810, *Pediococcus acidilactici* MI807) isolates were found to produce BLIS (sakacin 806, pediocin 808, 809, 810 and pediocin Ac807 respectively). Isolates of new bacteriocins producing LAB strains depend for subclass II. The results of inhibitory spectra studies and pH sensitivity analysis indicated that the BLIS under investigation were different from each other. These novel BLIS (sakacin 806, pediocin 808, pediocin 809, pediocin 810 and pediocin Ac807) have been tested for their antimicrobial activity against *B. subtilis*, one of the most important micro-organisms responsible for ropiness in bread;

furthermore their sensitivity to various baking enzymes has been examined (Narbutaite et al., 2008). Antimicrobial activity was tested using an overlay assay method; the results showed that the BLIS studied here were effective against *B. subtilis*. To our knowledge, this is the first report of BLIS-producing LAB isolated from sourdoughs which are active against *B. subtilis*.

Bacteriocins have gained importance as natural biopreservatives for the control of spoilage and pathogenic organisms in foods. Latest studies highlights the possibility of using LAB exhibiting antimicrobial activity against *B. subtilis* in sourdough bread making, a desirable characteristic when selecting for more competitive starters. The strains described here can have an impact when used as starter cultures for traditional sourdough fermentation by delaying spore germination and inhibiting the outgrowth of *B. subtilis*. This opens up the possibility of using such LAB on an industry scale. Future work will also focus on obtaining the amino acid sequences of the BLIS presented here.

4.2 Potential of lactic acid bacteria to degrade biogenic amines in different fermentation media

A variety of fermented foods especially protein-rich foods e.g. fermented vegetables, legume products, beers and wines contain biogenic amines (BAs) (Kalač et al., 2002). During the fermentation process protein breakdown products, peptides and amino acids, used by spoilage and also by the fermentation microorganisms represent precursors for BAs formation (Hernandez-Jover et al., 1997; Bodmer et al., 1999). BAs are formed through the decarboxylation of specific free amino acids by exogenous decarboxylases released from the microbial population associated with the raw material.

Some biogenic amines such as histamine (HIS), tyramine (TYR), putrescine (PUT) and cadaverine (CAD) are important for their physiological and toxicological effects on the human body. They may exert either psychoactive or vasoactive effects on sensitive humans. Histamine is physiologically the most important BA. Histamine has been found to cause the most frequent food-borne intoxications associated with BAs; it acts as a mediator and is involved in pathophysiological processes such as allergies and inflammations (Gonzaga et al, 2009). Tyramine can evoke nausea, vomiting, migraine, hypertention and headaches (Shalaby, 1996). Putrescine and cadaverine can increase the negative effect of other amines by interfering with detoxification enzymes that metabolize them (Stratton et al., 1991). The consumption of foods with high concentrations of BAs can induce adverse reactions such as nausea, headaches, rashes and changes in blood pressure (Ladero et al., 2010).

The main BAs associated with such fermented plant product as wine are HIS, TYR and PUT (Ancin-Azpilicueta et al., 2008). Their presence in wine is considered as marker molecules of quality loss, and some EU countries even have recommendations for the amount of histamine acceptable in wine which impacts on the import and export of wines to these countires. Most fermented foods, such as cheese, fermented sausages and beer, which are consumed more frequently than wines, have biogenic amine content (Fernandez et al., 2007). However, the precence of alcohol in wine may enhance the activity of amines because it inhibits monoamine oxidase enzymes. These enzymes depending for the detoxification system in the intestinal tract of mammals convert amines into non-toxic products, which are further excreted out of the organism.

Fermentation Processes Using Lactic Acid Bacteria Producing Bacteriocins for Preservation and Improving
Functional Properties of Food Products

37

Regarding fruits and vegetables relatively low levels of biogenic amines were found in fruit juice and canned fruit/vegetable samples. The same tendency has been noticed in other publications, but sometimes the results are controversial. Moret et al., (2005) showed that vegetables generally contained low levels in biogenic amines (0.1–9.6 mg kg^{-1}) while Kalač et al., (2002) found relatively high levels of the amines in vegetables (0.8–52.5 mg kg^{-1}). The polyamines PUT and spermidin (SPM) are practically ubiquitous in all vegetables at a few mg/100 g of fresh weight and TYR is less widespread in vegetables (Kalač et al., 2002; Moret et al., 2005). They are implicated in a number of physiological processes, such as cell division regulation, plant growth, flowering, fruit development, response to stress and senescence (Bouchereau et al., 2000). Moreover, although PUT, SPD and other biogenic amines are generated in low quantities in most canned vegetables/fruits, they are not the primary metabolic products produced by the fermenting organisms (Stratton et al., 1991).

With the exception of tempe (Saaid et al., 2009) and taucu, relatively low levels of biogenic amines are found in the soy bean products tested. Studies by Mower and Bhagavan (1989) showed higher level of TYR (450 mg kg^{-1}) in salted black beans. The quantitative analysis of fermented products prepared for wheat bread production revealed that tyramine (32.6–215.8 mg kg^{-1}), histamine (20.8–96.7 mg kg^{-1}), and putrescine (33.7–195.2 mg kg^{-1}) showed as being the major occurring BAs (Bartkiene et al., 2011). Since several varieties of molds, yeasts and lactic acid bacteria are involved in the fermentation processes of such products and the raw material (soy bean) contains considerable amounts of protein, the formation of various amines might be expected during the fermentation (Shalaby, 1996). Studies have shown that biogenic amines in fermented soy bean products are most likely formed by the lactic microflora that is active during fermentation (Kirschbaum et al., 2000). TYR and HIS have been found at various levels in such products (Stratton et al., 1991). The variability of biogenic amines levels in the commercial fermented soy bean products samples had been attributed to the variations in manufacturing processes; variability in the ratio of soy bean in the raw material, microbial composition, conditions and duration of fermentation (Shalaby, 1996).

Knowledge concerning the origin and factors involved in BAs production in fermented products e.g. wine is well documented, and recently several reviews on this topic have been published (Costantini et al., 2009; Moreno-Aribas and Polo, 2010). They are generated either as the result of endogenous decarboxylase-positive microorganisms in raw materials or by the growth of contaminating decarboxylase-positive microorganisms in fermented products. With regards to wine microorganisms, a large amount of literature is available on the production of BAs. Several research group support the view that biogenic amines are formed in winemaking mainly by lactic acid bacteria (LAB) due to the decarboxilation of the free amino acids (Constantini et al., 2006; Lucas et al., 2008). The levels of BAs usually increase during fermentation due to decarboxylase activity of the LAB used as starter culture. Low acid conditions, such as those occurring during fermentation, favour the decarboxylation of amino acids (De las Rivas et al., 2005). The levels of free amino acids usually increase in fermented products during fermentation due to the action of endogenous and exogenous proteases through proteolysis processes (Hughes et al., 2002). It is thought that proteolysis might provide the nutrient for spoilage microorganisms, leading to a promoted growth of those microorganisms (Riebroy et al., 2004).

In this context, recently published paper (Garcia et al., 2011) reports novel data about the presence of histamine-, tyrosine- and putrescine-degrading enzymatic activities of wine-associated LAB. Of particular interest are the results concerning the degradation of putrescine, since no such degrading ability of any food LAB has previously been reported. The isolates tested (42 strains *Oenococcus oeni*, 7 strains *Pediococcus parvulus*, 4 strains *P. pentosaceus*, 6 strains *Lactobacillus plantarum*, 9 strains *L. hilgardi*, 3 strains *L. zeae*, 7 strains *L. casei*, 5 strains *L. paracasei* and 2 strains Leuconostoc mesenteroides) belong to the principal species of wine LAB and other related ecosystems and were selected because they came from wine cellars that often suffer from the problem of BAs in their wines (Moreno-Aribas and Polo, 2010). In this study the most potent amine-degrading species detected were *L. plantarum*, *P. parvulus* and, in particular, *P. pentosaceus* and *L. casei*, in spite of the fact that strains of these last species have never be reported to degrade histamine, tyramine and/or putriscine. None of the strains were able to produce these BAs as they did no show the decarboboxylase activity necessary for the production of these compounds in wine. However, this potential for histamine, tyramine and/or putrescine degradation among wine LAB does not appear to be very frequent, since out of the 85 strains examined, only nine displayed noteworthy amine-degrading activity in culture media. Further studies using other LAB species and/or strains may enable more potent amine-degrading enzyme producers to be identified. However, it was observed that positive strains displayed amine-degrading activity against several biogenic amines simultaneoustly, in accordance with previous works that also reported the presence of either one or two amine oxidases in other food fermenting microorganisms, such as *Micrococcus varians* and *Staphylococcuscarnosus* (Leuschner et al., 1998).

The fact that active bacteria which were able to significantly reduce the concentration of BAs in the conditions used in the study came from different fermentation media such as young wine, wood- aged wines, sherry wines (Table 1), suggest that there are ecological niches for the isolation of potential amine-degrading bacteria.

Strains	Degradation, % [a,b]		
	Histamine	Tyramine	Putrescine
L. casei IFI-CA-52	54	55	65
L. hilgardi IFI-CA-41	n.e.	n.e.	20
L. plantarum IFI-CA-26	33	n.e.	24
L. plantarum IFI-CA-54	23	17	24
O. oeri IFI-CA-32	12	n.e.	16
P. parvulus IFI-CA-30	20	15	53
P. pentosaceous IFI-CA-30	10	12	49
P. pentosaceous IFI-CA-83	19	22	39
P. pentosaceous IFI-CA-86	n.e.	54	69

[a] Activity is expressed as a percentage of control without strain and according to HPLC quantitative biogenic amine results.
[b] Mean value (n=3); n.e. : no effect was observed.

Table 1. Percentage of degradation of the biogenic amines (histamine, tyramine and putrescine) by wine-associated LAB in culture media

Recently, homofermentative *Pediococcus acidilactici* were isolated from spontaneous rye sourdoughs and characterised as producing pediocin Ac807 with antimicrobial activity

Fermentation Processes Using Lactic Acid Bacteria Producing Bacteriocins for Preservation and Improving
Functional Properties of Food Products

39

against *Bacillus subtilis* (Digaitiene et al., 2005; Narbutaite et al., 2008). Since fermentation by using *P. acidilactici* could improve or modify flavour, taste, and texture of used plant additives for wheat bread production, the safety characteristics of fermented products rich in proteins are not always predictable. Therefore the BAs investigation in untreated whole lupine flours and fermented products of different lupines species (*Lupinus angustifolius* and *Lupinus luteus*) after spontaneous fermentation and fermentation by *P. acidilactici* has been carried out (Bartkiene et al., 2011). This study showed that the BAs levels were found to be lower after fermentation (by 17%) of *L. luteus* flour by *P. acidilactici*. Also the total amount of BAs was significantly reduced (25%) during spontaneous fermentation of *L. luteus* flour compared to non-treated samples. Opposite to the *L. luteus*, fermentation of *L. angustifolius* flour led to an increase of 20.5% and 44% of total amount of BAs in spontaneous and *P. acidilactici* sourdoughs, respectively. The different BAs in fermented products, which have been prepared using different kinds of lupine can be explained by the quality of the raw material and /or by formation during the fermentation process involving microorganisms and these results are in agreement with previous research (Silla Santos, 1996).

The presented results agree to the findings that the use of decarboxylase negative microorganisms, e.g. LAB as starter cultures could be an important factor to be considered in order to reduce the levels of BAs in fermented foods.

4.3 Possible approaches of LAB and enzymes to biodegradation of mycotoxins

Mycotoxins are secondary metabolites produced by a wide variety of filamentous fungi, including species from the genera *Aspergillus, Fusarium* and *Penicillium*. They cause nutritional losses and represent a significant hazard to the food and feed chain. Humans have long been exposed to mycotoxins by several different routes: directly, via foods of plant origin, including cereals from which bread and bakery products are derived; by air (both indoors and outdoors); or indirectly, through foods of animal origin. Many countries, therefore, have established measures to safeguard the health of consumers by establishing regulations for food and feed. The most economically important mycotoxins occurring in food and feed are *aflatoxins, ochratoxin A, patulin,* and the *Fusarium toxins* (zearalenon, trichothecenes, fumonisins etc.) (Chassy, 2010).

Principally, there are three possibilities to avoid harmful effect of contamination of food and feed caused by mycotoxins: (1) prevention of contamination, (2) decontamination of mycotoxin-containing food and feed, and (3) inhibition of absorption of mycotoxin content of consumed food into the digestive tract (Halász et al., 2009).

The theoretically soundest approach of prevention is doubtless to breed cereals and other food and feed plants for resistance to mould infection and consequently exclude mycotoxin production. Particularly in breeding wheat and corn, significant improvement of resistance has been achieved. The identification of microbial species (and genes coding enzymes degrading mycotoxins) allows transfer of these genes into plants and production of such enzymes by transgenic plants. In this way, the safety problems connected with the use of live microorganisms may be avoided.

Another practical approach to prevention of mycotoxin contamination is the inhibition of the growth of molds and their production of mycotoxins. First, optimal harvesting, storage and processing methods, and conditions may be successful in prevention of mold growth.

Although the primary goal is the prevention of mycotoxin contamination, mycotoxin formation appears to be unavoidable under certain adverse conditions.

Treatment of grains by some chemicals to prevent mycotoxin formation is also possible. Most of these compounds work by inhibiting fungal growth. For example, approximately one hundred compounds have been found to inhibit aflatoxin production. Two extensively studied inhibitors of aflatoxin synthesis are dichlorvos (an organophosphate insecticide) and caffeine. As reported by (Halász et al., 2009) some surfactants have been found to suppress the growth of *Aspergillus flavus* and aflatoxin synthesis.

When contamination cannot be prevented, physical and chemical decontamination methods have been employed with varying success in the past, principally for feed. Whichever decontamination strategy is used, it must meet some basic criteria:

- the mycotoxin must be inactivated or destroyed by transformation to non-toxic compounds;
- fungal spores and mycelia should be destroyed, so that new toxins are not formed;
- the food or feed material should retain its nutritive value and remain palatable;
- the physical properties of raw material should not change significantly; and
- it must be economically feasible / the cost of decontamination should be less than the value of contaminated commodity.

Partial removal of mycotoxin may be achieved by dry cleaning of the grain and in the milling process, as well. Milling led to a fractionation, with increased level of mycotoxin in bran and decreased level in flour. The majority of mycotoxins are heat-stable so heat treatment, usually applied in food technology, does not have significant effect on mycotoxin level.

Efforts were made in several countries to find an economically acceptable way of destruction of mycotoxins into non-toxic products using different chemicals such as alkali and oxidative agents. Although such treatment reduces nearly completely the mycotoxin concentration, these chemicals also cause losses of some nutrients and such treatment is too drastic for grain destined for food uses.

Although the different methods used at present have been to some extent successful, most methods have major disadvantages, starting with limited efficacy to losses of important nutrients and generally with high costs.

More recently, biological decontamination and biodegradation of mycotoxins with microorganisms or enzymes, have been used (He et al., 2010; Juodeikiene et al., 2011). In this case no harmful chemicals where used, so no significant losses in nutritive value and palatability of decontaminated food and feed occurred. Today, ruminants appear to be a promising potential source of microbes or enzymes for use in the biotransformation of mycotoxins.

One of the most frequently used strategies for biodegradation of mycotoxins includes isolation of microorganisms able to degrade the given mycotoxin and treatment of food or feed in an appropriate fermentation process. From the food safety point-of-view, fermentation with microorganisms commonly used in food production (fermentation with lactic acid bacteria, alcoholic fermentation, traditional fermentation of vegetable protein

Fermentation Processes Using Lactic Acid Bacteria Producing Bacteriocins for Preservation and Improving
Functional Properties of Food Products

41

used in South Asia, etc.) should be preferred. Knowledge of enzymes that take part in degradation of mycotoxins opens some new approaches: (1) the production of genetically modified species of microorganisms commonly used in food production and their use for production of enzymes mentioned above; or (2) the transfer of genes coding for these enzymes to transgenic plants and use the plants for production of mycotoxin degrading enzymes.

In staple food such as bread and bakery products in the flour sector, yeast and lactobacilli now play an important role. It thus stands to reason that the same microbes and enzymes are the first to have been considered for use as detoxifying or decontaminating agents. This type of biodegradation could therefore prove a useful strategy for partially overcoming the problem of some mycotoxins. Indeed, this already takes place in bread and in sourdough processes (Bartkiene et al., 2008); and OTA in food can also undergo biodegradation (Abrunhosa et al., 2010) and certain antagonistic yeast strains can substantially degrade OTA. This might offer new possibilities for reducing this mycotoxin in bread and bakery products and their raw materials (Patharajan et al., 2011). The use of enzymes or engineered micro-organisms (provided that these are allowed by legislation) as processing aids in the bread and bakery sector would also prove beneficial. Genetic engineering technologies will improve the efficiency with which enzymes can be produced from these organisms, and will allow the production of engineered organisms which have the target genes. They will additionally increase the availability and bioavailability, and will improve the quality of the end product.

4.3.1 Inhibition of mycotoxins biosynthesis by lactic acid bacteria

Several papers dealing with the inhibition of mycotoxin biosynthesis by LAB have focused on aflatoxins (Thyagaraja & Hosono, 1994). During cell lysis, it is possible that LAB releases molecules that potentially inhibit mould growth and therefore lead to a lower accumulation of their mycotoxins (Gourama & Bullerman, 1995). These "anti-mycotoxinogenic" metabolites could also be produced during LAB growth. Gourama (1991), using a dialysis assay, demonstrated the occurrence of a metabolite that inhibits aflatoxin accumulation in *Lactobacillus* cell-free extracts. It was suggested that this inhibition of aflatoxin biosynthesis was not the result of a hydrogen peroxide production or a pH decrease (Karunaratne et al., 1990). These findings were consistent with those of Gourama (1991), who suggested that inhibition of aflatoxin biosynthesis by *Lactobacillus* cell free supernatants was probably due to specific bacterial metabolites. Coallier-Ascah and Idziak (1985) reported a significant reduction of aflatoxin biosynthesis by *Lactobacillus* cell free supernatants and suggested that this inhibition was related to a heat stable, low-molecular-weight inhibitory compound. Although *Lactobacillus* spp. were found to delay aflatoxin biosynthesis, other lactic strains such as *Lc. lactis* were found to stimulate aflatoxin accumulation (Luchese and Harrigan, 1990).

4.3.2 Decontamination of mycotoxins using microorganisms by binding or degradation

Biological detoxification of mycotoxins works mainly *via* two major processes, sorption and enzymatic degradation, both of which can be achieved by biological systems. Live micro-organisms can absorb either by attaching the mycotoxin to their cell wall components or by active internalization and accumulation. Dead microorganisms too can absorb mycotoxins, and this phenomenon can be exploited in the creation of biofilters for fluid decontamination or probiotics (which have proven binding capacity) to bind and remove the mycotoxin from the intestine.

Another approach to the biological decontamination of mycotoxins involves their degradation by selected micro-organisms. Recently critical review on biological detoxification by (Dalié et al., 2009) summarized different and interesting aspects of the biological detoxification of mycotoxins.

Micro-organism detoxification can be performed in many different ways (Magan and Olsen, 2004):

- The entire organism can be used as a starter culture, as in the fermentation of beer, wine and cider, or in lactic acid fermentation of vegetables, milk and meat.
- The purified enzyme can be used in soluble or immobilized (biofilter) forms.
- The gene encoding the enzymatic activity can be transferred and overexpressed in a heterologous system; interesting candidates for this application include yeasts, probiotics and plants.

Biological methods have been applied for the biodegradation and decontamination of different mycotoxins.

Aflatoxins. As the first mycotoxins to be discovered, were also the first target in screening for microbial degradation. Several examples of the detoxification of the most common and important mycotoxins are reviewed. Almost 40 years ago, several species of micro-organisms – including yeasts, moulds, bacteria, actinomycetes and algae – were screened for detoxification activity; based on this studies only one isolate was found, *Flavobacterium auranotiacum*, which significantly removed aflatoxin from a liquid medium (Ciegler et al., 1966).

Later aflatoxins decontamination during fermentation was reported in several cases. About 50% reduction in aflatoxins B1 and G1 has been reported during an early stage of miso fermentation. It was attributed to the degradation of the toxin by micro-organisms. Significant losses of aflatoxin B1 and ochratoxin A were observed during beer brewing (Chu *et al.*, 1975). Detoxification of aflatoxin B1 occurred during the fermentation of milk by LAB and in dough fermentation during breadmaking. Digestive tract micro-organisms are able to reduce mycotoxin levels not only by binding and removal but also by detoxification.

Most data dealing with the effects of LAB on the accumulation of mycotoxins are related to aflatoxin-producing moulds. Wiseman and Marth (1981) revealed the existence of an amensalism relationship between *Lc. lactis* and *A. parasiticus*. When these authors added the spores of *A. parasiticus* to a 13-day-old culture of *Lc. lactis*, they observed the entire repression of aflatoxin production. When the fungal spore suspension and the lactic strain were inoculated simultaneously, an increase in aflatoxin production was observed. In contrast, Coallier-Ascah and Idziak (1985) showed an inhibition of aflatoxin accumulation when both microorganisms were simultaneously cultivated in Lab-Lemco tryptone broth (LTB). Addition of glucose to the cultivation medium during the conidiation phase of the mould did not restore the the production of aflatoxin.

Several LAB have been found to be able to bind aflatoxin B_1 *in vitro* (Kankaanpää et al., 2000; Gratz et al., 2004), with an efficiency depending on the bacterial strain (Shah & Wu, 1999). El-Nezami with co-workers (1998a) have evaluated the ability of five *Lactobacillus* to bind aflatoxins *in vitro* and have shown that probiotic strains such as *Lb. rhamnosus* GG and *Lb. rhamnosus* LC-705 were very effective for removing aflatoxin B_1, with more than 80% of the toxin trapped in a 20µg/ml solution (Haskard et al.,1998).

According to Coallier-Ascah and Idziak (1985), the inhibition of aflatoxin accumulation was not related to a pH decrease but rather to the occurrence of a low-molecular-weight metabolite produced by the LAB at the beginning of its exponential phase of growth. Inhibition of aflatoxin production by other LAB belonging to the genus *Lactobacillus* was also reported (Karunaratne et al., 1990). It was assumed that this inhibition resulted from the production of a metabolite different from hydrogen peroxide or organic acid (Gourama, 1991). Haskard wich co-wokers (2001) demonstrated that *Lb. rhamnosus* GG (ATCC 53103) and *Lb. rhamnosus* LC-705 (DSM 7061) were able to eliminate aflatoxin B_1 from the culture medium by a physical process.

Several studies have suggested that the antimutagenic and anti-carcinogenic properties of probiotic bacteria can be attributed to their ability to non-covalently bind hazardous chemical compounds such as aflatoxins in the colon (El-Nezami et al., 1998b; Gratz et al., 2004). Both viable and non-viable forms of the probiotic bacterium *Lactobacillus rhamnosus* GG effectively removed aflatoxin B1 from an aqueous solution (El-Nezami et al., 1998b). Since metabolic activation is not necessary, binding can be attributed to weak, non-covalent, physical interactions, such as association to hydrophobic pockets on the bacterial surface (Haskard et al., 2000). Coallier-Ascah and Idziak (1985) reported a significant reduction of aflatoxin biosynthesis by *Lactobacillus* cell free supernatants and suggested that this inhibition was related to a heat stable, low-molecular-weight inhibitory compound.

Ochratoxin-A (OTA) The major OTA producers in food and feed products are considered to be *A. alliaceus, A. carbonarius, A. ochraceus, A. steynii, A. westerdijkiae, P. nordicum* and *P. verrucosum* (Frisvad et al., 2006). These are mainly associated with agricultural crops pre-harvest, or in post harvest storage situations. Biological methods use microorganisms, which can decompose, transform or adsorb OTA to detoxify contaminated products or to avoid the toxic effects when mycotoxins are ingested. These are the technologies of choice for decontamination proposes because they present several advantages from being mediated by enzymatic reactions. For example, they are very specific, efficient, environmentally friendly, and they preserve nutritive quality.

Two pathways may be involved in OTA microbiological degradation (Abrunhosa et al., 2010). First, OTA can be biodegraded through the hydrolysis of the amide bond that links the L-β-phenylalanine molecule to the OTα moiety. Since OTα and L-β-phenylalanine are virtually non-toxic, this mechanism can be considered to be a detoxification pathway. Second, a more hypothetical process involves OTA being degraded via the hydrolysis of the lactone ring. In this case, the final degradation product is an opened lactone form of OTA, which is of similar toxicity to OTA when administered to rats. However, it is less toxic to mice and *Bacillus brevis*. Although this is hypothetical, it is likely to occur since microbiological lactonohydrolases, which undertake a similar transformation, are common. Several protozoal, bacterial, yeast and filamentous fungal species are able to biodegrade OTA. After success in clarifying the mechanism and degradation products of ochratoxin, three directions in recent research may be observed (1) possibilities of bacterial degradation, study of molds able to degrade this mycotoxin and identification and isolation of enzymes taking part in the degradation process.

Lactobacillus strains were demonstrated to eliminate 0.05 mg OTA/L added to culture medium - in particular, *L. bulgaricus*, *L. helveticus*, *L. acidophillus*, eliminated up to 94%, 72% and 46%, respectively, of OTA (Böhm et al., 2000); *L. plantarum*, *L. brevis* and *L. sanfrancisco* were reported to eliminate 54%, 50% and 37%, respectively, of 0.3 mg OTA/L after 24 h of incubation (Piotrowska and Zakowska, 2000). It is now generally accepted that OTA adsorption to the cells walls is the predominant mechanism involved in this OTA detoxification phenomenon by lactic acid bacteria (LAB). For example, adsorption effects were claimed by Turbic et al. (2002), who found that heat and acid treated cells from two *Lactobacillus rhamnosus* strains were more effective at removing OTA from phosphate buffer solutions than viable cells. The strains removed 36% to 76% in the buffer solution (pH 7.4) after 2 h at 37°C. Similarly, Piotrowska and Zakowska (2005) verified that *L. acidophilus* and *L. rhamnosus* caused OTA reductions of 70% and 87% of 1 mg OTA/L after five days at 37°C, and that significant levels of the OTA were present in the centrifuged bacteria cells. Other LAB (*L. brevis*, *L. plantarum* and *L. sanfranciscencis*) also produced smaller decreases on OTA (approximately 50%). Finally, Del Prete et al. (2007) tested 15 strains of oenological LAB in order to determine the *in vitro* capacity to remove OTA, and reported *Oenococcus oeni* as the most effective, with OTA reductions of 28%. The involvement of cell-binding mechanisms was confirmed as (i) up to 57% of the OTA absorbed by the cells was recovered through methanol extraction from the bacteria pellets; (ii) crude cell-free extracts were not able to degrade OTA; and (iii) degradation products were not detected. Nevertheless, some authors consider that metabolism may also be involved. For example, Fuchs et al. (2008) confirmed that viable cells of *L. acidophilus* removed OTA more efficiently then unviable. A *L. acidophilus* strain was able to decrease ≥95% the OTA in buffer solutions (pH 5.0) containing 0.5 and 1 mg OTA/L when incubated at 37°C for 4 h. In addition, a detoxification effect was also demonstrated since pre-incubation of OTA with this strain reduced OTA toxicity to human derived liver cells (HepG2) (Fuchs et al., 2008). Other *L. acidophilus* strains demonstrated only a moderate reduction in OTA contents suggesting that the effect was strain specific. In summary, some LAB adsorbs OTA by a strain specific cell-wall binding mechanism, although some undetected catabolism can also be involved. The detection of this OTA catabolism may only be possible with radiolabeled OTA.

The potential of LAB as mycotoxin decontaminating agents has been studied in different fermentation processes and reviewed (Shetty and Jespersen, 2006). The ochratoxin-A content, its fate during wine-making and possibilities of its degradation have been intensively studied. Overviews concerning presence and fate of this mycotoxin in grapes, wine and beer were published by Mateo et al. (2007) and Varga and Kozakiewicz (2006). Although the decrease of OTA content in liquid phase during vinification process is observed by the majority of researchers, reports are controversial regarding the mechanism of OTA removal. Is it a result of malolactic fermentation due to the action of lactic acid bacteria (Kozakiewicz et al., 2003), or is it adsorption to yeast cell walls (Binder et al., 2000)? Reports about the capacity of proteolytic enzymes to hydrolyze OTA can also be found.

Furthermore, although the results of these studies look very promising for reducing OTA contamination, studies on model systems do not guarantee the degradation of OTA *in situ*,

Fermentation Processes Using Lactic Acid Bacteria Producing Bacteriocins for Preservation and Improving
Functional Properties of Food Products

45

using food or feed. Further studies are needed to characterize the products of degradation and to investigate the activity of these bacteria in food and feedstuffs.

Patulin contamination of apple and other fruit-based foods and beverages is an important food safety issue due to the high consumption of these commodities. Patulin contamination is considered of greatest concern in apples and apple products; however, this mycotoxin has also been found in other fruits, such as pears, peaches, strawberries, blueberries, cherries, apricots and grapes as well as in cheese (Halász et al., 2009). The initial studies concerning degradation of patulin by actively fermenting yeasts were reported in the 1979 by Stinson et al. (1979). However, authors were not able to chemically characterize the products of degradation. More recently, Moss and Long (2002) reported that under fermentative conditions, the commercial yeast *Saccharomyces cerevisiae* transformed patulin into ascladiol.

In a recent study (Richelli et al., 2007) the ability of *Gluconobacter oxydans* to degrade patulin was investigated and the degradation products of this mycotoxin determined. More than 96% of patulin was degraded after twelve-hour treatment, due to change of chemical structure (opening of the pyran ring). The degradation product was confirmed to be ascladiol. The genus *Gluconobacter*, whose taxonomy is at present under worldwide study, is made up of five different species (Tanasupawat et al., 2004; Sievers et al., 1995) which have no health risk, and that are commonly used in food manufacturing. Apple juice inoculated with this bacterium and incubated for 3 days still tasted like juice and was drinkable. However, keeping in mind the toxicity of ascladiol and eventual unsatisfactory organoleptic properties of alcoholic apple (fruit) juice (apple wine), the use of this bacterium at the industrial level needs additional investigation.

In screenings for patulin detoxifying bacteria, has been isolated a bacterium from fermented sausage; it was identified as *Lactobacillus plantarum*, and it significantly reduced patulin levels via an intracellular enzyme (Halász et al., 2009).

Fusarium **toxins** Considerable amounts of the *Fusarium* mycotoxins zearalenone (ZEN) and its derivative α-zearalenol, were bound *in vitro* to the probiotic bacteria *L. rhamnosus* GG and *L. rhamnosus* LC705. Both heat-treated and acid-treated bacteria were capable of removing the toxins, indicating that binding, not metabolism is the mechanism by which the toxins are removed from the media (El-Nezami et al., 2002). Zearalenone was also degraded by a mixed bacterial culture. A few other microbial activities that transform zearalenone have been published but are protected by patents. Several micro-organisms have been found that can degrade DON and T-2. On the basis of morphological and phylogenetic studies, the degrader strain was classified as a bacterium belonging to the *Agrobacterium Rhizobium* group. Interactions between lactic strains and ZEN and its derivative, β-zearalenol were also investigated.

DON levels did not change during beer malting and the amount of trichothecene did not change during wine alcoholic fermentation. In contrast, trichothecene and iso-trichothecin were decomposed during alcoholic fermentation of grape juice. It was suggested that the yeast epihydroxylase might be involved.

A significant proportion (38–48%) of both toxins was trapped in the bacterial pellet and no degradation product of zearalenone or α-zearalenol was detected (El-Nezami et al., 2002), leading to the conclusion that binding and not metabolism was the mechanism by which the toxins were removed from the media. Similar results were obtained with other mycotoxins including ochratoxin A (Del Prete et al., 2007; Fuchs et al., 2008) and fumonisin B_1 and B_2 (Niderkorn et al., 2006).

Therefore, two specific processes such as binding and inhibition of biosynthesis may be involved in the interaction between LAB and the accumulation of some mycotoxins.

Concerning the mechanisms of action involved in the removal of fumonisins by LAB, Niderkorn (2007) suggested that peptidoglycans were the most plausible fumonisin binding sites. The quenching ability of LAB was increased when bacteria were killed using different physical and chemical treatments, while lysozyme and mutanolysin enzymes that target peptidoglycans partially inhibited it. It was also reported that tricarballylic acid chains found in fumonisin molecules played an important role in the binding process since hydrolysed fumonisin had less affinity for LAB, and free amine group inactivation had no effect on the binding process (Niderkorn, 2007). The same article attempted to explain the low affinity of fumonisin B_1 using a molecular modelling approach. In fact, an additional hydroxyl group in fumonisin B_1 could form a hydrogen bond with one of the tricarballylic acid chains, resulting in a spatial configuration where the tricarballylic acid chain is less available to interact with bacterial peptodoglycans.

Removal of fumonisins by LAB was ascribed to adhesion to cell wall components rather than covalent binding or metabolism, since the dead cells fully retained their binding ability. Peptidoglycans probably play a key rule in this binding process. Therefore, elucidating the differences between bacterial cell wall components of LAB strains might make it possible to select LAB species with the potential to act as biopreservative agents capable of reducing exposure from fumonisins that occur in food and feed.

5. Conclusions

The use of bacteriocins and/or bacteriocin-producing strains of LAB are of great interest as they are generally recognized as safe organisms and their antimicrobial products as biopreservatives. Several studies confirmed that microorganisms and enzymes could be a practical way to reduce the concentrations of some contaminants and to avoid the toxic effects via bioremediation. This opens up wide possibilities of using such biotechnological means on an food industry scale. Further experiments like utilization of the strains in fermentation processes or using the enzyme preprations can exhibit great outcomes. Great source of BLIS producing LAB strains could be traditional fermentation processes which should be more widely distributed as novel microorganisms crossing the borders. Further development of these strains for the biopreservation of food products requires an understanding of the mechanisms of action of the antimicrobial activity and of the decontamination of certain contaminants e.g. biogenic amines and mycotoxins. It seems that, according to results of the experiments to date, microorganisms are the main living organisms applicable for the biodegradation of these contaminants. Progress in this field of molecular biology techniques, antimicrobial LAB strains with multi-functional properties, including the degradation of mycotoxins, can be engineered to significantly improve the quality, safety and acceptability of plant foods. Further studies and knowledge of enzymes taking part in mycotoxin degradation allows production of these enzymes and their use for detoxification instead of microorganisms or additionally with LAB strains. Despite the intensive research in this field and of the numerous publications that confirm the ability of various microorganisms to degrade mycotoxins, lack of results achieved until now in the development of practical commercial technologies by using BLIS producing strains hampered progress. The majority of experiments were carried out in model systems and in laboratory conditions. The control of degradation products and the effects of detoxification on nutritive and sensory properties is in every case a decisive part of research and potential application. The use of antifungal LAB instead of chemical preservatives would enable the food industry to meet the request of consumers for natural products. Finally a practical technology should be developed and controlled from an economical point of view. Several studies confirmed that different environmental factors have strongly affected on the growth and bacteriocin production of several LAB strains that are mainly considered for food applications. However, it is desirable to continue the studies on this subject to select the most efficient factors and their combinations, which could be used on a production scale. Future work on LAB bacteriocin production, purification, obtaining the amino acid sequences of the BLIS to increase their activity is required.

6. References

Abrunhosa, L., Paterson, R.R.M., Venâncio, A. 2010. Biodegradation of ochratoxin A for food and Feed Decontamination. *Toxins 2*, 1078-1099.

Ancin-Azpilicueta, C., Gonzalez-Marco, A., Jimenez-Mareno, N. 2008. Current knowledge about the precense of biogenic amines in wine. Critical Reviews in *Food Science and Nutrition* 48, 257–275.

ANVISA – Brazilian Agency of Sanitary Surveillance. Food with health claims, new foods/ ingredients, bioactive compounds and probiotics. http://www.anvisa.gov.br/alimentos/comissoes/tecno_lista_alega.htm. Accessed May 28, 2010.

Axelsson, L. Katla, T., Bjornslett, M., Vincent, Eijsink, G.H., Holck, A. 1988. A system for heterologous expression of bacteriocins in *Lactobacillus sakei*. *FEMS Microbiology Letters* 168 (1), 137–143.

Bailey, C.P., Von Holy, A. 1993. *Bacillus* spore contamination associated with commercial bread manufacture. *Food Microbiology* 10 (4), 287–294.

Batish, V.K., Grover, S., Lal, R. 1989. Screening lactic starter cultures for antifungal activity. *Cultured Dairy Products Journal* 24, 21–25.

Castellano, P., Farías, M.E., Holzapfel, W., Vignolo, G. 2001. Sensitivity variations of *Listeria* strains to the bacteriocins, lactocin 705, enterocin CRL35 and nisin. *Biotechnology Letters* 23 (8), 605–608.

Chassy, B.M. 2010. Food safety risk and consumer health. *New Biotechnology* 27 (5), 534-544.

Cintas, L.M., Casaus, M.P., Herranz, C., Nes, I.F., Hernández, P.E. 2001. Review: Bacteriocins of lactic acid bacteria. *Food Science and Techology International* 7 (4), 281–305.

Clarke, C.I., Schober, T.J., Dockery, P., O'Sullivan, K., Arendt, E.K. 2004. Wheat sourdough fermentation: effects of time and acidification on fundamental rheological properties. *Cereal Chemistry 81* (3), 409–417.

Coallier-Ascah, J., Idziak E., 1985. Interaction between *Streptococcus lactis* and *Aspergillus flavus* on production of aflatoxin. *Applied and Environmental Microbiology* 49, 163–167.

Collins, N.E., Kirschner, L.M., Von Holy, A. 1991. Characterization of *Bacillus* isolates from ropey bread, bakery equipment and raw materials. *South African Journal of Science* 87, 62–66.

Condón, S. Palop, A., Raso, J., Sala, F.J. 1996. Influence of the incubation temperature after heat treatment upon the estimated heat resistance values of spores of *Bacillus subtilis*. *Letters in Applied Microbiology* 22 (2), 149–152.

Constantini, A., Cersosimo, M., Del Prete, V., Garcia-Moruno, E. 2006. Production biogenic amines by lactic acid bacteria: screening by PSR, thin chromatography and high-performance liquid chromatography of strains isolated from wine and must. *Journal of Food Protection* 69, 391–396.

Corsetti, A. De Angelis, M,, Dellaglio, F., Paparella, A., Fox, P.F., Settanni, L., Gobbetti, M. 2003. Characterization of sourdough lactic acid bacteria based on genotypic and cell-wall protein analyses. *Journal of Applied Microbiology* 94 (4), 641–654.

Corsetti, A., Settanni, L., Van Sinderen, D. 2004. Characterization of bacteriocin-like inhibitory substances (BLIS) from sourdough lactic acid bacteria and evaluation of their in vitro and in situ activity. *Journal of Applied Microbiology* 96 (3), 521–534.

Costantini, A., Vaudano, E., Prete, W.D., Danei, M., Garcia-Maruno, E., 2009. Biogenic amine production by contaminating bacteria found in starter preparations used in winemaking. *Journal of Agricultura and Food Chemistry* 57, 10664-10669.

Cotter, P.D., Hill, C., Ross, R.P. 2005. Bacteriocins: developing innate immunity for food. *Nature Reviews Microbiology* 3 (10), 777–788.

Cruz, A.G., Antunes, A.E.C., Sousa, A.L.O.P., Faria, J.A.F., Saad, S.M.I. 2009. Ice-cream as a probiotic food carrier. *Food Research International* 42, 1233–1239.

Cuozzo, S.A., Sesma, F., Palacios, J.M., de Ruíz Holgado, A.P., Raya, R.R. 2000. Identification and nucleotide sequence of genes involved in the synthesis of lactocin 705, a two-peptide bacteriocin from *Lactobacillus casei* CRL 705. *FEMS Microbiology Letters* 185 (2), 157–161.

Bartkiene, E., Juodeikiene, G., Vidmantiene, D., Viskelis, P., Urbonaviciene, D. 2011. Nutritional and quality aspects of wheat sourdough bread using *L. luteus* and *L. angustifolius* flours fermented by *Pedioccocus acidilactici*. *International Journal of Food Science and Technology* 46, 1724–1733.

Bartkiene, E., Juodeikiene, G., Vidmantiene, D. 2008. Evaluation of deoxynivalenol in wheat by acoustic method and impact of starter on its concentration during wheat bread baking process. *Food Chemistry and Technology* 42 (1), 5–12.

Binder, E.M, Heidler, D., Schatzmayr, G., Thimm, N., Fuchs, E., Schuh, M., Krska, R., Binder, J. 2000. Microbial detoxification of mycotoxins in animal feed. In *Mycotoxins and Phytotoxins in Perspective at the Turn of the Millenium. Proceedings of the 10-th International IUPAC Symposium on Mycotoxins and Phytotoxins;* De Koe, W.J., Samson, R.A., Van Egmond, H.P., Gilbert, J., Sabino, M., Eds; Garuja, Brazil, May 20–25, IUPAC, 271–277.

Bouchereau, A., Guenot, P., Larher, F. 2000. Analysis of amines in plant materials. *Journal of Chromatography* B 747, 49–67.

Bodmer, S., Imark, C., Kneubühl, M. 1999. Biogenic amines in foods: histamine and food processing. *Inflammation Research* 48, 296–300.

Fermentation Processes Using Lactic Acid Bacteria Producing Bacteriocins for Preservation and Improving
Functional Properties of Food Products

49

Bogovic-Matijasic, B., Rogelj, I. 1998. Bacteriocin complex of *Lactobacillus acidophilus* LF221 – production studies in MRS–media at different pH–values and effect against *Lactobacillus helveticus* ATCC 15009. *Process Biochemistry* 33, 345–352.

Bogovič-Matijašić, B. Rogelj, I,, Nes, I.F., Holo, H. 1998. Isolation and characterization of two bacteriocins of *Lactobacillus acidophilus* LF221. *Applied Microbiology and Biotechnology* 49 (5), 606–612.

Böhm, J., Grajewski, J., Asperger, H., Rabus, B., Razzazi, E. 2000. Study on biodegradation of some trichothecenes (NIV, DON, DAS, T-2) and ochratoxin A by use of probiotic microorganisms. *Mycological Research* 16, 70–74.

Ciegler, A., Lillehoj, B., Peterson, H.H. 1966. Microbial detoxification of aflatoxin. *Journal of Applied Microbiology* 14, 934–939.

Dalié, D.K.D., Deschamps, A.M., Richard-Forget, F., 2009. A review: Lactic acid bacteria – Potential for control of mould growth and mycotoxins. *Food Control* 21 (4), 370–380.

De Las Rivas, B., Marcobal, A., Muñoz, R. 2005. Improvedmultiplex-PCR method for the simultaneous detection of foodbacteria producing biogenic amines. *FEMS Microbiology Letters* 244, 367–372.

Delgado, A., Brito, D., Peres, C., Arroyo-L'opez, F.N., Garrido-Fern'andez, A. 2005. Bacteriocin production by *Lactobacillus pentosus* B96 can be expressed as a function of temperature and NaCl concentration. *Food Microbiology* 22, 521–528.

Delgado, A., Arroyo-L'opez, N.F., Brito D., Peres, C., Fevereiro, P., Garrido-Fern'andez, A. 2007. Optimum bacteriocin production by *Lactobacillus plantarum* 17.2b requires absence of NaCl and apparently follows a mixed metabolite kinetics. *Journal of Biotechnology* 130, 193–201.

Delves-Broughton, J. 1990. Nisin and its uses as a food preservative. *Food Technology* 44, 100-117.

Deraz, S.F. Karlsson, E.N., Khalil, A.A., Mattiasson, B. 2007. Mode of action of acidocin D20079, a bacteriocin produced by the potential probiotic strain, *Lactobacillus acidophilus* DSM 20079. *Journal of Industrial Microbiology and Biotechnology* 34 (5), 373–379.

De Vuyst, L., Vancanneyt, M. 2007. Biodiversity and identification of sourdough lactic acid bacteria. *Food Microbiology* 24 (2), 120–127.

Del Prete, V., Rodriguez, H., Carrascosa, A.V., Rivas, B.D.L., Garcia-Moruno, E., Munoz, R. 2007. In vitro removal of ochratoxin A by wine lactic acid bacteria. *Journal of Food Protection* 70, 2155–2160.

Digaitiene, A., Hansen, Å., Juodeikiene, G., Josephsen, J. 2005. Microbial population in Lithuanian spontaneous rye sourdoughs. *Ecology and Technology* 5 (77), 193–198.

Driks, A. 2002. Overview: Development in bacteria: spore formation in *Bacillus subtilis*. *Cellular and Molecular Life Sciences* 59 (3), 389–391.

El-Nezami, H.S., Kankaanpää, P., Salminen, S., Ahokas, J. 1998a. Ability of dairy strains of lactic acid bacteria to bind a common food carcinogen, aflatoxin B1. *Food and Chemical Toxicology* 36, 321–326.

El-Nezami, H.S., Kankaanpää, P., Salminen S., Ahokas, J. 1998b. Physiochemical alterations enhance the ability of dairy strains of lactic acid bacteria to remove aflatoxin from contaminated media. *Journal of Food Protection* 61, 446–448.

El-Nezami, H.S., Polychronaki, N., Salminen, S., Mykkänen H. 2002. Binding rather metabolism may explain the interaction of two food-grade *Lactobacillus* strains with zearalenone and its derivative α-zearalenol. *Applied and Environmental Microbiology* 68, 3545–3549.

Eichenberger, P., Fujita, M., Jensen, S.T., Conlon, E.M., Rudner, D.Z., Wang, S.T., Ferguson, C., Haga, K., Sato, T., Liu, J.S., Losick, R. 2004. The program of gene transcription for a single differentiating cell type during sporulation in *Bacillus subtilis*. *Public Library of Science* 2 (10), 328.

Eijsink, V.G., Axelsson, L., Diep, D.B., Håvarstein, L.S., Holo, H., Nes, I.F. 2002. Production of class II bacteriocins by lactic acid bacteria; an example of biological warfare and communication. *Antonie Van Leeuwenhoek* 81 (1-4), 639–654.

Ennahar, S., Aoude-Werner, D., Sorokine, O., Van Dorsselaer, A., Bringel, F., Hubert, J.C., Hasselmann, C. 1996. Production of pediocin AcH by *Lactobacillus plantarum* WHE 92 isolated from cheese. *Applied and Environmental Microbiology* 62 (12), 4381-4387.

Errington, J. 2003. Regulation of endospore formation in *Bacillus subtilis*. *Nature Reviews Microbiology* 1 (2), 117–126.

FAO-WHO 2002. Food and agriculture organization of the United Nations, World Health Organization. Report of a joint FAOWHO working group on drafting guidelines for the evaluation of probiotics in food. Ontario. ftp://ftp.fao.org/docrep/fao/009/a0512e/a0512e00.pdf

Fernandez, M., Linares, D.M., Rodriguez, A., Alvarez, M.A. 2007. Factors affecting tyramine production in *Enteroccus durans* IPLA 655. *Applied Microbiology and Biotechnology* 73, 1400-1406.

Frisvad, J.C., Thrane, U., Samson, R.A., Pitt, J.I. 2006. Important mycotoxins and the fungi which produce them. *Advances in Food Mycology* 571, 3–31.

Fuchs, S., Sontag, G., Stidl, R., Ehrlich, V., Kundi, M., Knasmuller, S. 2008. Detoxification of patulin and ochratoxin A, two abundant mycotoxins, by lactic acid bacteria. *Food and Chemical Toxicology* 46, 1398–1407.

Gálvez, A., Valdivia, E., Martínez-Bueno, M., Maqueda, M. 1990. Induction of autolysis in *Enterococcus faecalis* S-47 by peptide AS-48. *Journal of Applied Bacteriology* 69, 406–413.

Galvez, A., Lopez, R.L., Abriouel, H. 2008. Application of bacteriocins in the control of food-borne pathogenic and spoilage bacteria. *Critical Reviews in Biotechnology* 28, 125–152.

Gänzle, M.G. 1998. Useful Properties of *Lactobacilli* for application as protective cultures in food. *PhD thesis*. University of Hohenheim, Germany.

Garcia-Ruiz, A., Gonzalez-Rompinelli, E.M., Bartolome, B., Moreno-Arribas, M.V. 2011. Potential of wine–associated lactic acid bacteria to degrade biogenic amines. *International Journal of Microbiology* 148, 115–120.

Gillor, O., Etzion, A., Riley, M.A. 2008. The dual role of bacteriocins as anti- and probiotics. *Applied Microbiology and Biotechnology* 81 (4), 591–606.

Gobbetti, M. *et al.* 2005. Biochemistry and physiology of sourdough lactic acid bacteria. *Trends in Food Science and Technology* 16 (1-3), 57–69.

Gonzaga, V.E., Lescano, A.G., Huaman, A.A., Salmon-Mulanovich, G., Blazes, D.I. 2009. Histamine levels in fish from markets in Lima, Peru. *Journal of Food Protection* 72, 1112-1115.

Fermentation Processes Using Lactic Acid Bacteria Producing Bacteriocins for Preservation and Improving
Functional Properties of Food Products

51

Gourama, H. 1997a. Inhibition of growth and mycotoxin production of *Penicillium* by *Lactobacillus* species. *Lebensmittel- Wissenschaft und-Technologie* 30, 279–283.

Gourama, H., Bullerman, L.B. 1997b. Anti-aflatoxigenic activity of *Lactobacillus casei* pseudoplantarum. *International Journal of Food Microbiology* 34, 131–143.

Gourama, H., Bullerman, L.B. 1995. Inhibition of growth and aflatoxin production of *Aspergillus flavus* by *Lactobacillus* species. *Journal of Food Protection* 58, 1249–1256.

Gourama, H. 1991. Growth and aflatoxin production of *Aspergillus flavus* in the presence *Lactobacillus* species. *Ph.D. thesis*, University of Nebraska-Lincoln.

Gratz, S., Mykkänen, H., Ouwehand, A.C, Juvonen, R., Salminen S., El-Nezami H.S. 2004. Intestinal mucus alters the ability of probiotic bacteria to bind aflatoxin B1 *in vitro*. *Applied and Environmental Microbiology* 70, 6306–6308.

Grossman, A.D., Losick, R. 1988. Extracellular control of spore formation in *Bacillus subtilis*. *Proceedings of the National Academy of Sciences* 85 (12), 4369–4373.

Harlander, S.K. 1993. Bacteriocins of Lactic Acid Bacteria, Academic Press, San Diego, CA, pp. 233–247.

Halász, A., Lásztity, R., Abonyi, T., Bata, A. 2009. Decontamination of mycotoxin-containing food and feed by biodegradation. *Food Reviews International* 25, 284–298.

Hammes, W.P., Gänzle, M.G. 1998. Sourdough breads and related products. In Wood, B.J.B. Eds. *Microbiology of Fermented Foods (Volume 1)*. London: Chapman & Hall, UK., pp. 199– 216.

Hansen, Å.S. 2004. Sourdough bread. In Hui et al. Eds., *Handbook of Food and Beverage Fermentation Technology*, Marcel Dekker Inc., Florida, USA, pp. 729–755.

Harris, L.J. Daeschel, M.A., Stiles, M.E, Klaenhammer, T.R. 1989. Antimicrobial activity of lactic acid bacteria against *Listeria monocytogenes*. *Journal of Food Protection* 52 (6), 384–387.

Haskard, C.A., El-Nezami, H.S., Peltonen, K.D., Salminen, S., Ahokas, J.T. 1998. Sequestration of aflatoxin B1 by probiotic strains: Binding capacity and localization. *Revue de Medecine Veterinaire* 149, 571.

Haskard, C.A., Binnion, C., Ahokas, J. 2000. Factors affecting the sequestration of aflatoxin by *Lactobacillus rhamnosus* strain GG. *Chemico-Biological Interactions* 128, 39–49.

Haskard, C.A., El-Nezami, H.S., Kankaanpää, P.E., Salminen, S., Ahokas, J.T. 2001. Surface binding of aflatoxin B1 by lactic acid bacteria. *Applied and Environmental Microbiology* 67, 3086–3091.

Hassan, Y.I., Bullerman, L.B. 2008. Antifungal activity of Lactobacillus paracasei ssp. tolerans isolated from a sourdough bread culture. *International Journal of Food Microbiology* 121 (1), 112–115.

He, J., Zhou, T., Young, J.C., Boland, G.J., Scott, P.M. 2010. Chemical and biological transformations for detoxification of trichothecene mycotoxins in human and animal food chains: a review. *Trends in Food Science & Technology* 21 (2), 67–76.

Heng, N.C.K., Wescombe, P.A., Burton, J.P., Jack, R.W., Tagg, J.R. 2007. The Diversity of bacteriocins in Gram-positive bacteria. In Riley, M.A., Chavan, M., Eds., *Bacteriocins: ecology and evolution*. Springer Berlin Heidelberg, New York, USA, pp. 45–93.

Hernandez-Jover, T., Izquierdo-Pulido, M., Veciana-Nogues, M.T., Marine-Font, A., Vidal-Carou, M.C. 1997. Biogenic amines and polyamine contents in meat and meat products. *Journal of Agricultural Food Chemistry* 45, 2098–2102.

Holzapfel, W.H., Geisen, R., Schillinger, U. 1995. Biological preservation of foods with reference to protective cultures, bacteriocins and food-grade enzymes. *International Journal of Food Microbiology* 24, 343–362.

Hoover, D.G., Chen, H. 2005. Bacteriocins with potential for use in foods. In Davidson, P.M., Sofos, J.N., Branen, A.L. Eds., *Antimicrobials in Food (3rd edition)*. Taylor & Francis Group, LLC, FL, USA, pp. 389–428.

Hoover, D.G., Dishart, K.J., Hermes, M.A. 1989. Antagonistic effect of *Pediococcus* spp. against *Listeria monocytogenes*. *Food Biotechnology* 3 (2), 183–196.

Hugas, M. Pagés, F., Garriga, M., Monfort, J.M. 1998. Application of the bacteriogenic *Lactobacillus sakei* CTC494 to prevent growth of *Listeria* in fresh and cooked meat products packed with different atmospheres. *Food Microbiology* 15 (6), 639–650.

Hughes, M.C., Kerry, J.P., Arendt, E.K., Kenneally, P.M., McSweeney, P.L.H., O'Neill, E.E. 2002. Characterization of proteolysis during the ripening of semi-dry fermented sausages. *Meat Science* 62, 205–216.

Jack, R.W., Tagg, J.R., Ray, B. 1995. Bacteriocins of Gram-positive bacteria. *Microbiological Reviews* 59, 171–200.

Jiménez-Díaz, R., Rios-Sanchez, R.M., Desmazeaud, M., Ruiz-Barba, J.L., Piard, J.C. 1993. Plantaricin S and T, two new bacteriocins produced by *Lactobacillus plantarum* LPCO10 isolated from a green olive fermentation. *Applied and Environmental Microbiology* 59, 1416–1424.

Johnsen, L. Fimland, G., Eijsink, V., Nissen-Meyer, J. 2000. Engineering increased stability in the antimicrobial peptide pediocin PA-1. *Applied and Environmental Microbiology* 66 (11), 4798–4802.

Joshi, V.K., Sharma, S., Ranaet, N.S. 2006. Bacteriocin from lactic acid fermented vegetables. *Food Technology and Biotechnology* 44 (3), 435–439.

Juodeikiene, G., Salomskiene, J., Eidukonyte, D., Vidmantiene, D., Narbutaite, V., Vaiciulyte-Funk, L. Impact of novel fermented products on the base of extruded wheat material on the quality of wheat bread, *Food Technology and Biotechnology, 2011* (Article in Press).

Juodeikiene, G., Basinskiene, L., Vidmantiene, D., Makaravicius, T., Bartkiene, E. Benefits of β-xylanase for wheat biomass conversion to bioethanol. *Journal of the Science of Food and Agriculture*, 2011 Jul 11. (Article in Press).

Ladero, V., Calles- Enríquez, M., Fernández, M., Alvarez, M.A. 2010. Toxicological effects of dietary biogenic amines. *Current Nutrition and Food Science* 6, 145–156.

Larsen, A.G., Nørrung, B. 1993. Inhibition of *Listeria monocytogenes* by bavaricin A, a bacteriocin produced by *Lactobacillus bavaricus* MI401. *Letters in Applied Microbiology* 17 (3), 132– 134.

Leal-Sánchez, M.V., Jiménez-Díaz, R., Maldonado- Barragán, A., Garrido-Fernández, A., Ruiz-Barba, J.L., 2002. Optimization of bacteriocin production by batch fermentation of *Lactobacillus plantarum* LPCO10. *Applied and Environmental Microbiology* 68, 4465–4471.

Leal, M.V., Baras, M., Ruiz-Barba, J.L., Floriano, B., Jiménez-Díaz, R. 1998. Bacteriocin production and competitiveness of *Lactobacillus plantarum* LPCO10 in olive juice broth, a culture medium obtained from olives. *International Journal of Food Microbiology* 43, 129–134.

Leer, R.J., Van der Vossen, J.M.B.M., Van Giezen, M., Van Noort Johannes, M., Pouwels, P.H. 1995. Genetic analysis of acidocin B, a novel bacteriocin produced by *Lactobacillus acidophilus*. *Microbiology* 141 (7), 1629–1635.

Leroy, F., Verluyten, J., Messens, W., De Vuyst, L. 2002. Modeling contributes to the understanding of the different behaviour of bacteriocin-producing strains in a meat environment. *International Dairy Journal* 12, 247–253.

Leuschner, R.S., Heidel, M., Hammes, W.P. 1998. Histamine and tyramine degradation by food fermenting microorganisms. *International Journal of Food Microbiology* 39, 1–10.

Luchese, R.H., Harrigan, W.F. 1990. Growth of and aflatoxin production by *Aspergillus parasiticus* when in the presence of either *Lactococcus lactis* or lactic acid and at different initial pH values. *Journal of Applied Bacteriology* 69, 512–519.

Lucas, P.M., Gaisse, O., Lonvaud-Funel, A. 2008. High frequency of histamine producing bacteria in the enological environment and instability of the histidine decarboxylase production phenotype. *Applied and Environmental Microbiology* 74, 811–817.

Kalač, P., Šavel, J., Križek, M., Pelikánová, T., Prokopová, M. 2002. Biogenic amine formation in bottled beer. *Food Chemistry* 79, 431–434.

Kalač, P., Švecova, S., Pelikánová, T. 2002. Levels of biogenic amines in typical vegetable products. *Food Chemistry* 77, 349–351.

Kankaanpää, P., Tuomola, E., El-Nezami, H., Ahokas, J., Salminen, S.J. 2000. Binding of aflatoxin B1 alters the adhesion properties of *Lactobacillus rhamnosus* strain GG in Caco-2 model. *Journal of Food Protection* 63, 412–414.

Katina, K., Arendt, E., Liukkonen, K. H., Autio, K., Flander, L., Poutanen, K. 2005. Potential of sourdough for healthier cereal products. *Trends in Food Science and Technology* 16, 104–112.

Katina, K., Sauri, M., Alakomi, H.L., Mattila-Sandholm, T. 2002. Potential of lactic acid bacteria to inhibit rope spoilage in wheat sourdough bread. *Lebensmittel-Wissenschaft und-Technologie* 35 (1), 38–45.

Karunaratne, A., Wezenberg, E., Bullerman, L.B. 1990. Inhibition of mold growth and aflatoxin production by *Lactobacillus* spp. *Journal of Food Protection* 53, 230–236.

Kim, W.S., Hall, R.J., Dunn, N.W. 1997. The effect of nisin concentration and nutrient depletion on nisin production of *Lactococcus lactis*. *Applied Microbiology and Biotechnology* 50, 429– 433.

Kingamkono, R., Sjögren, E., Svanberg, U., Kaijser, B. 1994. pH and acidity in lactic-fermenting cereal gruels – effects on viability of enteropathogenic microorganisms. *World Journal of Microbiology and Biotechnology* 10, (6), 664–669.

Kirschbaum, J., Rebscher, K., Bruckner, H. 2000. Liquid chromatographic determination of biogenic amines in fermented foods after derivatization with 3,5-dinitrobenzoyl chloride. *Journal of Chromatography A* 881, 517–530.

Klaenhammer, T.R. 1993. Genetics of bacteriocins produced by lactic acid bacteria. *FEMS Microbiology Reviews* 12, 39–86.

Kozakiewicz, Z., Battilani, P., Cabanes, I., Venancio, A., Mule, G., Tjamos, E. Making wine safer. In *Meeting the Mycotoxin Menace*, van Egmond, H., van Osenbruggen, T., Lopez Garcia, R., Visconti, A., Eds., Wageningen Academic Publisher: Wageningen, 2003 pp.131–140.

Kramer, J., Gilbert, R. *Bacillus cereus* and other *Bacillus* sp. In Doyle, M.P., Eds., *Foodborne Bacterial Pathogens*. Marcel Dekker Inc., New York, USA, 1989, pp. 22–70.

Kunst, F. *et al.* 1997. The complete genome sequence of the Gram-positive bacterium *Bacillus subtilis*. *Nature* 390 (6657), 249–256.

Mao, Y., Muriana, P.M., Cousin, M.A. 2001. Purification and transpositional inactivation of lacticin FS92, a broad-spectrum bacteriocin produced by *Lactococcus lactis* FS92. *Food Microbiology* 18 (2), 165–175.

Marín, S., Guynot, M.E., Neira, P., Bernadó, M., Sanchis, V., Ramos, A.J. 2002. Risk assessment of the use of sub-optimal levels of weak-acid preservatives in the control of mould growth on bakery products. *International Journal of Food Microbiology* 79 (3), 203–211.

Marugg, J.D., Gonzalez, C.F., Kunka, B.S. Ledeboer, A.M., Pucci, M.J., Toonen, M.Y., Walker, S.A., Zoetmulder, L.C., Vandenbergh, P.A. 1992. Cloning, expression, and nucleotide sequence of genes involved in production of pediocin PA-1, and bacteriocin from *Pediococcus acidilactici* PAC1.0. *Applied and Environmental Microbiology* 58 (8), 2360–2367.

Martınez-Anaya, M.A. 1996. Enzymes and bread flavor. *Journal of Agricultural and Food Chemistry* 44, 2469–2480.

Martínez-Cuesta, M.C., Requena, T., Peláez, C. 2006. Cell membrane damage induced by lacticin 3147 enhances aldehyde formation in *Lactococcus lactis* IFPL730. *International Journal of Food Microbiology* 109 (3), 198–204.

Martinez, R.C.R., De Martinis, E.C.P. 2006. Effect of *Leuconostoc mesenteroides* 11 bacteriocin in the multiplication control of *Listeria monocytogenes*. *Ciznia e Tecnologia de Alimentos* 26 (1), 52–55.

Mateo, R., Medina, A., Mateo, E.M., Mateo, F., Jimenez, M. 2007. An overview of ochratoxin A in beer and wine. *International Journal of Food Microbiology 119* (1–2), 79–83.

Moreno-Aribas, M.V., Polo M.C. 2010.Wine Chemistry and Biochemistry. In *Mycotoxins in food Detection and control: Biological decontamination of mycotoxins*, Magan, N. and Olsen, M., Eds., Springer New York., Woodhead Publishing Ltd and CRC Press LLC, pp. 2006–211.

McKay, L.L., Baldwin, K.A. 1990. Application for biotechnology: present and future improvements in lactic acid bacteria. *FEMS Microbiology reviews 7*, 3–14.

McAuliffe, O., Hill, C., Ross, R.P. 1999. Inhibition of *Listeria monocytogenes* in cottage cheese manufactured with a lacticin 3147-producing starter culture. *Journal of Applied Microbiology* 86 (2), 251–256.

Mendoza, F., Maqueda, M., Gálvez, A., Martínez-Bueno, M., Valdivia, E. 1999. Antilisterial activity of peptide AS-48 and study of changes induced in the cell envelope properties of an AS-48-adapted strain of *Listeria monocytogenes*. *Applied and Environmental Microbiology* 65, 618–625.

Mentes, Ö., Ercan, R., Akçelik, M. 2007. Inhibitor activities of two *Lactobacillus* strains, isolated from sourdough, against rope-forming *Bacillus* strains. *Food Control* 18 (4), 359–363.

Messens,W., De Vuyst, L. 2002. Inhibitory substances produced by lactobacilli isolated from sourdoughs – a review. *International Journal of Food Microbiology* 72, 31–43.

Messens, W., Neysens, P., Vansieleghem, W., Vanderhoeven, J., De Vuyst, L. 2002. Modeling growth and bacteriocin production by *Lactobacillus amylovorus* DCE 471 in response to temperature and pH values used for sourdough fermentations. *Applied and Environmental Microbiology* 68 (3), 1431–1435.

Miller, K.W., Schamber, R., Osmanagaoglu, O., Ray, B. 1998. Isolation and characterization of pediocin AcH chimeric protein mutants with altered bactericidal activity. *Applied and Environmental Microbiology* 64 (6), 1997–2005.

Minei, C.C., Gomes, B.C., Ratti, R.P., D'Angelis, C.E.M., De Martinis, E.C.P. 2008. Influence of peroxyacetic acid and nisin and coculture with *Enterococcus faecium* on *Listeria monocytogenes* biofilm formation. *Journal of Food Protection* 71, 634–638.

Morency, H., Mota-Meira, M., LaPointe, G., Lacroix, C., Lavoie, M.C. 2001. Comparison of the activity spectra against pathogens of bacterial strains producing a mutacin or a lantibiotic. *Canadian Journal of Microbiology* 47 (4), 322–331.

Moreno, I., Lerayer, A.L.S., Baldini, V.L.S., Leitao, M.F.F. 2000. Characterization of bacteriocins produced by *Lactococcus lactis* strains. *Brazilian Journal of Microbiology* 31, 184–192.

Moret, S., Smela, D., Populin, T., Conte, L.S. 2005. A survey on free biogenic amine content of fresh and preserved vegetables. *Food Chemistry* 89, 355–361.

Moretro, T., Aassen, I.M., Storro, I., Axelsson, L. 2000. Production of sakacin P by *Lactobacillus sakei* in a completely defined medium. *Journal of Applied Microbiology* 88, 536–545.

Mørtvedt, C.I., Nes, I.F. 1990. Plasmid-associated bacteriocin production by a *Lactobacillus sake* strain. *Journal of General Microbiology* 136 (8), 1601–1607.

Moss, M.O., Long, M.T. 2002. Fate of patulin in the presence of the yeast *Saccharomyces cerevisiae*. *Food Additives & Contaminants* 19, 387–399.

Mota-Meira, M., Morency, H., Lavoie, M.C. 2005. In vivo activity of mutacin B-Ny266. *Journal of Antimicrobial Chemotherapy* 56 (5), 869–871.

Mower, H. F., Bhagavan, N.V. 1989. Tyramine content of Asian and Pacific foods determined by high performance liquid chromatography. *Food Chemistry* 31, 251–257.

Narbutaite, V., Fernandez, A., Horn, N., Juodeikiene, G., Narbad, A. 2008. Influence of baking enzymes on antimicrobial activity of five bacteriocin-like inhibitory substances produced by lactic acid bacteria isolated from Lithuanian sourdoughs. *Letters in Applied Microbiology* 47 (6), 555–560.

Nes, I. F., Holo, H. 2000. Class II antimicrobial peptides from lactic acid bacteria. *Biopolymers* 55 (1), 50–61.

Nes, I. F., Diep, D.B., Håvarstein, L.S., Brurberg, M.B., Eijsink, V., Holo, H. 1996. Biosynthesis of bacteriocins in lactic acid bacteria. *Antonie Leeuwenhoek* 70, 113–128.

Neysen, P., De Vuyst, L. 2005. Kinetic and modeling of sourdough lactic bacteria. *Trends in Food Science & Technology* 16, 95–103.

Niderkorn, V., Boudra, H., Morgavi, D.P. 2006. Binding of *Fusarium* mycotoxins by fermentative bacteria *in vitro*. *Applied and Environmental Microbiology* 101, 849–856.

Niderkorn, V. 2007. Activites de biotransformation et de séquestration des fusariotoxines chez les bactéries fermentaires pour la détoxification des ensilages de maïs. *PhD thesis*, Blaise Pascal University, France.

Oppegård, C., Fimland, G., Thorbæk, L., Nissen-Meyer, J. 2007. Analysis of the two-peptide bacteriocins Lactococcin G and Enterocin 1071 by site-directed mutagenesis. *Applied and Environmental Microbiology* 73 (9), 2931–2938.

Östman, E.M., Nilsson, M., Elmstahl, H., Molin, G., Bjorck, I. 2002. On the effect of lactic acid on blood glucose and insulin responses to cereal products: mechanistic studies in healthy subjects and in vitro. *Journal of Cereal Science* 36 (3), 339–346.

Parada, J.L., Caron, C.R., Bianchi, A., Medeiros, P., Soccol, C. R. 2007. Bacteriocins from lactic acid bacteria: purification, properties and use as biopreservatives. *Brazilian Archives of Biology and Technology* 50, 521–542.

Patharajan, S., Reddy, K.R.N., Karthikeyan, V., Spadaro, D., Lore, A., Gullino, M.L., Garibaldi, A. 2011. Potential of yeast antagonists on invitro biodegradation of ochratoxin A. *Food Control* 22, 290–296.

Pattison, T.L., Lindsay, D., von Holy, A. 2004. Natural antimicrobials as potential replacements for calcium propionate in bread. *South African Journal of Science 100* (7–8), 342–348.

Pepe, O., Blaiotta, G., Moschetti, G., Greco, T., Villani, F. 2003. Rope-producing strains of *Bacillus* spp. from wheat bread and strategy for their control by lactic acid bacteria. *Applied and Environmental Microbiology* 69 (4), 2321–2329.

Piggot, P.J., Hilbert, D.W. 2004. Sporulation of *Bacillus subtilis*. *Current Opinion in Microbiology* 7 (6), 579–586.

Piotrowska, M., Zakowska, Z. 2000. The biodegradation of ochratoxin A in food products by lactic acid bacteria and baker's yeast. In *Progress in Biotechnology (Food Biotechnology)*; Bielecki, S., Tramper, J., Polak, J., Eds., Elsevier, Amsterdam, The Netherlands, Volume 17, pp. 307–310.

Piotrowska, M., Zakowska, Z. 2005. The elimination of ochratoxin A by lactic acid bacteria strains. *Polish Journal of Microbiology* 54, 279–286.

Phillips, Z.E., Strauch, M.A. 2002. *Bacillus subtilis* sporulation and stationary phase gene expression. *Cellular and Molecular Life Sciences* 59 (3), 392–402.

Powell, J.E., Todorov, S.D., van Reenen, C.A., Dicks, L.M.T., Witthuhn, R.C. 2006. Growth inhibition of *Enterococcus mundtii* in Kefir by *in situ* production of bacteriocin ST8KF. *Le Lait* 86, 401–405.

Rasch, M., Knøchel, S. 1998. Variations in tolerance of *Listeria monocytogenes* to nisin, pediocin PA-1 and bavaricin A. *Letters in Applied Microbiology* 27 (5), 275–278.

Ravel, J., Fraser, C.M. 2005. Genomics at the genus scale. *Trends in Microbiology* 13 (3), 95–97.

Richelli, A., Baruzzi, F., Solfrizzo, M., Morea, M., Fanizzi, F.P. 2007. Biotransformation of patulin by *Gluconobacter oxydans*. *Applied and Environmental Microbiology* 73, 785–792.

Riebroy, S., Benjakul, S., Visessanguan, W., Kijrongrojana, K., Tanaka, M. 2004. Some characteristics of commercial Som-fug produced in Thailand. *Food Chemistry* 88, 527–535.

Fermentation Processes Using Lactic Acid Bacteria Producing Bacteriocins for Preservation and Improving
Functional Properties of Food Products

57

Röcken, W. 1996. Applied aspects of sourdough fermentation. *Advances in Food Sciences* 18 (5–6), 212–216.

Rogers, L.A. 1928. The inhibitory effect of *Sreptococcus lactis* on *Lactobacillus bulgaricus*. *Journal of Bacteriology* 16, 321–325.

Rosenkvist, H., Hansen, Å. 1995. Contamination profiles and characterization of *Bacillus* species in wheat bread and raw materials for bread production. *International Journal of Food Microbiology* 26 (3), 353–363

Røssland, E., Andersen Borge, G.I., Langsrud, T., Sørhaug. T. 2003. Inhibition of *Bacillus cereus* by strains of *Lactobacillus* and *Lactococcus* in milk. *International Journal of Food Microbiology* 89 (2–3), 205–212.

Ross, R.P., Morgan, S., Hill, C. 2002. Preservation and fermentation: past, present and future. *International Journal of Food Microbiology* 79, 3–16.

Ross, R.P., Stanton, C., Hill, C., Fitzgerald, G.F., Coffey, A. 2000. Novel cultures for cheese improvement. *Trends in Food Science and Technology* 11 (3), 96–104.

Roy, U., Batish, V.K., Grover, S., Neelakantan, S. 1996. Production of antifungal substance by *Lactococcus lactis* subsp. *lactis* CHD-28.3. *International Journal of Food Microbiology* 32, 27–34.

Ryan, L.A.M., Dal Bello, F., Arendt, E.K. 2008. The use of sourdough fermented by antifungal LAB to reduce the amount of calcium propionate in bread. *International Journal of Food Microbiology* 125 (3), 274–278.

Saaid, M., Saad, B., Hashim, N.H., Ali, M.A.S., Saleh, M.I. 2009. Determination of biogenic amines in selected Malaysian food. *Food Chemistry* 113, 1356–1362.

Sahl, H.G., Bierbaum., G. 1998. Lantibiotics: biosynthesis and biological activities of uniquely modified peptides from Gram-positive bacteria. *Revista de Microbiologia* 52, 41–79.

Sadeghi, A. 2008. The secrets of sourdough: A review of miraculous potentials of sourdough in bread shelf life. *Biotechnology* 7 (3), 413–417.

Settani, L., Massitti, O., Van Sinderen, D., Corsetti, A. 2005. In situ activity of a bacteriocin – producing *Lactococcus lactis* strain. Influence on the interactions between lactic acid bacteria during sourdough fermentation. *Journal of Applied Microbiology* 99, 670–681.

Schillinger, U., Lücke, F.K. 1989. Antibacterial activity of *Lactobacillus sake* isolated from meat. *Applied and Environmental Microbiology* 55 (8), 1901-1906.

Schleifer, K.H., Kraus, J., Dvorak, C., Kilpper-Bälz, R., Collins, M.D., Fischer, W. 1985. Transfer of *Streptoccus lactis* and related streptoccus to the genus of *Lactococcus* gen nov. *Systematic and Applied Microbiology* 6, 183–195.

Sievers, M., Garerth, I.C., Becsh, C., Ludwig, W., Teuber, M. 1995. Phylogenetic position of *Gluconobacter*species as a coherent cluster from all *Acetobacter* species on the basis of 16S ribosomal RNA sequences. *FEMS Microbiology Letters* 126, 123–126.

Silla Santos, M.H. 1996. Biogenic amines: their importance in foods. *International Journal of Food Microbiology* 29, 213–231.

Silveira, T.F.V., Vianna, C.M.M., Mosegui, G.B.G. 2009. Brazilian legislation for functional foods and the interface with the legislation for other food and medicine classes: contradictions and omissions. *Physis Revista de Saúde Coletiva* 19, 1189–1202.

Simova, E.D., Beshkova, D.B., Dimitrov, Z.P. 2009. Characterization and antimicrobial spectrum of bacteriocins produced by lactic acid bacteria isolated from traditional Bulgarian dairy products. *Journal of Applied Microbiology* 106 (2), 692–701.

Şimşek, Ö., Çon, A.H., Tulumoğlu, Ş. 2006. Isolating lactic starter cultures with antimicrobial activity for sourdough processes. *Food Control* 17 (4), 263–270.

Shah, N., Wu, X. 1999. Aflatoxin B$_1$ binding abilities of probiotic bacteria. *Bioscience and Microflora* 18, 43–48.

Shalaby, A.R. 1996. Significance of biogenic amines to food safety and human health. *Food Research International* 29, 675–690.

Shetty, P.H., Jespersen, L. 2006. *Saccharomyces cerevisiae* and lactic acid bacteria as potential mycotoxin decontaminating agents. *Trends in Food Science & Technology* 17, 48–55.

Sorokulova, I.B., Reva, O.N., Smirnov, V.V., Pinchuk, I.V., Lapa, S.V., Urdaci, M.C. 2003. Genetic diversity and involvement in bread spoilage of *Bacillus* strains isolated from flour and ropy bread. *Letters in Applied Microbiology* 37 (2), 169–173.

Stinson, E.E., Osman, S.F., Bills, D.D. 2006. Water soluble products from patulin during alcoholic fermentation of apple juice. *Journal of Food Science* 44 (3), 788–789.

Stratton, J.E., Hutkins R.W., Taylor S.I. 1991. Biogenic amines in cheese and other fermented foods: a review. *Journal of Food Protection* 54, 460–470.

Stauffer, C.E. 1994. Enzymes used in bakery products. Fundamentals of enzymes. *AIB Tech Bull* XVI, 1–6.

Stiles, M.E. 1996. Biopreservation by lactic acid bacteria. *Antonie van Leuwenhoek* 70, 331–345.

Svanberg, U., Sjögren, E., Lorri, W., Svennerholm, A.M., Kaijser, B. 1992. Inhibited growth of common enteropathogenic bacteria in lactic-fermented cereal gruels. *World Journal of Microbiology and Biotechnology* 8 (6), 601–606.

Suomalainen, T.H., Mäyrä-Makinen, A.M. 1999. Propionic acid bacteria as protective cultures in fermented milks and breads. *Lait* 79 (1), 165–174.

Tanasupawat, S., Thawai, C., Yukphan, P., Moonmangmee, D., Itoh, T., Adachi, O., Yamada, Y. 2004. *Gluconobacter thailandicus* sp. nov., an acetic acid bacterium in the alpfa-Proteobacteria. *The Journal of General and Applied Microbiology* 50, 159–167.

Thomas, L.V., Clarkson, M.R., Delves-Broughton, J. In *Natural food antimicrobial systems*, Thomas, L.V., Clarkson, M.R. and Delves-Broughton, J., Eds., CRC Press, A.S. Naidu, USA, 2000, pp. 463–524.

Thomas, L.V., Delves-Broughton, J. 2001. New advances in the application of food preservative nisin. *Recent Advances in Food Science* 2, 11–22.

Thompson, J.M., Dodd, C.E.R., Waites, W.M. 1993. Spoilage of bread by *Bacillus. International Biodeterioration & Biodegradation* 32 (1-3), 55–66.

Todorov, S., Gotcheva, B., Dousset, X., Onno, B., Ivanova, I. 2000. Influence of growth medium on bacteriocin production in *Lactobacillus plantarum* ST31. *Biotechnology & Biotechnological Equipment* 14, 50–55.

Todorov, S.D., Dicks, L.M.T. 2004. Effect of medium components on bacteriocin production by *Lactobacillus pentosus* ST151BR, a strain isolated from beer produced by the fermentation of maize, barley and soy flour. *World Journal of Microbiology and Biotechnology* 20, 643–650.

Fermentation Processes Using Lactic Acid Bacteria Producing Bacteriocins for Preservation and Improving
Functional Properties of Food Products

59

Todorov, S.D., Van Reenen, C.A., Dicks, L.M.T. 2004. Optimization of bacteriocin production by *Lactobacillus plantarum* ST13BR, a strain isolated from barley beer. *Journal of General and Applied Microbiology* 50, 149–157.

Todorov, S.D., Dicks, L.M.T. 2004. Influence of growth conditions on the production of a bacteriocin by *Lactococcus lactis* subsp. *lactis* ST34BR, a strain isolated from barley beer. *Journal of Basic Microbiology* 44, 305–316.

Todorov, S.D., Dicks L.M.T. 2005a. Effect of growth medium on bacteriocin production by *Lactobacillus plantarum* ST194BZ, a strain isolated from boza. *Food Technology and Biotechnology* 43, 165–173.

Todorov, S.D., Dicks L.M.T. 2005b. *Lactobacillus plantarum* isolated from molasses produces bacteriocins active against Gram-negative bacteria. *Enzyme and Microbial Technology* 36, 318–326.

Todorov, S.D., Dicks L.M.T. 2006a. Effect of medium components onbacteriocin production by *Lactobacillus plantarum* strains ST23LD and ST341LD, isolated from spoiled olive brine. *Research in Microbiology* 161, 102–108.

Todorov, S.D., Dicks L.M.T. 2006b. Medium components effecting bacteriocin production by two strains of *Lactobacillus plantarum* ST414BZ and ST664BZ isolated from boza. *Biologia* 61, 269–274.

Todorov, S.D., Nyati, H., Meincken, M., Dicks, L.M.T. 2007a. Partialcharacterization of bacteriocin AMA-K, produced by *Lactobacillus plantarum* AMA-K isolated from naturally fermented milk from Zimbabwe. *Food Control* 18, 656–664.

Todorov, S.D., Powell, J.E., Meincken, M., Witthuhn R.C., Dicks L.M.T. 2007b. Factors affecting the adsorption of *Lactobacillus plantarum* bacteriocin bacST8KF to *Enterococcus faecalis* and *Listeria innocua*. *International Journal of Dairy Technology* 60, 221–227.

Todorov, S.D. 2008. Bacteriocin production by *Lactobacillus plantarum* AMA-K isolated from Amasi, a Zimbabwean fermented milk product and study of adsorption of bacteriocin AMA-K to *Listeria* spp. *Brazilian Journal of Microbiology* 38, 178–187.

Turbic, A., Ahokas, J.T., Haskard, C.A. 2002. Selective *in vitro* binding of dietary mutagens, individually or in combination, by lactic acid bacteria. *Food Additives & Contaminants* 19, 144–152.

Thyagaraja, N., Hosono, A. 1994. Binding properties of lactic acid bacteria from 'Idly' towards food-borne mutagens. *Food and Chemical Toxicology* 32, 805–809.

Zotta, T., Parente, E., Ricciardi, A. 2009. Viability staining and detection of metabolic activity of sourdough lactic acid bacteria under stress conditions. *World Journal of Microbiology and Biotechnology* 25 (6), 1119–1124.

Valerio, F. Favilla, M., De Bellis, P., Sisto, A., de Candia, S., Lavermicocca, P. 2009. Antifungal activity of strains of lactic acid bacteria isolated from a semolina ecosystem against *Penicillium roqueforti*, *Aspergillus niger* and *Endomyces fibuliger* contaminating bakery products. *Systematic and Applied Microbiology* 32 (6), 438–448.

Valerio, F., De Bellis, P., Lonigro, S.L., Visconti, A., Lavermicocca, P. 2008. Use of *Lactobacillus plantarum* fermentation products in bread-making to prevent *Bacillus subtilis* ropy spoilage. *International Journal of Food Microbiology* 122 (3), 328–332.

Varga, J., Kozakiewicz, Z. 2006. Ochratoxin-A in grapes and grape-derived products. *Trends in Food Science & Technology* 1, 72–81.

Verellen, T.L.J., Bruggeman, G., Van Reenen, C.A., Dicks, L.M.T., Vandamme, E.J. 1998. Fermentation optimization of plantaricin 423, a bacteriocin produced by *Lactobacillus plantarum* 423. *Journal of Fermentation and Bioengineering* 86, 174–179.

Verluyten, J., Leroy, F., De Vuyst, L. 2004a. Influence of complex nitrogen source on growth of and curvacin A production by sausage isolate *Lactobacillus curvatus* LTH1174. *Applied and Environmental Microbiology* 70, 5081–5088.

Viljoen, C.R., von Holy, A. 1997. Microbial populations associated with commercial bread production. *Journal of Basic Microbiology* 37 (6), 439–444.

Volavsek, P.J.A., Kirshner, L.A.M., von Holy, A. 1992. Accelerated methods to predict the rope-inducing potential of bread raw materials. *South African Journal of Science* 87, 99–102.

Wiedemann, I., Breukink, E., van Kraaij, C., Kuipers, O.P., Bierbaum, G., de Kruijff, B., Sahl, H.G. 2001. Specific binding of nisin to the peptidoglycan precursor lipid II combines pore formation and inhibition of cell wall biosynthesis for potent antibiotic activity. *The Journal of Biological Chemistry* 276 (3), 1772–1779.

Wiseman, D.W., Marth, E.H. 1981. Growth and aflatoxin production by *Aspergillus parasiticus* when in the presence of *Streptococcus lactis, Mycopathologia* 73, 49–56.

Worobo, R.W., Van Belkum, M.J., Sailer, M., Roy, K.L., Vederas, J.C., Stiles, M.E. 1995. A signal peptide secretion-dependent bacteriocin from *Carnobacterium divergens*. *Journal of Bacteriology* 177 (11), 3143–3149.

Quintavalla, S., Parolari, G. 1993. Effects of temperature, aw and pH on the growth of *Bacillus* cells and spores: a response surface methodology study. *International Journal of Food Microbiology* 19 (3), 207–216.

Total Recycle System of Food Waste for Poly-L-Lactic Acid Output

Kenji Sakai[1], Pramod Poudel[1,2] and Yoshihito Shirai[3]

[1]Graduate School of Bioresource and Bioenvironmental Sciences,
Faculty of Agriculture, Kyushu University, Fukuoka,
[2]National College (NIST), Department of Microbiology,
Tribhuvan University, Kathmandu,
[3]Graduate School of Life Science and Systems Engineering,
Kyushu Institute of Technology, Kitakyushu, Fukuoka,
[1,3]Japan
[2]Nepal

1. Introduction

1.1 Impacts of food waste

Food waste is defined as wholesome edible material intended for human consumption arising at any point in the food supply chain that is instead discarded, lost, degraded or consumed by pests. The average consumer in Europe and North-America throws away ca. 100 kg of food per year according to a new report published by the UN's Food and Agriculture Organization (Annual report of FAO, 2011 & Parfitt et al., 2010). The study centers on food loss and food waste during the whole supply chain from production to consumption and finds that around "one-third of the edible parts of food produced for human consumption gets lost or wasted globally" representing about 1.3 billion ton per year. Around 20% of about 50 metric tons of waste that is generated annually in Japan is high moisture content refuse from kitchens and the food industries. The social, economic and environmental impacts of food waste are enormous. Such wastes readily decompose, generate odors, and sometimes cause illnesses. Municipal solid wastes including food waste are usually incinerated or land filled which ultimately generates many problems such as liberation of harmful compounds like dioxin and furans (Addink & Olie, 1995). Incineration facilities can be damaged by temperature fluctuations when food waste with high water content is burned in semi continuous process. In addition, it is difficult to recover energy from such waste incineration processes because the heating value of food waste is low (Harrison et al., 2000). This requires frequent and periodic collection and treatment of waste i.e. irrespective of their values. When excess food waste is disposed of in a landfill, it decomposes and is a significant source of methane gas, which is highly effective at trapping heat in the atmosphere than CO_2 (Camobreco et al., 1999). Annually, food waste in the United States accounted for slightly more than 100 metric tons of methane originating from landfills. At the European level, the overall environmental impact is at least 170 metric tons of CO_2 emitted annually.

In this regard, the significance is considered as an important concept for aiming at the formation of the recycle-oriented society. Accordingly, untreated food waste contributes to excess consumption of freshwater and fossil fuels which, along with methane and CO_2 emissions from decomposing food, impacts global climate change. The prompt implementation of total recycling system can play a beneficial role in the utilization of municipal waste.

The design of this system can be conducted considering not only to the environmental impacts and energy increase in the recycling but also to the best economical efficiency as sustainable bio-based materials. So, management of municipal solid waste including kitchen waste via microbiological processes improves these wastes and reduces the need for both landfill space and fuel used in waste incineration. Direct composting and methane fermentation, which produce fertilizers and biogas, are alternative ways to reuse food waste but these processes have been applied only in rural areas. On the other hand, it has been found that municipal food waste is nutritious substratum for natural lactic acid bacteria (Sakai et al., 2000b). This finding indicated another reuse route of food waste, suitable for urban areas.

	No.	Amount/Place [Kg/y]	Total Amount [t/y]	Impurity [%]	Food waste [t/y]	Total sugar [t/y]	Glucose [t/y]
Large scale retail store (1)	28	437.5	12.3	15.6	10.4	1.5	1.3
Large scale retail store (2)	124	290.0	36.0	13.1	31.3	2.2	1.8
Convenience store	299	17.6	5.3	8.2	4.9	1.6	1.1
Hotel (>100 room)	29	160.0	4.6	1.8	4.5	0.5	0.4
Hospital (>100 bed)	64	143.5	9.2	0	9.2	1.1	0.8
Department	10	542.7	5.4	12.5	4.7	0.5	0.5
Total			72.8		65.0	7.4	5.9

Table 1. Food waste recycling generated in Kitakyushu-City, Japan.

According to the report of Shirai & Sakai (2006), food waste collected from each town sectors of Kitakyushu-city of Japan, as shown in Table 1 above, accumulated 7.4 tons of sugar per year (7.4 t/y) out of which consisted 80% glucose after the treatment with very low concentration of enzyme. Further, the overall glucose generated from the food waste from house kitchen is shown in Table 2 below.

Day	Site	Food composition (%)				Water (%)	Total Sugar (g/kg wet waste)	Glucose (g/kg wet waste)
		Cereal	Fish & Meat	Vegetables	Fruits			
2003/1/21	1	17.6	10.5	55.5	16.4	76.2	115	89
2003/1/21	2	21.5	9.1	32.7	36.7	75.3	131	78
2003/1/21	3	10.0	13.7	76.3	0	72.7	93	64
2003/1/22	4	15.8	8.4	42.5	33.3	75.9	108	104
2003/1/22	5	12.4	5.1	37.9	44.6	76.5	120	76
2003/1/24	1	24.0	12.6	23.1	40.3	76.5	148	71
2003/1/24	2	20.4	28.0	30.6	21.0	80.5	108	55
2003/1/24	3	11.1	11.0	57.2	20.7	73.8	139	86
2003/1/25	4	11.0	11.0	57.4	20.5	75.2	139	77
2003/1/25	5	13.0	6.4	47.2	33.4	72.4	191	117
Average		15.7	11.6	46.0	25.1	75.5	129	83

Table 2. Composition of food refuse wasted from house kitchen at Kitakyushu area.

2. Bio-economy system

Today's industrial economy is largely dependent on petroleum oil which provides the basis of most of our energy and chemical feedstock. There is increasing concern over the impact of these traditional manufacturing processes on the environment. Therefore, considering to the resource materials' exhaustion, we need to substantially reduce our dependence in the petroleum feedstock by establishing a bio-based economy.

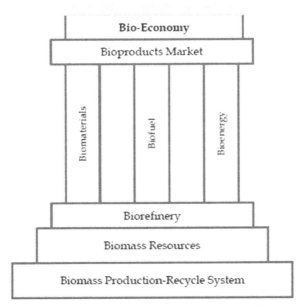

Fig. 1. Schematic Presentation of Sustainable Bio-Economy System (Revised from Kamm et al., 2005).

The bio-economy is the sustainable production and conversion of biomass for a range of food, pharmaceuticals, fiber and industrial products, and energy. In it, the renewable biomass encompasses any biological material to be used as raw material (KBBE, 2010). It helps to:

increase the scientific understanding of biomass resources and improve the tailoring of those resources; improve sustainable systems to develop, harvest and process biomass resources; improve efficiency and performance in conversion and distribution processes and technologies for the development of bio-based products; create the regulatory and market environment necessary for sustainable development and the use of bio-based products (Fig. 1). Bio-based products are virtually similar to their petroleum-based counterparts but they are manufactured from renewable resources (Kamm et al., 2005).

Generally biomass resources are strategic plant biomass rich in sugar (corn, rice, cassava, cane, beet etc) and oil (oil palm). To establish bio-economy, it is prerequisite to avoid confliction with food, cultivation field and deforestation. There are also important biomass resources in residues from agriculture and forestry including both wet and dry waste materials, for instance, sewage sludge and municipal solid waste. One of the pillars of bio-economy system is the systematic conversion of biomass resources i.e. bio-refinery (Fig. 1). Similarly, the increased production of bio-fuels, especially biodiesel from the transesterification of fats and oils from oil plants (Palm, Soybean, Jatropha etc), is making glycerin a cheap organic material. Basically, conversion of these biomass resources to useful sustainable products includes two general pathways: thermo-chemical and bio-chemical conversion pathways. Briefly, biochemical conversion pathways use microorganisms to convert biomass resources into methane, hydrogen gas, and organic acid or simple alcohols usually in combination with some mechanical or chemical pre-treatment step. Substantial research effort has been expended to make this a raw material for various organic chemicals. Not the least of these is material that can be used in various thermoplastic and thermo-set polymers. Equally important, succinic acid, a biomass derived product posits its large potential for a variety of applications. This dicarboxylic acid can be converted to a huge amount of green chemical of industrial value, such as polyesters (derived from succinic acid and butandiol) which is used for soft plastics.

3. Poly-L-lactic acid

3.1 Lactic acid

Lactic acid has both hydroxyl and carboxyl groups with one chiral carbon atom existing in two stereoisomers L- and D-lactic acid, and it is widely used in the food, pharmaceutical, and general chemical industries (Sakai et al., 2001). The L form differs from the D form in its effect on polarized light. For L-lactic acid, the plane is rotated in a clockwise (dextro) direction; whereas, the D form rotates the plane in an anticlockwise (laevo) direction. Basically, the chemical synthesis only produces the racemic mixture of the L (+) and D (-) enantiomers, while microbial fermentation using biomass resources has the advantage of producing optically pure L(+)- or D(-)-lactic acid (Hafvendahl & Hagerdal, 2000). Among basic compounds from biomass, lactic acid is relatively unique C-3 compound obtained from C-6 glucose without any oxidation-reduction of the carbon atoms. Lactic acid can be polymerized to form the biodegradable and recyclable polyester poly-lactic acid which is considered a potential substitute for plastics manufactured from petroleum (Ohara & Sawa, 1994). No doubt, we are subsequently focusing on the production of lactic acid with high optical purity from food waste (kitchen refuse).

Optical purity is measured as;

$$\text{Optical purity (\%)} = ([L] - [D]) \times 100 / ([L] + [D])$$

where [L] denotes to the concentration of L-lactic acid and [D] to that of D-lactic acid.

We found that food waste collected from commercial sectors such as retail store, convenience store, college and university contained high amount of total sugars (129g/kg) as shown in Table 1 and 2 (Shirai & Sakai, 2006). They are mainly starch and can easily be converted to glucose enzymatically (82g/kg). Subsequently, for the production of lactic acid generating glucose from the food refuse was subjected to the fermentation using *Lactobacillus rhamnosus* which produces high amount of L-lactic acid (Sakai et al., 2004a, Fig.2). Considerably, the rate of lactic acid production was more than 85% which was satisfactory with the highest yield.

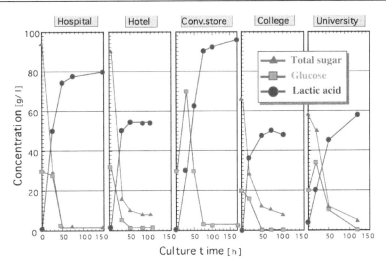

Fig. 2. Lactic acid production profile from different commodities

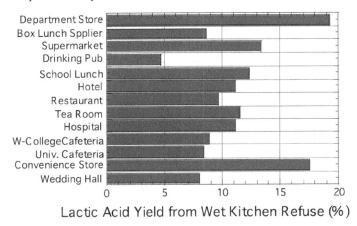

Lactic Acid Yield from Wet Kitchen Refuse (%)

Fig. 3. Lactic acid yield using various kitchen refuse from commercial sectors

3.2 Poly-lactic acid from food waste

Poly lactic acid (PLA) is thermoplastic aliphatic polyester synthesized from L- or D-Lactic acid (Fig.4). Highly optical pure L- or D-lactic acid is necessary to obtain high crystalline poly-lactic acid which leads to the high strength, chemical and heat resistances of the polymers. PLA polymers range from amorphous glassy polymers with a glass transition of 58°C to semi-crystalline/ highly crystalline products with crystalline melting points ranging from 130°C to 180°C.

We propose a novel recycling system for municipal food waste that combines fermentation and chemical processes to produce high-quality poly-L-lactate (PLLA) biodegradable plastics (Fig. 5). The process consists of removal of endogenous D- or L-lactic acid from minced food waste by a *Propionibacterium*, L-lactic acid fermentation under semisolid conditions, L-lactic acid purification via butyl esterification, and L-lactic acid polymerization via LL-lactide. The total design of the process enables a high yield of PLLA with high optical activity (i.e., a high proportion of optical isomers) and novel recycling of all materials produced at each step with energy savings and minimal emissions. Approximately, 50% of the total carbon was removed mostly as L-lactic acid and 100 kg of collected food waste yielded 7.0 kg PLLA. The physical properties of the PLLA yielded in this manner were comparable to those of PLLA generated from commercially available L-lactic acid (Table 4). Evaluation of the process is also discussed from the viewpoints of material and energy balances and environmental impacts (Fig. 5).

Although the ester bond of poly-L-lactate (PLLA) is susceptible to some enzymes, including proteinases and lipases, and PLLA has been recognized as a biodegradable plastic (Sakai et al., 2001), its biodegradation in soil is rather slow and it depends on morphology and thickness (Miyazaki & Harano, 2001). Therefore, PLLA may better be developed as a chemically recyclable plastic with an appropriate collection system for the used materials but not as a single-use plastic (Nishida et al., 2004).

Monomers and Dimers	Melting Point (°C)	Poly-lactic acid	Melting point (°C)
L-lactic acid	25.8	PLLA/PDLA Stereo-complex	245
DL-lactic acid	16.8	PLLA	245
LL-lactide	97.8	PLLA	175
Meso-lactide	52.0	PDLA	175
DL-lactide (Co-crystal)	124.0	PLDLA	50

Table 3. Melting point for Lactic acid and its Polymers

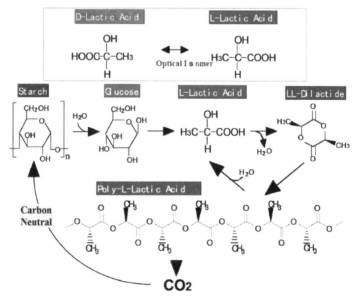

Fig. 4. Poly- L-lactic acid from starch

The proposed PLLA process has an energy advantage over even the general poly-lactide process because the feedstock is totally food waste. In the process with corn starch, nearly 30% of gross fossil energy use goes into producing and processing corn to provide dextrose to feed the lactic acid fermentation. Since the feedstock to the proposed PLLA process is food waste that must otherwise be disposed of, the only upstream fossil energy allocated to the production of PLLA would be that required for collection of the separated waste (Sakai, 2004b, 2007).

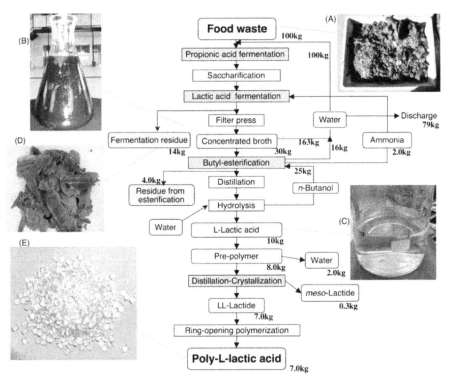

Fig. 5. Process outline of PLLA production from food waste (A), Food waste (B), concentrated broth after lactic acid fermentation (C), Purified L-lactic acid (D), Fermentation residue (E), Pellets of PLLA.

Optical purity	97.5%
Average molecular weight	200kDa
Melting point	175°C
Glass transition temperature	58°C

Table 4. Characteristics of PLLA produced from collected food waste.

The material balance and energy requirements of the total process are summarized in Table 5. The overall experimental process yielded 68.8 g PLLA from 1 kg food waste (1.0 kg PLLA/14.6 kg food waste). This means that 34% of total carbon in the food waste was recovered as PLLA. In comparison, the first commercial PLLA plant operated by Cargill Dow Polymers reportedly requires gross fossil process energy of 39.5 MJ/kg (Vink et al., 2003). Meanwhile, the process energy required for production of bottle grade polyethylene terephthalate (PET) and high-density polyethylene (HDPE) using petrochemicals is 27 MJ/kg and 23 MJ/kg respectively (Boustead, 2002). Furthermore, the process was designed to have low environmental impact. The fermentation residue is rich in nitrogen (C/N=6.5; concentrations of N, P and K were 75, 2.6, and 0.7 mg/g dry matter respectively) reduced in weight to 14% of the untreated food waste, and the precipitated residue produced at the esterification step contains high concentrations of phosphorus and potassium (C/N = 7.7; concentrations of N, P and K were 39, 28, and 23 mg/g dry matter respectively). These stable residues were confirmed to be useful fertilizers (Mori et al., 2008). Condensed water, ammonia, and butanol were reused during the process. Consequently, nearly all materials are converted to valuable resources or recycled in the process. As the production energy required is comparable to that required in the PLLA process using maize, we have been trying to improve the process especially to reduce energy required at the process of lactic acid fermentation as described below.

Besides recycling process of municipal food waste using mesophile (L. rhamnosus), the prompt utilization of its biomass as a feed additive for the animals was also proceeded to fulfill the zero emission concept (Umeki, 2004, 2005).

Items	Content and Yield		Carbon yield
	(Kg/Kg wet waste)	(Kg/Kg dry waste)	(%)
Dry material[a]	0.215	1.0	-
Carbon content[a]	0.101	0.470	100
Total soluble sugar[a,b]	0.143	0.665	-
Lactic acid in culture filtrate[c]	0.118	0.549	47
Purified L-lactc acid[c]	0.099	0.459	37
PLLA[c]	0.069	0.320	34
Fermentation residue[d]	0.14	0.101	27
Esterification residue[e]	0.04	0.038	16

[a] Average of 20 samples from 15 companies.
[b] Average concentration in saccharified samples.
[c] PLLA was experimentally produced from three representative culture filtrate samples. Average yield was calculated using efficiencies of each step (purified L-lactic acid from culture filtrate, 78.7%; PLLA from purified L-lactic acid, 91.9%).
[d] Representative data: water contents of fermentation residue and esterification residue were 38% and 6.4%, respectively.

Table 5. Product yield and carbon balance

4. Microorganisms for lactic acid production (MLAP)

4.1 Lactic acid bacteria (LAB)

The term lactic acid bacteria (LAB) means bacterial group that produces lactic acid as the major metabolite, and is used in different meaning from microorganism for lactic acid production (MLAP): they are gram-positive, acid tolerant, non-sporulating, non-respiring rod or cocci with low-GC content, able to produce L-type, D-type, or, L/D lactic acid as the major metabolic end product (more than 50%). Maximum growth temperature of it is up to 43°C. The core genera of LAB are *Lactobacillus, Leuconostoc, Pediococcus, Lactococcus,* and *Streptococcus* as well as the more peripheral *Aerococcus, Carnobacterium, Enterococcus,*

Oenococcus, Teragenococcus, Vagococcus, and *Weisella* belonging to the order *Lactobacillales.* *Lactobacillus rhamnosus* has been reported for L-lactic production from kitchen refuse (Sakai et al., 2004b). Similarly, Oh et al., (2005) used strains of *Enterococcus faecalis* for the lactic acid production from sterilized wheat hydrolysates. On the other hand, microorganism which produces   high amount L-lactic acid and is used for industrial lactic acid production (MLAP) distributed in more variety of genera in bacteria, yeast, and fungi.

4.2 Non-LAB

As non-LAB, *Rhizopus oryzae, R. microsporus, Bacillus subtilis, B. coagulans* has been used for L-lactic acid production (Miura et al., 2003, Ohara et al., 1996 & Sakai et al., 2006c). Particularly, optically active L-lactic acid production from *Rhizopus oryzae* strains is significant (Miura et al., 2003). Industrial production of L-lactic acid using *Rhizopus* sps. has several advantages over using lactic acid bacteria (LAB).

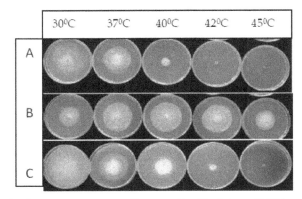

Fig. 6. Effect of incubation temperature on the growth of isolates. A) *Rhizopus oryzae* (TISTR 3514), B) *Rhizopus microsporus* (TISTR 3518) and C) *Rhizopus oryzae* (TISTR 3523).

The fungus only produces L-lactic acid, while LAB frequently produces the D-isomers as well. Therefore, the optical purity of L-lactic acid produced from the fungus is relatively higher than that from LAB. L-lactic acid production has been reported in only the *R. oryzae* group. In addition, variety of studies on construction of lactic acid-producing *Escherichia coli* and *Saccharomyces cerevisiae* by genetic engineering have been reported (Sakai, 2008).These strains would be promising for the industrial production under strictly closed sterilized fermentation using certain purified substrate sugar. According to Kitpreechavanich et al., (2008), a thermotolerant *Rhizopus* strain which is capable of producing L-lactic acid from starch substrate was identified as *R. microsporus* (Fig. 6).

4.3 Thermophilic/thermotolerant bacteria MLAP

The term 'thermophilic' has been progressively more restricted to organisms which can grow or form products at temperatures between 45°C and 70°C with optimal 60°C (Madison et al., 2009). Dijkhuizen & Arfan (1990), reported that thermotolerant organisms grow at temperature between 35°C and 60°C with optimal 50°C -55°C. Thermophilic bacteria are common in soil, compost and volcanic habitats and have a limited species composition (Zeikus, 1979).

Meantime, we have found that several thermotolerant/thermophilic bacterial species in Bacillaceae are able to produce certain amount of optically active L-lactic acid (Table 7). Compared to *Lactobacilli* and *Lactococci*; *Bacillus* species generally show interesting microbial properties. Most of them are basically aerobic and they form spores under certain environmental conditions. They do not produce D-lactic acid. Some of them show growth limitation at temperatures around 70°C. Some species produce polysaccharide-hydrolyzing enzymes such as amylase, chitinase, or xylanase. Many strains ferment glycerol, D-galactose, D-fructose, D-xylose, sucrose, cellobiose as well as starch which are constituent sugars in food and agricultural waste. Therefore, not only the characteristics of this bacterium are quite suitable for the bioconversion of starch from food waste but also it would be applicable to other agricultural wastes.

Thermotolerant strain *Bacillus licheniformis* has been explored for the L-lactic acid production from standard kitchen refuse under open condition i.e. 40g/*l* L-lactic acid with 97% optical activity and 2.5g/*l*.h productivity (Sakai & Yamanami 2006b). Moreover, thermophilic bacterium *Bacillus coagulans* is quite useful for producing optically active L-lactic acid from non-steriled kitchen refuse (Sakai & Ezaki, 2006c). The *B. coagulans* selectively grew at 55°C under open condition, while *Lactobacillus plantarum*, which is a major species in natural fermentation of kitchen refuse under mesophilic condition, suppressed its growth. Temperature and growth relations in different temperature classes of *B. coagulans* and *L. plantarum* are shown in Fig. 7.

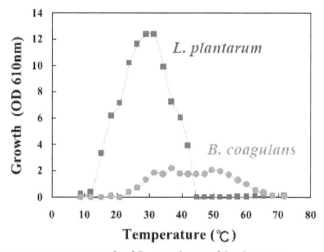

Fig. 7. Effect of temperature on growth of *B. coagulans* and *L. plantarum*

5. Open fermentation for total recycle of food waste

5.1 Merits of open fermentation

Nonsterile open fermentation has various merits over conventional sterilized and closed fermentations. For example, it requires no facilities for sterilization and no steam for autoclaving. Thus, nonsterile open fermentation of kitchen refuse could be implemented at on-site storage facilities for municipal food waste before the waste is transported to centralized processing plants. Because autoclaving is avoided, substrate sugars and other nutritional constituents required for lactic acid fermentation remain intact. The Maillard reaction, for instance, not only decreases the amount of available sugars and amino acids but also produces unfavorable furfural compounds that inhibit bacterial growth. In addition, food waste is unsuitable for filter sterilization or separate autoclaving of substrate from other medium constituents. Nonsterile open fermentation avoids these complications; however, the optical purity of accumulated lactic acid from such fermentation at room temperature is low (Sakai & Ezaki, 2006c). This type of natural lactic acid fermentation also occurs during the collection and storage of municipal kitchen refuses (Sakai et al., 2004b). On the other hand, the thermophilic bacterium *Bacillus coagulans* is useful for producing optically active L-lactic acid from kitchen refuse under nonsterile condition (Heriban et al., 1993).

5.2 Mesothermal recycle of food waste

During the investigation of open fermentation at atmospheric temperature, we found that naturally-existed mesophile *Lactobacillus plantarum* preferentially proliferated and selectively accumulated lactic acid in non-sterile kitchen refuse (food waste) under pH swing control (intermittent pH adjustment) (Table 6). Despite the reproducible and selective proliferation of the species, this strain produced both L- and D-lactic acid with nearly equal racemic body ratio. As optically inactive lactic acid is not suitable for high-quality of PLA, we tried to improve the optical activity by inoculating *L. rhamnosus* or *Lactococcus lactis* which are L-lactic acid producing LAB. But this kind of open fermentation also resulted proliferation of naturally existed *L. plantarum* and accumulation of lactic acid with low optical activity. In comparison, Frederico et al., (1994) also reported that *L. plantarum* accumulated low amount of lactic acid during the fermentation of fruit juice under sterilized condition. As shown in Table 6 (Run 1-1 to 1-5), the amount of accumulated lactic acid varied according to the intervals of pH adjustment, and maximum accumulation was observed with pH adjustment of 6hour (6h) or 12h.

Run[a]	Adjusted pH	Interval (hour)[b]	Productivity (g/l.h)[c]	Accumulation (g/l)[d]	Selectivity (%)[e]
1-1	7	0	1.05	19	83
1-2	7	6	0.70	44	92
1-3	7	12	0.58	45	94
1-4	7	24	0.4	31	94
1-5	7	_[f]	0.25	13	87
2-6	3	_[g]	0.04	2.0	-
2-7	5	6	0.42	32	96
2-8	7	6	0.65	45	94
2-9	10	6	0.58	45	92

[a] MKR samples of runs 1-1 to 1-5 and runs 2-6 to 2-9 were differently prepared.
[b] Interval of intermittent pH adjustment.
[c] Average production rate of lactic acid to reach maximum concentration.
[d] Maximum concentration of lactic acid accumulated.
[e] Ratio of accumulation of lactic acid to total organic acids
[f] MKR paste was adjusted at pH 7.0 initially and incubated without pH adjustment.
[g] No pH change was observed

Table 6. Effect of intermittent pH adjustment on accumulation of lactic acid during the open fermentation of MKR paste by mesophile.

5.3 Molecular monitoring of bacteria during recycle of food waste

From the very nature of a thing, non-sterilized fermentation process generally proceeds under a mixed culture condition. We have repeatedly isolated and identified the microbial structure during the course of open fermentation of kitchen refuse. Meantime, we cultivated, purified and characterized several microbial isolates, which counts laborious and time-consuming, and only predominant cultivable species can be identified. Therefore, we applied 16SrRNA-targeted fluorescence *in-situ* hybridization (FISH) to analyze the microbial population during open lactic acid fermentation (Sakai et al., 2004a, Fig. 8). For this, we designed probes for monitoring non-sterilized open fermentation of kitchen refuse such as a LAB group specific probe (LAC722) and a *B. coagulans* specific probe (Bcoa191). Similarly, specificity of Bcoa191 probe for *B. coagulans* in whole-cell hybridization of the new probe was confirmed *B. coagulans*, and differentiated the species from other bacteria as shown is Fig. 9 (Sakai & Ezaki, 2006c).

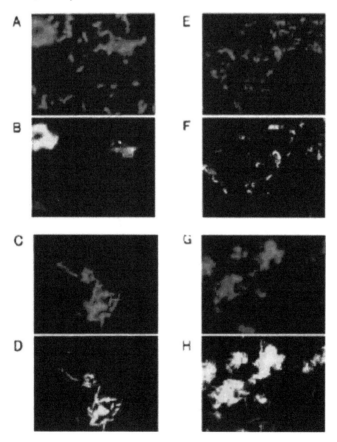

Fig. 8. Typical FISH staining during open fermentation of kitchen refuse. Samples at time zero (A, B, C, D), or 48 hours (C, D, G, H), without (A, B, C, D) or with (E, F, G, H) inoculated seed culture stained with rhodamine-EUB338 (A, C, E, G) or FITC-LAC722(L) (B, D, F, H), (Sakai et al., 2004a).

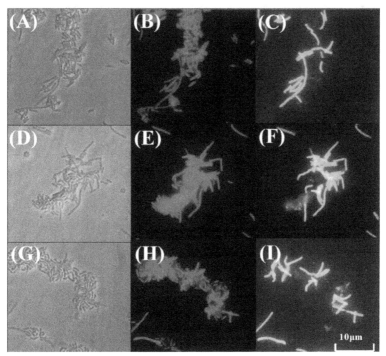

Fig. 9. Differential staining of *B. coagulans* using new probe Bcoa191 in 16S-Fluorescence In Situ Hybridization (FISH). *B. coagulans* cells were mixed with *L. plantarum* (A-C), L. *rhamnosus* (D-F), or *E. coli* (G-I). The mixed-cell samples were subjected to 16S-FISH, and the photomicrographs of phase contrast microscopic observation (A, D, G) and fluoro-microscopic observation for rhodamine-EUB338 (B, E, H) or FITC-Bcoa191 (C, F, I) are shown (Sakai & Ezaki, 2006c).

5.4 Thermotolerant MLAP in total recycle of food waste

As shown in Table 6, the L-lactic production rate and optical purity of mesophilic lactic acid bacteria was low. We, furthermore, tried to use the thermotolerant *Bacillus* species for the total utilization of food waste for PLA production and its biomass utilization. Production of lactic acid by some *Bacillus* species, including *Bacillus coagulans*, *Bacillus stearothermophilus*, *Bacillus subtilis* and *Bacillus licheniformis*, had already been reported (Bischoff et al., 2010, Heriban et al., 1993; Ohara & Yahata, 1996; Sakai & Yamanami, 2006b).

Recently, we isolated and identified novel thermotolerant *Bacillus* species from the mixed culture system. We, subsequently, used these strains for L-lactic acid production from the food waste. During the total utilization of food waste, the conditions for the fermentation of food waste were optimized as described previously (Sakai, 2006a, 2006e). Interestingly, novel thermotolerant strains *B. soli* U4-3 & U4-6 and *B. subtilis* N3-9 produced high amount of L-lactic acid within 6 hours of fermentation at 50°C with cent percent optical purity. L-lactic acid production profile is shown in Table 7 below. Meantime, L-lactic acid produced was further used for the PLA production which is one of the instances in total recycle of food waste.

Isolate No.	Species	L-lactic acid (g/l)	Yield/g (%)	Optical Activity (%)
N1-3	*Bacillus coagulans*	20.6	69.0	99.8
N1-4	*B. coagulans*	25.1	61.1	99.9
N1-7	*B. niacini*	11.7	46.4	99.5
N1-12	*B. coagulans*	28.6	60.2	100
N2-1	*B. coagulans*	29.1	64.0	99.4
N2-10	*B. subtilis*	31.5	68.9	99.0
N3-6	*B. subtilis*	32.6	82.7	99.7
N3-9	*B. subtilis*	35.1	74.0	100
U4-3	*B. soli*	30.3	70.8	100
U4-6	*B. soli*	29.3	85.5	100
N5-7	*B. subtilis*	28.4	61.0	99.3
N5-8	*B. subtilis*	23.3	55.3	99.0

Table 7. L-lactic acid production by thermotolerant *Bacillus* strains isolated from high-temperature and Aerobic fermenter.

In general, for the commercial production of poly-L-lactic acid plastic from biomass wastes, a feasible fermentation process to produce optically active L-lactic acid is required (Sakai, 2004a, 2004b, 2006d). By using collected kitchen refuse, saccharified liquid containing 117g/l soluble sugar was obtained (Table 8). This figure is fairly representative of collected kitchen refuse (Table 2). Following the incubation with B. *coagulans* at 55°C, pH 6.5, 86g/l L-lactic acid with 97% optical purity was produced under non-sterile conditions. The yields of total lactic acid from total carbon and total sugar were 53% and 98% respectively. These figures are comparable to those achieved by *L. rhamnosus* incubation using sterilized collected kitchen refuse (Fig.10).

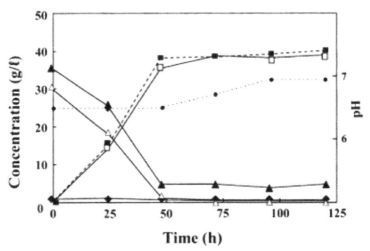

Fig. 10. Open fermentation of MKR using *B*. coagulans under constant pH 6.5 at 55°C under open culture conditions. The changes in the concentrations of total lactic acid (closed squares), L-lactic acid (open squares), D-lactic acid (closed diamonds), total sugar (closed triangles), and glucose (open triangles) are represented along with pH change (close circles).

Parameters	Closed fermentation with *Lactobacillus rhamnosus*		Open fermentation with *Bacillus coagulans*	
	Initial	Final	Initial	Final
Total sugar (g/l)	74	14	117	31
Total lactic acid (g/l)	3	61	2	86
Optical purity (%)	1.4	95	2.9	97
Carbon yield (%)	-	38	-	53
Sugar yield (%)	-	97	-	98

Table 8. Summary of open and closed fermentation of kitchen refuse using mesophile and thermophile.

6. Conclusions and future prospective

The majority of the worldwide industrial economics are now largely dependent on petroleum oil which provide basis for most all of our energy and chemical feedstock. Meanwhile, there is increasingly concern over the impact of these traditional manufacturing processes or the environment, i.e. the effect of CO_2 emissions on global warming as well as exhaustion of fossil resources. In order to maintain the world population in terms of food, fuel, and organic chemicals, we need to substantially reduce our dependence on petroleum feedstock by establishing a bio-based economy.

Principally, production and harvest of biomass plant is neither self-sustained nor environmentally friendly. It is a harvesting-out process of nutritious compound from field. Food waste and wastewaters, further, are unavoidably produced to pollute environment. So that, the total system design for recycle of all elements, not only carbon as neutral but also including nitrogen, potassium and phosphorus, is important for sustainable biomass production. Cascade utilization of biobased-products and recycle of biomaterials in a waste stream and wastewater, is another key technology for carbon sequestration and for the sustainable production-utilization system like metabolic network in human body.

We human beings are keeping our body function to be active by taking into the energy and chemicals as food. At the same time, we continuously use over half of total energy at our liver and pancreas, organs working in catabolism cleaning up our blood and recovering metabolites to maintain our body functions healthy. Treatment and utilization of waste materials may be compared with recycle of biomolecules via venous blood stream. In this context, our society has to further enrich the quality and quantity of 'venous industry' to treat waste and recover resources from them sustaining our society to be healthy.

Here, we present a total recycle system of food waste via chemical production with energy and facility savings and minimal emissions from waste materials. It should be further investigated to trait by improving the leading case study in 'Bio-economy system'. The challenge of the next decade will be to develop zero-emission bio-based environmentally friendly products from geographically distributed feedstock and worldwide generated food waste by simultaneous reduction of pollution indeed.

7. References

Addink, R. & Olie, K. (1995). Mechanisms of Formation and Destruction of Polychlorinated Dibenzo-p-dioxins and Dibenzofurans in Heterogeneous Systems. *Environmental Science and Technology,* 29, 1425-1435.

Bischoff, K. M.; Liu, S.; Hughes, S. R. & Rich, J. O. (2010). Fermentation of corn fiber and hydrolysate to lactic acid by the moderate thermophile *Bacillus coagulans*. *Biotechnology Letters,* 32, 823-828.

Boustead, I. (2002). *Eco-profiles of the European plastics industry: Polyethylene terephthalate.* Prepared for the European Centre for Plastics in the Environment, Brussels. www.apme.org

Camobreco, V.; Ham, R.; Barlaz, M.; Repa, E.; Felker, M.; Rousseau, C. & Rathle, J. (1999). Life-cycle inventory of a modern municipal solid waste landfill, *Waste Management and Research,* 17, 394–408.

Dijkhuizen, L. & Arfan, N. (1990). Methanol metabolism in thermotolerant methlotropic *Bacillus* species. *Federation of European Microbiological Societies, Microbiology reviews,* 87, 215-220.

Frederico, V. P.; Henry, P. F. & David, F. O. (1994). Kinetics and Modeling of Lactic Acid Production by *Lactobacillus plantarum*. *Applied and Environmental Microbiology,* 60, 2627-2636.

Harrison, K. W.; Dumas, R. D.; Barlaz, M. A. and Nishtala, S. R. (2000). A life-cycle inventory model of municipal solid waste combustion. *Journal of the Air and Waste Management* Association, 50, 993–1003.

Heriban, V.; Sturdik, E.; Zalibara, L. & Matus, P. (1993). Process and metabolic characteristics of *Bacillus coagulans* as a lactic acid producer. *Letters in Applied Microbiology,* 16, 243–246.

Hofvendahl, K. & Hagerdal, B.H. (2000). Factors affecting the fermentative lactic acid production from renewable resources. *Enzyme Microbial Technology,* 26, 87–107.

Kamm, B.; Gruber, P.R. & Kamm, M. (2005). *Biorefineries-Industrial Processes and Products,* Wiley-VCH, ISBN 3-527-31027-4, Germany

KBBE (2010). *The Knowledge Based Bio-Economy in Europe: Achievements and Challenges.* http://www.bio economy.net/reports/files/KBBE_2020_BE_presidency.pdf

Kitpreechavanich, V.; Maneeboon, T.; Kayano, Y. & Sakai, K. (2008). Comparative Characterization of L-Lactic Acid-Producing Thermotolerant *Rhizopus* Fungi. *Journal of Bioscience and Bioengineering,* 106, 541–546.

Madigan, M.T.; Martino, J.M.; Dunlap, P.V. & Clark, D.P. (Ed.). (2009). *Brock Biology of Microorganisms.* Pearson International, pp. 159, ISBN 0-321-5365-0, NewYork, USA

Miura, S.; Arimura, T.; Hosino, M.; Kojima, L.D. & Okabe, M. (2003). Optimization and scale-up of L-lactic acid fermentation by mutant strain Rhizopus sp.MK. *Journal of Bioscience and Bioengineering,* 96, 65-69.

Miyazaki, H. & Harano, K. (2001). Degradation test on biodegradable plastics polymers into composting. In: 2000 *Annual Research Report of Oita Industrial Research Institute.* Ono-Kousoku, Oita, Japan: Oita Industrial Research Institute, pp. 154–157

Mori, M.; Kuribayashi, M.; Nakamura, M.; Nishimura, Y.; Shira, Y. & Sakai, K. (2008). Material Balance in the Process of producing Poly-L-lactic acid from Municipal Food waste and Recycling the Byproduct as Organic Nitrogen Fertilizer for Rice Cultivation. *The Japan Society of Waste Management Exports,* 29, 400-408.

Nishida, H.; Fan, Y. ; Shirai, Y.; Tokiwa, Y. & Takeshi Endo T. (2004).Thermal degradation behaviour of poly(lactic acid) stereo complex. *Polymer Degradation and Stability*, 86, 197-208.

Oda, Y.; Park, B.; Moon, K. & Tonomura, K. (1997). Recycling of bakery wastes using an amylolytic lactic acid bacterium. *Bioresource Technology*, 60, 101–106.

Oh, H.; Wee, Y.J.; Yun, J.S.; Han, S.H.; Jung, S. & Ryu, H.W. (2005). Lactic acid production from agricultural resources as cheap raw materials. *Journal of Bioscience and Bioengineering*, 96, 1492–1498.

Ohara, H. & Sawa, S. (1994). Purification process of L-lactic acid for bio-degradable poly-L-lactate. In Biodegradable plastics and polymers, edited by Doi Y. & Fukuda K. Amsterdam: Elsevier

Ohara, H. & Yahata, M. (1996). L-Lactic acid production by *Bacillus* sp. in anaerobic and aerobic culture. *Journal of Fermentation and Bioengineering*, 81, 272–274.

Parfitt, J.; Barthel, M. & Macnaughton, S. (2010). Food waste within food supply chains: quantification and potential for change to 2050. *Philosophical Transactions of Royal Society*, 365, 3065-3081

Sakai, K.; Kudoh, E.; Wakayama, M. & Moriguchi, M. (2000a). Molecular analysis of microbial community in an activated sludge enriched by inorganic-nitrite medium. *Microbes Environment*, 15, 103–112.

Sakai, K.; Murata, Y.; Yamazumi, H.; Tau, Y.; Mori, M.; Moriguchi, M. & Shirai, Y. (2000b). Selective proliferation of lactic acid bacteria and accumulation of lactic acid during open fermentation of kitchen refuse with intermittent pH adjustment. *Food Science Technology*, 6, 140–145.

Sakai, K.; Kawano, H.; Iwami, A.; Nakamura, M. & Moriguchi, M. (2001). Isolation of a thermophilic poly-l-lactide degrading bacterium from compost and its enzymatic characterization. *Journal of Bioscience and Bioengineering*, 92, 298–300.

Sakai, K.; Mori, M.; Fujii, A.; Iwami, Y.; Chukeatirote, E. & Shirai, Y. (2004a). Fluorescent in situ hybridization analysis of open lactic acid fermentation of kitchen refuse using rRNA targeted oligonucleotide probes. *Journal of Bioscience and Bioengineering*, 98, 48–56.

Sakai, K.; Taniguchi, M.; Miura, S.; Ohara, H.; Matsumoto, T. & Shirai, Y. (2004b). Making plastics from garbage: a novel process for poly-L-lactate production from municipal food waste. *Journal of Industrial Ecology*, 7, 63–73.

Sakai, K.; Fujii, N. & Chukeatirote, E. (2006a). Racemization of L-lactic acid in pH-swing open fermentation of kitchen refuse by selective proliferation of *Lactobacillus plantarum*. *Journal of Bioscience and Bioengineering*, 102, 227–232.

Sakai, K. & Yamanami, T. (2006b). Thermotolerant *Bacillus licheniformis* TY7 Produces Optically active L-lactic acid from kitchen refuse under open condition. *Journal of Bioscience and Bioengineering*, 102, 132–134.

Sakai, K.; & Ezaki, Y. (2006c). Open L-lactic acid fermentation of food refuse using thermophilic *Bacillus coagulans* and fluorescence in situ hybridization analysis of microflora. *Journal of Bioscience and Bioengineering*, 101, 457–463.

Sakai, K.; Oue K.; Umeki, M.; Mori, M. & Mochizuki, S. (2006d). Species-specific FISH analysis of the cecal microflora of rat administered lactic acid bacterial cells. World Journal of Microbiology and Biotechnology, 22, 493–499.

Sakai, K.; Ezaki, Y.; Tongpim, S. & Kitpreechavanich,V. (2006e). High-Temperature L-Lactic Acid Fermentation of Food Waste under Open Condition and Its FISH Analysis of Its Micro Flora. *Kasetsart Journal (Natural Science)*, 40, 35-39.

Sakai, K. (2007). Establishment of carbondioxide- free composting of municipal food waste with producing poly L-lactic acid plastic. *Soil microorganisms*, 61, 103-110.

Sakai, K. (2008). Production of Optically Active Lactic acid by using Non-LAB Microorganisms. White Biotechnology; *The front of Energy and Material Development*, Ohara, H. & Kimura Y. CMC publisher, pp. 98-108 ISSN, Tokyo, Japan (in Japanese)

Shirai, Y. & Sakai, K. (2006). *Utilization of Food refuse, Hand book of Food Engineering*. Asakura Shotan, pp. 460, ISBN 4-354-43091-K, Tokyo, Japan (in Japanese)

Umeki, M.; Oue K.; Mori M.; Mochizuki S. & Sakai K. (2005). Fluorescent *In Situ* Hybridisation Analysis of Cecal Microflora in rats simultaneously administrated Lactobacillus rhamnosus KY-3 and Cellobiose. *Food Science and Technology Research*, 11, 168-170.

Umeki, M.; Oue, K.; Mochizuki, S.; Shirai, Y. & Sakai, K. (2004). *Effect of Lactobacillus rhamnosus* KY-3 and Cellobiose as Synbiotics on Lipid Metabolism in Rats. *Journal of Nutritional Science and Vitaminology*, 50, 330-334.

United Nations Food and Agriculture Organization (FAO), (2011). Global food losses and food waste. International Congress, Rome.
http://www.fao.org/fileadmin/user_upload/ags/publications/GFL_web.pdf

Vink, E. T. H.; Rabago, K. R.; Glassner, D. A. & Gruber, P. R.(2003). Applications of life cycle assessment to Nature Works polylactide (PLA) production. *Polymer Degradation and Stability*, 80, 403–419.

Zeikus, J. G. (1979). Thermophilic bacteria: ecology, physiology and technology. *Enzyme Microbial Technology*, vol. 1, IPS Business Press

4

Making Green Polymers Even Greener: Towards Sustainable Production of Polyhydroxyalkanoates from Agroindustrial By-Products

José G. C. Gomez[1], Beatriz S. Méndez[2], Pablo I. Nikel[2,3],
M. Julia Pettinari[2], María A. Prieto[4] and Luiziana F. Silva[1]
[1]Institute of Biomedical Sciences, University of São Paulo
[2]Department of Biological Chemistry, Faculty of Sciences,
University of Buenos Aires and National Council for Research (CONICET),
[3]Institute for Research in Biotechnology, University of San Martín,
[4]Department of Environmental Biology, Centro de Investigaciones Biológicas,
[1]Brazil
[2,3]Argentina
[4]Spain

1. Introduction

This review addresses recent achievements on the development of energy-saving and environmentally-friendly bioprocesses for the synthesis of polyhydroxyalkanoates (PHAs), a kind of non-petrochemical bioplastics. Different cutting-edge strategies developed in order to achieve bioprocesses with enhanced sustainability will be described. These are mainly based on the use of cheap substrates concomitantly lowering energy consumption levels, thus diminishing the environmental impact of PHAs production. We will also cover studies that shed light on the physiology of PHA-producing microorganisms by means of metabolic flux analysis, and also those that analyzed polymer modifications aimed at modifying physico-chemical properties. An overview of the applications of PHAs, including novel functionalized varieties, will conclude this review.

1.1 Environmental preservation

Microbial fermentations and many other industrial processes mostly rely on two fossil fuels (petroleum and gas) as sources of energy. This biased strategy has contributed to global climate change by emitting large amounts of carbon dioxide to the atmosphere and, as a collateral consequence, has favored the generation of an extended range of difficult-to-dispose-of goods by petrochemical industries. Accumulation of microscopic plastic debris at sea is particularly alarming, as well as the exponentially increasing need of landfill for municipal solid waste disposal. This particular situation has renewed the interest in strategies based on energy-saving bioprocesses. These strategies propose the replacement of petroleum by renewable resources and the manufacture of non-petrochemical goods, such as bioplastics, in order to reduce the pollution phenomena.

1.2 Bioplastics

The name biopolymers is currently used for polymers that are either synthesized by living organisms or produced from substrates obtained from living organisms. Examples of the first kind of biopolymers are naturally occurring polymers such as cellulose, starch, and PHAs. Among the second kind, there are poly(lactic acid), that can be synthesized from biologically-obtained lactic acid, or even polyethylene, when it is produced from ethylene obtained from bio-ethanol. Bioplastics are biopolymers with plastic properties. Bioplastics synthesized by living organisms are generally biodegradable; and chemically synthesized polymers, especially those derived from petroleum, are generally non-biodegradable, while those that are "bio-based" (*i.e.*, obtained using a biologically produced substrate such as bio-ethanol), have several degrees of biodegradability. Polyethylene and polypropylene, whether bio-based or not, are considered non-biodegradable, even when there have been claims of slow degradation of these polymers in nature (Corti et al., 2010). There are exceptions to the relationship between biological origin and biodegradability, as not all biopolymers are biodegradable, and not all biodegradable polymers are biopolymers. There are some plastics obtained from non-biological processes that can also be biodegraded, such as poly(ε-caprolactone) and the petroleum derived polymer poly(butadiene adipate-*co*-terephthalate) (Queiroz & Collares-Queiroz, 2009); and there are also polymers synthesized by microorganisms that are not biodegradable, such as polythioesthers, obtained by the polymerization of mercaptoalkanoic acids by PHA synthases (Steinbüchel, 2005).

Currently, there are many different biodegradable bioplastics. Among these, we found blends containing natural polymers, such as starch and cellulose; and polymers synthesized chemically from different substrates, such as poly(lactic acid), poly(ε-caprolactone), and others (Rehm, 2010). Starch can be blended with other compounds to obtain polymers which could be used for several applications, but this material is quickly damaged in contact with water. Poly(lactic acid) is not normally degraded by microorganisms, but it is easily hydrolyzed and can be composted. PHAs are natural bioplastics produced by many bacteria from different substrates. In sharp contrast to the other bioplastics mentioned above, these polymers are totally biodegradable, as all microorganisms that naturally accumulate PHAs can degrade them. Moreover, PHAs can also be degraded by many other microorganisms, both bacteria and fungi, under either aerobic or anaerobic conditions.

These polymers are synthesized naturally by a wide variety of bacterial species as a reserve for carbon and energy. Nowadays, PHAs continue to attract increasing industrial interest as renewable, biodegradable, biocompatible, and extremely versatile thermoplastics (Steinbüchel & Lütke-Eversloh, 2003; Suriyamongkol et al., 2007). PHAs are the only water-proof thermoplastic materials available that are fully biodegraded both in aerobic and anaerobic environments. Two classes of PHAs are distinguished according to their monomer composition: short-chain length (SCL) PHAs and medium-chain length (MCL) PHAs. SCL-PHAs are polymers of 3-hydroxyacid monomers with a chain length of three to five carbon atoms, such as poly(3-hydroxybutyrate) (PHB, the most common PHA); whereas MCL-PHAs contain 3-hydroxyacid monomers with six to sixteen carbon atoms. All of them are optically active R-($-$) compounds. This versatility is partly due to the wide substrate range of the PHA-synthesizing enzymes, and gives PHAs an extended spectrum of associated properties which is a clear advantage *vis-à-vis* to other bioplastics. Around 200 different monomer constituents were found in the polymers analyzed so far (Steinbüchel & Lütke-Eversloh, 2003).

Making Green Polymers Even Greener: Towards Sustainable Production of Polyhydroxyalkanoates from
Agroindustrial By-Products

81

1.3 Environmental Issues

Current high-yield bioprocesses for the synthesis of PHAs require fully aerobic conditions, which means that they are high energy-consuming processes. The environmental impact of replacing oil-derived plastics with biopolymers has been the subject of several studies, among them, those regarding bacterial PHB production in bioreactors (Gerngross, 1999). A complete life cycle assessment for PHB production from the cradle to the factory gate has been published by Harding et al. (2007). Those studies pointed out that, in spite of the fact that PHB is more environmentally friendly than oil-derived polymers, the great amount of energy required for its production must be taken into account when assessing its environmental impact. Similar results were obtained from research applied to the manufacturing of polymers obtained from transgenic plants (Zhong et al., 2009), or from agricultural substrates such as corn (Kim & Dale, 2005).

From these researches and other studies it was concluded that when the amount of energy used for sterilization, aeration, and agitation (both in the bioreactor and downstream processing), as well as the energy needed for the production of the agricultural feed-stocks to be used as carbon sources is considered, the environmental performance of PHAs equals that of petrochemical polymers. Different initiatives to overcome this problem are described below.

2. Towards an enhanced sustainable production

In the following sections, we will discuss several cutting-edge strategies intended to enhance the sustainability of PHA production processes (as summarized in Fig. 1). Even when they will discussed in a sequential fashion, beginning with the choice of suitable substrates up to the rational functionalization of polymer properties, it is important to mention that bioprocesses designed for PHA synthesis from agroindustrial by-products are subjected to continuous improvement.

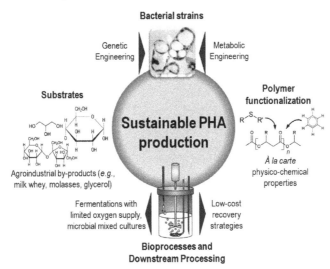

Fig. 1. Strategies used to enhance the sustainability of PHA production processes. Note that modifications to improve different steps in the process as a whole can be implemented in a cyclic, iterative fashion.

2.1 Substrates

It is widely accepted that the prize of the carbon source is one of the main factors affecting the cost of PHAs, influencing the sustainability of production processes. However, the choice of a suitable carbon source is not a clearcut issue. The use of industrial and agricultural by-products could require extensive purification, equalling or even surpassing the energy demand of cost-intensive agricultural feed-stocks.

Some early approaches to overcome this situation have integrated PHA and sugar production in a substrate- and energy-closed system (Nonato et al., 2001). Milk whey, a substrate which does not require extensive purification, was the most appropriate option for many other initiatives. Recently, glycerol has received attention as a potential carbon substrate due to its accumulation as a by-product of biodiesel synthesis.

2.1.1 Milk whey

About 80-90% of the processed milk volume is converted to whey during cheese and casein production by the dairy industry (Wong & Lee, 1998). Whey is rich in lactose, proteins, lipids, and lactic acid (Yang et al., 1994). After casein precipitation from raw milk, skimmed whey is produced, which is then concentrated and ultra-filtrated producing whey permeate (rich in lactose) and whey retentate (rich in proteins and containing a considerable amount of residual lactose).

Some components of whey retentate are useful in the pharmaceutical industry. Whey permeate contains *ca.* 81% of the original lactose in milk and is appropriate for biotechnological processes (Nath et al., 2008).

Young et al. (1994) evaluated for the first time the production of PHB from lactose by *Burkholderia cepacia*. Since then, many other isolated bacteria were evaluated for the production of PHB from lactose or milk whey (Nath et al., 2008). However, the cultures were always performed at low-cell-densities, thus hindering an appropriate evaluation of their economical relevance. The production of the copolymer poly(3-hydroxybutyrate-*co*-3-hydroxyvalerate) [P(HB-*co*-HV)] from milk whey was also demonstrated in cultures either supplemented or not supplemented with valeric (pentanoic) acid (Koller et al., 2005, 2008). Production of PHAs with different monomer compositions using milk whey as the main carbon source will allow their use in different applications.

Ralstonia eutropha[1] has been extensively established as the platform for PHA production *par excellence* (Reinecke & Steinbüchel, 2009). However, it is unable to hydrolyse lactose, and galactose (its hydrolysis product) is not metabolized. Koller et al. (2007) proposed an alternative process consisting of two steps for PHB production from milk whey: lactose was first converted to lactic acid by lactobacilli, and the resulting lactic acid was then used as carbon source by *R. eutropha* for PHA production. Marangoni et al. (2002) hydrolyzed milk whey in order to make glucose available for *R. eutropha*, but it should be considered that galactose would remain unused in the culture medium since it is not metabolized by *R. eutropha*, as mentioned. After the expression of genes encoding β-galactosidase and

---

[1]*R. eutropha* is currently known as *Cupriavidus necator*. In this review we will adopt the name *R. eutropha* which is most frequently used in the literature, including the announcement of its complete genome sequence (Pohlmann et al., 2006).

Making Green Polymers Even Greener: Towards Sustainable Production of Polyhydroxyalkanoates from Agroindustrial By-Products

83

galactokinase in *R. eutropha* it became able to use lactose, albeit at a very slow rate (Pries et al., 1990). Recently, the *lacZ* (encoding β-galactosidase) and *lacI* (encoding the *lac* operon repressor protein) genes from *Escherichia coli* were introduced in the genome of *R. eutropha* interrupting *phaZ1* (encoding an intracellular PHB depolymerase). Cell concentration reached values higher than 8 g · L^{-1} and the PHB content was about 20-25% (wt/wt), demonstrating the capability of this recombinant *R. eutropha* strain to use lactose (Povolo et al., 2010).

Lee and co-workers studied high-cell density cultures of recombinant *E. coli* for the production of PHB from milk whey (Ahn et al., 2000; Wong & Lee, 1998). The best results were reached when using a highly concentrated milk whey solution associated to an external membrane module to retain the cell mass inside the bioreactor. Under these working conditions, and after 36.5 h of cultivation, PHB volumetric productivities as high as 4.6 g · L^{-1} · h^{-1} were reached (*i.e.*, cell and PHB concentrations of 194 and 168 g · L^{-1}, respectively). These processes were scaled up (30 and 300 L), but the productivities attained were low (Park et al., 2002).

Kim (2000) also studied the production of PHB from milk whey by recombinant *E. coli* strains harboring the PHB biosynthetic genes from *R. eutropha*. After 35 h of cultivation, cell and PHB concentrations reached 55 and 32 g · L^{-1}, respectively; corresponding to a PHB volumetric productivity of 0.9 g · L^{-1} · h^{-1}. In a 24-h fed-batch process using milk whey and corn steep liquor as the main carbon and nitrogen sources, a recombinant *E. coli* strain (harboring the PHB biosynthetic genes from *Azotobacter* sp. strain FA8) reached cell and PHB concentrations of 70.1 and 51.1 g · L^{-1}, respectively, corresponding to a PHB volumetric productivity of 2.13 g · L^{-1} · h^{-1} (Nikel et al., 2006a).

PHB production processes from milk whey, based on fed-batch cultivation of recombinant *E. coli* strains and reaching high cell densities and PHB volumetric productivities, were established (Ahn et al., 2000; Nikel et al., 2006a; Wong & Lee, 1998), and are economically sound. However, further studies will be needed to scale up these processes keeping high productivities, cell concentration, and PHB content. Milk whey would satisfy the materials demands in processes for the production of other PHAs. However, taking into account the relevance of energy demands in these processes (Gerngross, 1999; Nonato et al., 2001), a renewable source must be considered to fulfill energy requirements and make the bioprocess truly sustainable.

2.1.2 Sugarcane molasses

Molasses is the residual syrup generated in sugar-refining mills after repeated sugar extraction by applying crystallization to sugarcane or sugar beet juice. At this point, sugar extraction is no longer economically viable, despite still having relatively high sucrose content. Low-grade molasses, inappropriate to be used in food or feed, has been suggested as substrate to produce PHAs (Solaiman et al., 2006).

A number of processes have been developed using molasses to produce PHA by bacteria. Most of the data provided on polymer content seem not to be competitive or sustainable at the moment. Beet and soy molasses have also been tested as alternative substrates (Solaiman et al., 2006). PHA production from molasses has been reported using Gram negative bacteria, such as recombinant *E. coli* and *Klebsiella* strains (Zhang et al., 1994), *R. eutropha*

(Oliveira et al., 2004), *Pseudomonas cepacia* (Çelik et al., 2005), or Gram positive bacteria such as *Bacillus* strains (Kulpreecha et al., 2009). In some cases, molasses was only used as additive [0.3 to 2.5% (wt/wt)] to grow *R. eutropha* in liquid or solid-state cultures along with other main substrates, reaching a maximum PHA content ranging from 26 to 39% (wt/wt) (Beaulieu et al., 1995; Oliveira et al., 2004). *P. cepacia* G13 accumulated PHA up to 70% (wt/wt) in culture media supplemented with 3% (wt/vol) beet molasses (Çelik et al., 2005).

Considering the use of sugarcane molasses as the main carbon source, Kulpreecha et al. (2009) tested *Bacillus megaterium* on sugarcane molasses and achieved a cell dry mass concentration of 72.7 g $\cdot L^{-1}$ in 24 h, with a PHB content of 42% (wt/wt); a good process that can still be improved since dissolved oxygen was a limiting factor. Brazil is currently one of the world leaders on sugarcane production (569 million tons in 2008-2009; Sawaya Jank, 2011). Initial attempts to use sugarcane molasses or high-test-molasses to produce PHAs were partially limited by its high nitrogen content (unpublished data).

Sugarcane molasses is no longer a waste material in Brazil but a by-product showing a good market value, and since 1970 it has been increasingly used on bio-ethanol production. The bagasse excess and sugarcane leaves are promising substrates to produce second-generation bio-ethanol. However, further developments are needed to solve the inability of yeasts to use the xylose fraction released from bagasse hydrolysis. Therefore, xylose and arabinose may be the new target by-products to be used in order to produce PHA in the integrated model of a sugar mill bio-refinery.

2.1.3 Glycerol

In the last years, a very important increase in the production of biodiesel has caused a sharp fall in the cost of glycerol, the main by-product of the biodiesel synthesis (da Silva et al., 2009; Solaiman et al., 2006). As a result, glycerol has become a very attractive substrate for bacterial growth. Additionally, because carbon atoms in glycerol are more reduced than in glucose or lactose, cells using glycerol are in a more reduced physiological state, favoring polymer synthesis. The use of glycerol for microbial PHA synthesis has been analyzed in natural PHA producers, such as *Methylobacterium rhodesianum* and *R. eutropha* (Borman & Roth, 1999), several *Pseudomonas* strains (Solaiman et al., 2006), the recently described *Zobellella denitrificans* (Ibrahim & Steinbüchel, 2009), and *Bacillus* sp. (Reddy et al., 2009), among others. Glycerol has also been investigated as a substrate for PHB synthesis in recombinant *E. coli* carrying the PHB biosynthetic genes from *Streptomyces aureofaciens* (Mahishi et al., 2003), and *Azotobacter* sp. strain FA8 (Nikel et al., 2008b).

PHAs obtained from glycerol were reported to have a significantly lower molecular weight (M_r) than polymers synthesized from other substrates, typically less than 1 MDa. In *Methylobacterium extorquens* and *R. eutropha*, PHB obtained from glycerol, ethanol, or methanol had a lower M_r than that obtained from other substrates (such as succinate, glucose, and fructose), and the M_r of the polymer was shown to decrease with increasing glycerol concentrations (Taidi et al., 1994). This effect was further analyzed and attributed to chain termination caused by glycerol (Madden et al., 1999). In studies performed using different *Pseudomonas* strains, the M_r of the polymers obtained, such as PHB produced by *P. oleovorans* and MCL-PHA synthesized by *P. corrugata,* was also observed to decrease with increasing glycerol concentrations [from 1% to 5% (wt/vol)] (Ashby et al., 2005). A recent study performed using *R. eutropha* describes PHB obtained from commercial glycerol and from waste glycerol with a M_r of 957 and 786 kDa, respectively, less than half of that of PHB obtained from glucose (Cavalheiro et al., 2009). In contrast, in a recent report describing P(HB-*co*-HV) accumulation in a *Bacillus* strain, similar M_rs, lower than 700 kDa, were observed for the polymer obtained from the two carbon sources (Reddy et al., 2009). A low M_r is undesirable for industrial processing of the polymer, so the results available in the literature pointed to a drawback in the use of glycerol as a substrate for the microbial production of PHAs. However, based on recent results obtained with recombinant *E. coli*, it has been proposed that it is possible to obtain PHB from glycerol with M_rs similar to those of the polymer obtained from glucose or lactose by using adequate bacterial strains and culture conditions (de Almeida et al., 2010).

2.2 Strains

As stated before, industrial synthesis of PHAs must improve sustainability in order to reach an appropriate production cost and diminish environment damage. The manipulation of natural PHA producers and recombinant strains to achieve high PHA production has been the subject of many studies (reviewed in Aldor & Keasling, 2003; Jung et al., 2010; Keshavarz & Roy, 2010; Madison & Huisman, 1999; Steinbüchel, 2001), and will not be considered in this review. In spite of the fact that several bacterial species are currently being used in biotechnological processes, *E. coli* remains as the "workhorse" of industrial developments. This species has been the selected host for genetic techniques devised to introduce the PHA biosynthetic genes, improve their expression, provide suitable quality and concentration of substrates to the PHA synthase, as well as to modify the host strains to improve their performance in the bioreactor (Li et al., 2007).

In this section, we will focus on recent studies using different *E. coli* mutant strains and metabolic flux analysis with the objective of increasing sustainability in PHAs synthesis processes.

2.2.1 Modification of host strains

When they are grown in bioreactors, all microorganisms, including PHB-producing recombinant *E. coli* strains, are subjected to extreme (and often oscillating) conditions, such as shear forces, extreme aeration (either low or high), pH, and growth temperatures chosen to obtain maximum product yield. These extreme conditions often lead to membrane debilitation, cell filamentation, or protein precipitation. A strategy used to avoid filamentation was to over-express the gene encoding FtsZ (involved in cell division) in *E. coli* harboring the *pha* genes from *R. eutropha*, thus improving the polymer productivity from 2.08 g $\cdot$ L^{-1} $\cdot$ h^{-1} in the wild-type strain up to 3.4 g $\cdot$ L^{-1} $\cdot$ h^{-1} in the filamentation-suppressed derivative strain and consequently enhancing the process sustainability (Wang & Lee, 1997).

Current processes for the synthesis of PHAs require fully aerobic conditions, which mean that they are high energy-consuming processes. Most of the energy requirement is needed to fulfill aeration and agitation inside the culture vessel of the bioreactor. Therefore, strategies aimed towards improving the respiratory capacity of the host strain under micro-aerobic growth conditions were developed to reduce aeration needs. *Vitreoscilla* haemoglobin is supposed to facilitate intracellular oxygen transfer and assimilation, and the gene encoding this protein was introduced in PHB-producing *E. coli* improving the growth and polymer yield, simultaneously avoiding the need of pure oxygen supplementation to achieve high-cell-density cultures (Horng et al., 2010, 2011).

Another approach was based on the control of the redox state. In *E. coli*, the two-component signal transduction system ArcAB modulates, the expression of many operons according to the redox state of the environment (Lynch & Lin, 1996). The main targets for repression by the phosphorylated regulator are the genes that encode the enzymes involved in aerobic respiration and oxidative bioreactions, such as those of the tricarboxylic acid cycle. *E. coli arc* mutants are unregulated for aerobic respiration, and the genes encoding components of the tricarboxylic acid cycle are fully expressed under micro-aerobic growth conditions.

As a consequence, the pool of reducing equivalents is elevated and could be funneled into reduced bioproducts such as PHB. This approach enabled the increase of PHB content up to 35% (wt/wt) in an *arcA* mutant strain grown in a semi-synthetic medium with gentle (75 rpm) agitation, conditions in which no PHB was accumulated by the wild-type strain (Nikel et al., 2006b). Another global regulatory system manipulated to increase PHB synthesis is CreBC, a two-component signal transduction pair, where CreB is the regulator and CreC the sensor kinase. The *cre* regulon includes different genes, and some of them are involved in carbon metabolism (Avison et al., 2001). *E. coli* strain CT1061, an *arcA* and *creC* constitutive mutant, has enhanced carbon source consumption as well as a reducing intracellular environment (characterized by a high $NADH/NAD^+$ ratio, *ca.* 1 mol $\cdot$ mol^{-1}), making it adequate as a candidate host for reduced biochemical synthesis (Nikel et al., 2008a). Introduction of the PHB biosynthetic genes from *Azotobacter* sp. strain FA8 in *E. coli* CT1061 resulted in increased PHB yield from glucose- or glycerol-supplemented semi-synthetic media, associated to the highly reduced redox state in this strain (Nikel et al., 2006b).

Another approach, based on the same rationale, was the use of anaerobic promoters to achieve PHB production under micro- or anaerobic conditions. Among the promoters tested, the one for *E. coli* alcohol dehydrogenase was the most effective in promoting micro-aerobic synthesis of PHB (Wei et al., 2009).

2.2.2 Metabolic engineering of PHA biosynthesis

Industrial microorganisms have been traditionally developed *via* multiple rounds of random mutagenesis followed by selection of desired phenotypes. However, these techniques do not take into account important features of the bioprocess itself, *inter alia*, increased sustainability. Approaches for microbial synthesis of valuable bioproducts have increasingly evolved towards more systematic strategies. Metabolic Engineering is a multidisciplinary field defined as the directed improvement of product(s) synthesis or cellular properties through the rationale modification of specific biochemical reaction(s), or the introduction of new one(s), as well as manipulating regulatory cellular processes (Stephanopoulos, 1999). In connection with this concept, Synthetic Biology is a newly coined term which defines a group of methodologies aimed at creating novel functional parts, modules, systems, and, ultimately, novel (micro)organisms through the integrated use of biological techniques and mathematical methods traditionally employed in Engineering designs (Lee et al., 2010). Metabolic Engineering and Systems Biology are different from other cellular engineering strategies since their systematic approaches focus on understanding the whole metabolic network in the cell. As a consequence, they can be used as powerful tools to increase bioprocess sustainability by taking into account different cellular and process features at the

Making Green Polymers Even Greener: Towards Sustainable Production of Polyhydroxyalkanoates from
Agroindustrial By-Products

87

same time. Metabolic Engineering is characterized by a cyclic process involving evaluation of metabolic performance of cells, establishment of appropriate target(s) for genetic engineering, and implementation of genetic modification(s) (Nielsen, 2001). The use of analytical tools and metabolic models to study the performance of cells and to identify the appropriate target for genetic modification allows distinguishing Metabolic Engineering from classical genetic engineering and characterize it as a system approach (Nielsen & Jewett, 2008).

PHAs synthesis is an interesting target for Metabolic Engineering manipulation as both polymer assembly and accumulation take place *in vivo*, offering the chance to optimize different metabolic and cellular processes at the same time (Jung et al., 2010; Tyo et al., 2010). The simplest Metabolic Engineering strategy for PHA synthesis manipulation would be to choose the appropriate carbon source(s) supplied to the bacterial host to control and direct carbon flux through relevant precursors and polymer biosynthesis enzymes. This strategy has traditionally been exploited to modulate polymer composition by varying the feed ratio of different substrate precursors (Lütke-Eversloh et al., 2001, Marangoni et al., 2002). Additionally, knowledge of the metabolic network operation under PHA-producing conditions would enable the rational streamlining of catabolic pathways to harness the greatest possible amount of carbon source for polymer synthesis. Knowing the distribution of fluxes is an important way to improve PHA production process towards efficient (and sustainable) polymer accumulation. Intracellular fluxes are quantitative descriptors which can be used to choose appropriate targets for modification of the metabolic network activity, increasing the formation of a desired product (*e.g.*, PHAs).

In silico genome scale analysis of metabolic models were also implemented to identify potential targets for manipulation and strain improvement of efficient PHA producers. Using this approach, Lim et al. (2002) identified *zwf* and *gnd* (encoding glucose-6-phosphate dehydrogenase and 6-phosphogluconate dehydrogenase, respectively) as relevant targets for manipulation in recombinant *E. coli* to redirect catabolic fluxes towards the pentose phosphate pathway, resulting in a high $NADPH/NADP^+$ ratio that favored PHA accumulation [up to 41% (wt/wt)]. Another study dealing with *in silico* metabolic analysis of PHB-accumulating *E. coli* strains showed that the Entner-Doudoroff pathway represents an important contribution to PHB synthesis (Hong et al., 2003), a fact also evidenced in proteomic analysis (Han et al., 2001). These studies clearly pointed towards the fact that choosing the adequate mutant background through systematic analysis of metabolic networks allowed the enhancement of PHB production processes.

A breakthrough in Metabolic Engineering is related to the emergence of ^{13}C-labeling methodologies to study the efficiency of complex metabolic networks. As the labeled substrate proceeds through the metabolic network, the pools of downstream metabolites become labeled and, at steady state, the fraction of labeled substrate in a given pool can be used to calculate the flux through that pathway. ^{13}C-based metabolic flux analysis uses the labeling information in proteinogenic amino acids to infer the labeling patterns of the respective precursor metabolites from central carbon metabolism (Sauer, 2006). The labeling information can be determined either by gas chromatography-mass spectrometry or nuclear magnetic resonance spectroscopy. The resulting labeling information is used as additional constraints for metabolic network models that utilize the biochemical stoichiometry, the substrate uptake, product secretion, and biomass formation rates to compute the intracellular flux distribution. Two alternative labeling information interpretation methods are used: comprehensive isotopomer modeling (Wiechert, 2001), and net-flux calculation utilizing results from metabolic flux ratio analysis (Fischer et al., 2004).

Metabolic networks are not the only targets for rational design of sustainable PHA production processes. In fact, regulatory circuits within the cell can be manipulated in order to obtain a desirable phenotype. Signal transduction pathways are involved in intercellular interactions and communication of extracellular conditions to the interior of the cell. The final outcome of such a signaling pathway is often the activation of specific transcription factor(s) that, in turn, control(s) gene expression. As stated before, in *E. coli* aerobic and anaerobic respiration, as well as fermentation pathways, are switched on and off by the ArcAB system, enabling bacterial cells to optimize energy generation according to the oxygen levels in the surrounding medium, and CreBC is responsive to the carbon source used and oxygen availability. Metabolic flux analysis based on ^{13}C-labelling showed that both ArcAB and CreBC systems have a deep impact on central metabolic pathways of *E. coli* under micro-aerobic growth conditions (Nikel et al., 2009), offering valuable information for rationale modification of regulatory networks aimed at polymer (and other bioproducts) synthesis. These results highlighted the idea that manipulation of the genes encoding global regulators could provide a relevant tool for the modulation of central metabolism and reducing power availability for biotechnological purposes, rather than manipulating the genes directly involved in the metabolic pathway of interest.

2.3 Bioprocesses and downstream processing

During the bioprocess conducing to PHAs production, energy is needed for the generation of steam used for sterilization, aeration and agitation in the reactor, and downstream processing. Several strategies which aimed to enhance both the polymer yield and the process sustainability by means of diminishing energy consumption were developed. Bacterial growth in the reactor was the target of these attempts, which were specially centered on two key aspects: (*i*) the growth of recombinant *E. coli* (facultative aerobe) under conditions not fully aerobic, thus decreasing aeration and agitation needs, and (*ii*) the development of mixed cultures, which circumvents sterilization.

Carlson et al. (2005) observed that recombinant *E. coli* DH5α carrying the *pha* genes from *R. eutropha* can support PHB accumulation in anaerobiosis when grown in rich media. The authors also developed a theoretical model of the biochemical network to interpret the experimental results and to study the metabolic capabilities of *E. coli* under anaerobic conditions. One of the few reports in the scientific literature on fed-batch cultivation in micro-aerobiosis describes a process for the synthesis of PHB developed under these conditions using glycerol as substrate and the concomitant synthesis of a valuable by-product, bio-ethanol, during micro-aerobic PHB accumulation. Micro-aerobic fed-batch cultures allowed a 2.57-fold increase in volumetric productivity when compared with batch cultures, attaining a PHB content of 51% (wt/wt) (Nikel et al., 2008b). In this work, the authors introduced the *pha* genes from *Azotobacter* sp. strain FA8 into an *arcA creC* mutant of *E. coli*, unregulated for redox control and carbon catabolism. In a fed-batch aerobic cultivation of a recombinant *E. coli* it was also reported that a PHB content of 80% (wt/wt) was obtained with oxygen limitation and a small increase in agitation using milk whey as the main carbon source (Kim, 2000).

Making Green Polymers Even Greener: Towards Sustainable Production of Polyhydroxyalkanoates from
Agroindustrial By-Products

89

An alternative to fed-batch processes to produce PHA from waste materials is the use of open microbial mixed cultures (MMCs). MMCs are microbial populations, often with unknown composition, selected by the operational conditions imposed on the biological system (currently referred to as "feast and famine", or aerobic dynamic feeding) resulting on polymer accumulation not induced by nutrient limitation. This system reduces bioreactor and operation costs, including sterilization, and is suitable for the use of agroindustrial wastes with unknown or variable composition (Serafim et al., 2008). Studies using sugarcane molasses in MMCs showed that by controlling the concentration of the influent substrate in the bioreactor, 88% of the working microorganisms accumulated PHA up to 74.5% (wt/wt) (Albuquerque et al., 2010), corresponding to a PHA concentration of $ca.$ 5.1 g $\cdot$ L^{-1}. MMC have been extensively studied, including the implementation of different strategies to manipulate the polymer monomer composition (Albuquerque et al., 2011). MMCs allow the use of already existing wastewater treatment plants to produce PHA but require long operation periods, on the opposite of some existing processes. The choice of one or another operational mode ($i.e.$, fed-batch or MMC) as a sustainable process depends on the scenario of each region.

As stated before, PHB and related copolymers are produced in Brazil in a bioprocess facility integrated into a sugarcane mill. The energy necessary for the production process is provided by waste biomass. Carbon dioxide emissions to the environment are photosynthetically assimilated by the sugarcane crop and liquid wastes are recycled to the cane fields (Nonato et al., 2001).

Considering downstream processing, the recovery of PHAs usually demand a considerable energy input for centrifugation and cell disruption (Harding et al., 2007). Several strategies have been used to diminish the downstream processing costs and the toxic effects of organic solvents traditionally used for polymer solubilization (Berger et al., 1989). The methods based on non-PHA cell mass dissolution are considered a smart alternative (Kapritchkoff et al., 2006; Martínez et al., 2011). These methods, extensively reviewed by Jacquel et al. (2008), utilize alkali, enzymes, slightly acid solutions, and different pre-treatments. Among the recent achievements in this area, there is a new method based on dissolution of non-PHA cell mass by protons in aqueous solution and the crystallization of PHAs (Yu & Chen, 2006). By applying these conditions, high purity (97.9%) and high recovery yield (98.7%) were obtained.

An eventual breakthrough in polymer recovery could be the generation of a suitable mutant of $Alcanivorax$ $borkumensis$ characterized by the extracellular deposition of MCL-PHA when grown on alkanes, allowing the recovery of the polymer from the culture medium (Sabirova et al., 2006).

2.4 Tailor made polymers

Microbiologists have the skills to engineer bacteria for the production of tailored polymeric reserve materials (Hunter, 2010). Since the discovery that some bacteria can incorporate 3-hydroxyalkanoates bearing functional groups from related substrates (Lenz et al., 1992), research has led to structural diversification of PHAs by modulated processes during biosynthesis and chemical modifications (Hazer & Steinbüchel, 2007). Holmes et al. (1984) described the controlled synthesis of P(HB-co-HV) in $R.$ $eutropha$, in which the 3-hydroxyvalerate fraction in the polymer could be controlled by the concentration of propionate in the growth medium. After the discovery of poly(3-hydroxyoctanoate-*co*-3-hydroxyhexanoate) in octane-grown *Pseudomonas oleovorans* (de Smet et al., 1983), the range of different constituents of PHAs expanded rapidly, and *ca*. 200 different PHA monomers have been identified (Steinbüchel & Lütke-Eversloh, 2003). However, the most commonly applied route for tailoring PHAs is their *in situ* functionalization by biosynthetically producing side chains with terminal double bonds followed by chemistry (revised in Scholz, 2010). PHAs with terminal double bonds were first described by Lageveen et al. (1988) and received a lot of follow-up research (Fritzsche et al., 1990; Hartmann et al., 2006; Park et al., 1998). In Pseudomonads, PHAs that are formed from glycerol, gluconate, or related sugars have a different composition with respect to PHAs obtained from fatty acids. Whereas the latter PHAs have 3-hydroxyoctanoate as the main constituent, sugar-grown cells accumulate PHAs in which 3-hydroxydecanoate is the main constituent, along with small amounts of unsaturated monomers (Huijberts et al., 1992).

The resulting tailor-made structural and material properties have positioned PHAs well to contribute to the manufacturing of second and third generation biomaterials for medical applications, which require a variety of tailor-made chemical architectures, physical properties, and surface characteristics (Chen, 2009; Escapa et al., 2011). Bacterial copolyesters with vinyl groups have attracted attention because the unsaturated terminal group is highly reactive when compared to other terminal groups. The evaluation of different plant oils as carbon source for PHA production by *Pseudomonas* spp. revealed the possibility of tailored synthesis of these polymers containing variable molar fractions of unsaturated monomers in a sustainable way (Silva-Queiroz et al., 2009). Some studies described the biosynthesis of alkyl esters substituted MCL-PHA (Scholz et al., 1994), as well as PHAs containing sulphur-groups in the side chains, comprising either thiophenoxy functional groups (Takagi et al., 1999), or thioester groups (Ewering et al., 2002). Moreover, biopolymers with thioester linkages in the polymer backbone, containing 3-mercaptopropionate or 3-mercaptobutyrate in addition to 3-hydroxybutyrate as the monomer constituents, were isolated from *R. eutropha* (Lütke-Eversloh et al., 2002). Molecular biology strategies designed to increase the production of MCL-PHA in *Pseudomonas* was firstly described in *P. putida* U (García et al., 1999). The existence in the genome of this strain of several sets of iso-enzymes encoding genes similar to those belonging to the *fad* regulon from *E. coli* from the β-oxidation of fatty acids have been described (Olivera et al., 2001a, 2001b). Engineered strains carrying mutations in the *fadA-fadB* genes had a strong intracellular accumulation of biopolyesters. Furthermore, the application of this strategy resulted in an over-accumulation of functionalized MCL-PHAs bearing aromatic side groups (Olivera et al., 2001b).

Similarly, the existence of several sets of *fad* genes in the model microorganism *P. putida* KT2440 has been mentioned in the literature, which is in agreement with the huge metabolic versatility of this strain (Nelson et al., 2002). When the *fadA* and *fadB* genes were knocked-out in its derived strain *P. putida* KT2442, PHAs with a higher fraction of long chain length monomers than the wild type, or even containing monomers with thioester-groups were produced (Escapa et al., 2011; Ouyang et al., 2007). Interestingly, terminal oxo- or thio-ester groups could undergo trans-esterifications reactions (Escapa et al., 2011).

3. Applications

The versatile copolymer P(HB-*co*-HV) was initially manufactured as shampoo bottles and other cosmetic containers (Hocking & Marchessault, 1994). Later on, pens, cups, and

Making Green Polymers Even Greener: Towards Sustainable Production of Polyhydroxyalkanoates from
Agroindustrial By-Products

91

packaging elements (*e.g.*, films) made with PHAs also appeared in the market. PHAs are biocompatible and for this reason they have also attracted attention as raw material to be used in medical devices (Wu et al., 2009). Being composed by R-$(-)$ monomers, PHAs are a source of chiral compounds with a high demand from the pharmaceutical industries (Chen & Wu, 2005). However, the manufacture of PHAs is carried out at small facilities and, as a consequence, it lacks the economic benefit of a large scale production (Chanprateep, 2010). A complete description of the goods produced as prototypes or already traded is presented by Philip et al. (2007).

4. Future research

- The technical potential substitution of plastic applications (thermoplastic and thermosets) and man-made fibers (*e.g.*, staple fibers and filaments) by bio-based plastics have been estimated based on their typical physical properties. The potential of biobased plastics for replacement of petrochemical plastics is 90%, corresponding to 240 million tons per year. PHA would respond for *ca.* 30 million tons (Akaraonye et al., 2010). Realizing this potential represents a great challenge, especially in a sustainable way.
- Bacterial growth in bioreactors needs an *ad fundum* understanding of microbial physiology and regulatory processes in order to select cultivation conditions aimed at an enhanced energy-saving process. All the attempts to grow PHA microbial producers under low oxygen supply provide an interesting starting point for these processes, but polymer yields are lower than those obtained under aerobic conditions. Additional process development and optimization are needed to achieve high PHA volumetric productivities and polymer content.
- The use of industrial and agricultural by-products is certainly needed for sustainability. However, high amounts of energy are still needed for production, extraction, and purification of PHAs. Hence, the definition of renewable energy sources will be also quite important.
- Metabolic-Engineering driven approaches should be a relevant tool to establish processes allowing to reach PHA yields close to the theoretical maximum from a given carbon source. Considering the relevance of carbon source on PHAs production cost, it will be important to explore the full metabolic potential of microbial cells.
- The great diversity of monomers detected as PHAs constituents is certainly the feature determining their great potential for technical replacement of petrochemical thermoplastics. Therefore, directed evolution of enzymes involved in PHA biosynthesis and Metabolic Engineering approaches of bacterial hosts will be the driving force to establish bioprocesses for the controlled production of PHAs with monomer composition *à la carte* and hence suitable for a number of applications. The potential of technical replacement could even be increased as the outcome of intensive scientific and technological work to explore the diversity of PHAs composition.
- Systems-level analysis of metabolic, signaling, and regulatory networks makes it possible to comprehensively understand global physiological processes taking place in PHA-accumulating *E. coli* strains. New targets and strategies for the improvement of PHA production will certainly be developed in the next future, including tailor-made PHAs with desired monomer compositions and M_rs. Ideally, and in order to design a completely sustainable PHA production process, strains developed using these system-

based approaches should be further metabolically engineered to produce PHAs up to a sufficiently high polymer content with high productivity from the most inexpensive carbon source through fine-controlled fermentation schemes.

Despite these great challenges, the current scenario is highly promising for the development of sustainable PHA production bioprocesses which could fulfill our needs for biopolymers applications.

5. Acknowledgments

This work was supported by the Ibero-American Programme for Science, Technology, and Development (CYTED). The authors are members of a CYTED network.

6. References

Ahn, W.S.; Park, S.J. & Lee, S.Y. (2000). Production of Poly(3-Hydroxybutyrate) by Fed-Batch Culture of Recombinant *Escherichia coli* with a Highly Concentrated Whey Solution. *Applied and Environmental Microbiology*, Vol. 66, No. 8, (August 2000), pp. 3624-3627, ISSN 0099-2240

Akaraonye, E.; Keshavarz, T. & Roy, I. (2010). Production of Polyhydroxyalkanoates: The Future Green Materials of Choice. *Journal of Chemical Technology and Biotechnology*, Vol. 85, No. 6, (June 2010), pp. 732-743, ISSN 1097-4660

Albuquerque, M.G.E.; Concas, S., Bengtsson, S. & Reis, M.A.M. (2010). Mixed Culture Polyhydroxyalkanoates Production from Sugar Molasses: The Use of a 2-Stage CSTR System for Culture Selection. *Bioresource Technology*, Vol. 101, No. 18, (September 2010), pp. 7112-7122, ISSN 0960-8524

Albuquerque, M.G.E.; Martino, V., Pollet, E., Avérous, L. & Reis, M.A.M. (2011). Mixed Culture Polyhydroxyalkanoate (PHA) Production from Volatile Fatty Acid (VFA)-Rich Streams: Effect of Substrate Composition and Feeding Regime on PHA Productivity, Composition and Properties. *Journal of Biotechnology*, Vol. 151, No. 1, (January 2011), pp. 66-76, ISSN 0168-1656

Aldor, I.S. & Keasling, J.S. (2003). Process Design for Microbial Plastic Factories: Metabolic Engineering of Polyhydroxyalkanoates. *Current Opinion in Biotechnology*, Vol. 14, No. 5, (October 2003), pp. 475-483, ISSN 0958-1669

Ashby, R.D.; Solaiman, D.K.Y. & Foglia, T. (2005). Synthesis of Short-/Medium-Chain-Length Poly(Hydroxyalkanoate) Blends by Mixed Culture Fermentation of Glycerol. *Biomacromolecules*, Vol. 6, No. 4, (July 2005), pp. 2106-2112, ISSN 1525-7797

Avison, M.B.; Horton, R.E., Walsh, T.R. & Bennett, P.M. (2001). *Escherichia coli* CreBC Is a Global Regulator of Gene Expression That Responds to Growth in Minimal Media. *Journal of Biological Chemistry*, Vol. 276, No. 29, (July 2001), pp. 26955-26961, ISSN 0021-9258

Beaulieu, M.; Beaulieu, Y., Mélinard, J., Pandian, S. & Goulet, J. (1995). Influence of Ammonium Salts and Cane Molasses on Growth of *Alcaligenes eutrophus* and Production of Polyhydroxybutyrate. *Applied and Environmental Microbiology*, Vol. 61, No. 1, (January 1995), pp. 165-169, ISSN 0099-2240

Berger, E.; Ramsay, B.A., Ramsay, J.A., Chavarie, C. & Braunegg, G. (1989). PHB Recovery by Hypochlorite Digestion of Non-PHB Biomass. *Biotechnology Techniques*, Vol. 3, No. 4, (April 1989), pp. 227-232, ISSN 0951-208X

Borman, E.J. & Roth, M. (1999). The Production of Polyhydroxybutyrate by *Methylobacterium rhodesianum* and *Ralstonia eutropha* in Media Containing Glycerol and Casein Hydrolysates. *Biotechnology Letters*, Vol. 21, No. 12, (December 1999), pp. 1059-1063, ISSN 0141-5492

Carlson, R.; Wlaschin, A. & Srienc, F. (2005). Kinetic Studies and Biochemical Pathway Analysis of Anaerobic Poly-(R)-3-Hydroxybutyric Acid Synthesis in *Escherichia coli*. *Applied and Environmental Microbiology*, Vol. 71, No. 2, (February 2005), pp. 713-720, ISSN 0099-2240

Cavalheiro, J.M.B.T.; de Almeida, M.C.M.D., Grandfils, C. & da Fonseca, M.M.R. (2009). Poly(3-Hydroxybutyrate) Production by *Cupriavidus necator* Using Waste Glycerol. *Process Biochemistry*, Vol. 44, No. 5, (May 2009), pp. 509-515, ISSN 1359-5113

Çelik, G.Y.; Beyatli, Y. & Aslim, B. (2005). Determination of Poly-β-Hydroxybutyrate (PHB) in Sugarbeet Molasses by *Pseudomonas cepacia* G13 Strain. *Zuckerindustrie*, Vol. 130, No. 3, (March 2005), pp. 201-203, ISSN 0344-8657

Chanprateep, S. (2010). Current Trends in Biodegradable Polyhydroxyalkanoates. *Journal of Bioscience and Bioengineering*, Vol. 110, No. 6, (December 2010), pp. 621-632, ISSN 1389-1723

Chen, G.G.Q. & Wu, Q. (2005). Microbial Production and Applications of Chiral Hydroxyalkanoates. *Applied Microbiology and Biotechnology*, Vol. 67, No. 5, (June 2005), pp. 592-599, ISSN 0175-7598

Chen, G.G.Q. (2009). A Microbial Polyhydroxyalkanoates (PHA) Based Bio- and Materials Industry. *Chemical Society Reviews*, Vol. 38, No. 8, (August 2009), pp. 2434-2446, ISSN 0306-0012

Corti, A.; Muniyasamy, S., Vitali, M., Imam, S.H. & Chiellini, E. (2010). Oxidation and Biodegradation of Polyethylene Films Containing Pro-Oxidant Additives: Synergistic Effects of Sunlight Exposure, Thermal Aging and Fungal Biodegradation. *Polymer Degradation and Stability*, Vol. 95, No. 6, (June 2010), pp. 1106-1114, ISSN 0141-3910

da Silva, G.P.; Mack, M. & Contiero, J. (2009). Glycerol: A Promising and Abundant Carbon Source for Industrial Microbiology. *Biotechnology Advances*, Vol. 27, No. 1, (January-February 2009), pp. 30-39, ISSN 0734-9750

de Almeida, A.; Giordano, A.M., Nikel, P.I. & Pettinari, M.J. (2010). Effects of Aeration on the Synthesis of Poly(3-Hydroxybutyrate) from Glycerol and Glucose in Recombinant *Escherichia coli*. *Applied and Environmental Microbiology*, Vol. 76, No. 6, (March 2010), pp. 2036-2040, ISSN 0099-2240

de Smet, M.J.; Eggink, G., Witholt, B., Kingma, J. & Wynberg, H. (1983). Characterization of Intracellular Inclusions Formed by *Pseudomonas oleovorans* During Growth on Octane. *Journal of Bacteriology*, Vol. 154, No. 2, (May 1983), pp. 870-878, ISSN 0021-9193

Escapa, I.F.; Morales, V., Martino, V.P., Pollet, E., Avérous, L., García, J.L. & Prieto, M.A. (2011). Disruption of β-Oxidation Pathway in *Pseudomonas putida* KT2442 to Produce New Functionalized PHAs with Thioester Groups. *Applied Microbiology and Biotechnology*, Vol. 89, No. 5, (March 2011), pp. 1583-1598, ISSN 0175-7598

Ewering, C.; Lütke-Eversloh, T., Luftmann, H. & Steinbüchel, A. (2002). Identification of Novel Sulfur-Containing Bacterial Polyesters: Biosynthesis of Poly(3-Hydroxy-S-Propyl-ω-Thioalkanoates) Containing Thioether Linkages in the Side Chains. *Microbiology*, Vol. 148, No. 5, (May 2002), pp. 1397-1406, ISSN 1350-0872

Fischer, E.; Zamboni, N. & Sauer, U. (2004). High-Throughput Metabolic Flux Analysis Based on Gas Chromatography-Mass Spectrometry Derived ^{13}C Constraints. *Analytical Biochemistry*, Vol. 325, No. 2, (February 2004), pp. 308-316, ISSN 0003-2697

Fritzsche, K.; Lenz, R.W. & Fuller, R.C. (1990). Production of Unsaturated Polyesters by *Pseudomonas oleovorans*. *International Journal of Biological Macromolecules*, Vol. 12, No. 2, (April 1990), pp. 85-91, ISSN 0141-8130

García, B.; Olivera, E.R., Miñambres, B., Fernández-Valverde, M., Cañedo, L.M., Prieto, M.A., García, J.L., Martínez, M. & Luengo, J.M. (1999). Novel Biodegradable Aromatic Plastics from a Bacterial Source. Genetic and Biochemical Studies on a Route of the Phenylacetyl-CoA Catabolon. *Journal of Biological Chemistry*, Vol. 274, No. 41, (October 1999), pp. 29228-29241, ISSN 0021-9258

Gerngross, T.U. (1999). Can Biotechnology Move Us toward a Sustainable Society? *Nature Biotechnology*, Vol. 17, No. 6, (June 1999), pp. 541-544, ISSN 1087-0156

Han, M.J.; Yoon, S.S. & Lee, S.Y. (2001). Proteome Analysis of Metabolically Engineered *Escherichia coli* Producing Poly(3-Hydroxybutyrate). *Journal of Bacteriology*, Vol. 183, No. 1, (January 2001), pp. 301-308, ISSN 0021-9193

Harding, K.G.; Dennis, J.S., von Blottnitz, H. & Harrison, S.T.L. (2007). Environmental Analysis of Plastic Production Processes: Comparing Petroleum-Based Polypropylene and Polyethylene with Biologically Based Poly-β-Hydroxybutyric Acid Using Life Cycle Analysis. *Journal of Biotechnology*, Vol. 130, No. 1, (May 2007), pp. 57-66, ISSN 0168-1656

Hartmann, R.; Hany, R., Pletscher, E., Ritter, A., Witholt, B. & Zinn, M. (2006). Tailor-Made Olefinic Medium-Chain-Length Poly[(R)-3-Hydroxyalkanoates] by *Pseudomonas putida* GPo1: Batch *versus* Chemostat Production. *Biotechnology and Bioengineering*, Vol. 93, No. 4, (March 2006), pp. 737-746, ISSN 1097-0290

Hazer, B. & Steinbüchel, A. (2007). Increased Diversification of Polyhydroxyalkanoates by Modification Reactions for Industrial and Medical Applications. *Applied Microbiology and Biotechnology*, Vol. 74, No. 1, (February 2007), pp. 1-12, ISSN 0175-7598

Hocking, P.J. & Marchessault, R.H. (1994). Biopolyesters, In: *Chemistry and Technology of Biodegradable Polymers*, G.J.L. Griffin, (Ed.), pp. 48-96, Blackie Academic & Professional, ISBN 0-7514-0003-3, Glasgow, United Kingdom

Holmes, P.A.; Collins, S.H. & Wright, L.F. (1984). 3-Hydroxybutyrate Polymers. *U.S. Patent 4,477,654*, (October 1984)

Hong, S.H.; Park, S.J., Moon, S.Y., Park, J.P. & Lee, S.Y. (2003). *In Silico* Prediction and Validation of the Importance of the Entner-Doudoroff Pathway in Poly(3-Hydroxybutyrate) Production by Metabolically Engineered *Escherichia coli*. *Biotechnology and Bioengineering*, Vol. 83, No. 7, (September 2003), pp. 854-863, ISSN 1097-0290

Horng, Y.T.; Chang, K.C., Chien, C.C., Wei, Y.H., Sun, Y.M. & Soo, P.C. (2010). Enhanced Polyhydroxybutyrate (PHB) Production *via* the Coexpressed *phaCAB* and *vgb* Genes Controlled by Arabinose P_{BAD} Promoter in *Escherichia coli*. *Letters in Applied Microbiology*, Vol. 50, No. 2, (February 2010), pp. 158-167, ISSN 1472-765X

Horng, Y.T.; Chien, C.C., Wei, Y.H., Chen, S.Y., Lan, J.C., Sun, Y.M. & Soo, P.C. (2011). Functional *cis*-Expression of *phaCAB* Genes for Poly(3-Hydroxybutyrate) Production by *Escherichia coli*. *Letters in Applied Microbiology*, Vol. 52, No. 5, (May 2011), pp. 475-483, ISSN 1472-765X

Huijberts, G.N.; Eggink, G., de Waard, P., Huisman, G.W. & Witholt, B. (1992). *Pseudomonas putida* KT2442 Cultivated on Glucose Accumulates Poly(3-Hydroxyalkanoates) Consisting of Saturated and Unsaturated Monomers. *Applied and Environmental Microbiology*, Vol. 58, No. 2, (February 1992), pp. 536-544, ISSN 0099-2240

Hunter, P. (2010). Can Bacteria Save the Planet? *EMBO Reports*, Vol. 11, No. 4, (April 2010), pp. 266-269, ISSN 1469-221X

Ibrahim, M.H.A. & Steinbüchel, A. (2009). Poly(3-Hydroxybutyrate) Production from Glycerol by *Zobellella denitrificans* MW1 *via* High-Cell-Density Fed-Batch Fermentation and Simplified Solvent Extraction. *Applied and Environmental Microbiology*, Vol. 75, No. 19, (October 2009), pp. 6222-6231, ISSN 0099-2240

Jacquel, N.; Lo, C.W., Wei, Y.H., Wu, H.S. & Wang, S.S. (2008). Isolation and Purification of Bacterial Poly(3-Hydroxyalkanoates). *Biochemical Engineering Journal*, Vol. 39, No. 1, (April 2008), pp. 15-27, ISSN 1369-703X

Jung, Y.K.; Lee, S.Y. & Tam, T.T. (2010). Towards Systems Metabolic Engineering of PHA Producers, In: *Plastics from Bacteria: Natural Functions and Applications*, G.G.Q. Chen, (Ed.), pp. 63-84, Springer-Verlag, ISBN 978-3-642-03286-8, Berlin, Germany

Kapritchkoff, F.M.; Viotti, A.P., Alli, R.C.P., Zuccolo, M., Pradella, J.G.C., Maiorano, A.E., Miranda, E.A. & Bonomi, A. (2006). Enzymatic Recovery and Purification of Polyhydroxybutyrate Produced by *Ralstonia eutropha*. *Journal of Biotechnology*, Vol. 122, No. 4, (April 2006), pp. 453-462, ISSN 0168-1656

Keshavarz, T. & Roy, I. (2010). Polyhydroxyalkanoates: Bioplastics with a Green Agenda. *Current Opinion in Microbiology*, Vol. 13, No. 3, (June 2010), pp. 321-326, ISSN 1369-5274

Kim, B.S. (2000). Production of Poly(3-Hydroxybutyrate) from Inexpensive Substrates. *Enzyme and Microbial Technology*, Vol. 27, No. 10, (December 2000), pp. 774-777, ISSN 0141-0229

Kim, S. & Dale, B.E. (2005). Lifecycle Assessment Study of Biopolymer (Polyhydroxyalkanoates) - Derived from No-Tilled Corn. *The International Journal of Life Cycle Assessment*, Vol. 10, No. 3, (May 2005), pp. 200-210, ISSN 0948-3349

Koller, M.; Bona, R., Braunegg, G., Hermann, C., Horvat, P., Kroutil, M., Martinz, J., Neto, J., Pereira, L. & Varila, P. (2005). Production of Polyhydroxyalkanoates from Agricultural Waste and Surplus Materials. *Biomacromolecules*, Vol. 6, No. 2, (March 2005), pp. 561-565, ISSN 1525-7797

Koller, M.; Hesse, P., Bona, R., Kutschera, C., Atlić, A. & Braunegg, G. (2007). Potential of Various Archae- and Eubacterial Strains as Industrial Polyhydroxyalkanoate Producers from Whey. *Macromolecular Bioscience*, Vol. 7, No. 2, (February 2007), pp. 218-226, ISSN 1616-5195

Koller, M.; Bona, R., Chiellini, E., Grillo-Fernandes, E., Horvat, P., Kutschera, C., Hesse, P. & Braunegg, G. (2008). Polyhydroxyalkanoate Production from Whey by *Pseudomonas hydrogenovora*. *Bioresource Technology*, Vol. 99, No. 11, (July 2008), pp. 4854-4863, ISSN 0960-8524

Kulpreecha, S.; Boonruangthavorn, A., Meksiriporn, B. & Thongchul, N. (2009). Inexpensive Fed-Batch Cultivation for High Poly(3-Hydroxybutyrate) Production by a New Isolate of *Bacillus megaterium*. *Journal of Bioscience and Bioengineering*, Vol. 107, No. 3, (March 2009), pp. 240-245, ISSN 1389-1723

Lageveen, R.G.; Huisman, G.W., Preusting, H., Ketelaar, P., Eggink, G. & Witholt, B. (1988). Formation of Polyesters by *Pseudomonas oleovorans*: Effect of Substrates on Formation and Composition of Poly-(R)-3-Hydroxyalkanoates and Poly-(R)-3-

Hydroxyalkenoates. *Applied and Environmental Microbiology*, Vol. 54, No. 12, (December 1988), pp. 2924-2932, ISSN 0099-2240

Lee, S.Y.; Kim, H.U., Yun, H., Sohn, S.B., Kim, J.S., Palsson, B.Ø., Herrgård, M.J. & Portnoy, V.A. (2010). Systems Biology, Genome-Scale Models, and Metabolic Engineering, In: *The Metabolic Pathway Engineering Handbook - Tools and Applications*, C.D. Smolke, (Ed.), pp. 15.1-15.11, CRC Press, ISBN 978-1-4200-7765-0, Boca Raton, Florida, United States of America

Lenz, R.W.; Kim, Y.B. & Fuller, R.C. (1992). Production of Unusual Bacterial Polyesters by *Pseudomonas oleovorans* through Cometabolism. *FEMS Microbiology Letters*, Vol. 103, No. 2-4, (December 1992), pp. 207-214, ISSN 1574-6968

Li, R.; Zhang, H. & Qi, Q. (2007). The Production of Polyhydroxyalkanoates in Recombinant *Escherichia coli*. *Bioresource Technology*, Vol. 98, No. 12, (September 2007), pp. 2313-2320, ISSN 0960-8524

Lim, S.J.; Jung, Y.M., Shin, H.D. & Lee, Y.H. (2002). Amplification of the NADPH-Related Genes *zwf* and *gnd* for the Oddball Biosynthesis of PHB in an *E. coli* Transformant Harboring a Cloned *phbCAB* Operon. *Journal of Bioscience and Bioengineering*, Vol. 93, No. 6, (October 2002), pp. 543-549, ISSN 1389-1723

Lütke-Eversloh, T.; Bergander, K., Luftmann, H. & Steinbüchel, A. (2001). Biosynthesis of Poly(3-Hydroxybutyrate-co-3-Mercaptobutyrate) as a Sulfur Analogue to Poly(3-Hydroxybutyrate) (PHB). *Biomacromolecules*, Vol. 2, No. 3, (August 2001), pp. 1061-1065, ISSN 1525-7797

Lütke-Eversloh, T.; Fischer, A., Remminghorst, U., Kawada, J., Marchessault, R.H., Bögershausen, A., Kalwei, M., Eckert, H., Reichelt, R., Liu, S.J. & Steinbüchel, A. (2002). Biosynthesis of Novel Thermoplastic Polythioesters by Engineered *Escherichia coli*. *Nature Materials*, Vol. 1, No. 4, (December 2002), pp. 236-240, ISSN 1476-1122

Lynch, A.S. & Lin, E.C.C. (1996). Responses to Molecular Oxygen, In: *Escherichia coli and Salmonella: Cellular and Molecular Biology*, F.C. Neidhardt, R. Curtiss III, J.L. Ingraham, E.C.C. Lin, K.B. Low, B. Magasanik, W.S. Reznikoff, M. Riley, M. Schaechter, H.E. Umbarger, (Eds.), pp. 1526-1538, ASM Press, ISBN 1-5558-1084-5, Washington, D.C., United States of America

Madden, L.A.; Anderson, A.J., Shah, D.T. & Asrar, J. (1999). Chain Termination in Polyhydroxyalkanoate Synthesis: Involvement of Exogenous Hydroxy-Compounds as Chain Transfer Agents. *International Journal of Biological Macromolecules*, Vol. 25, No. 1-3, (June 1999), pp. 43-53, ISSN 0141-8130

Madison, L.L. & Huisman, G.W. (1999). Metabolic Engineering of Poly(3-Hydroxyalkanoates): From DNA to Plastic. *Microbiology and Molecular Biology Reviews*, Vol. 63, No. 1, (March 1999), pp. 21-53, ISSN 1092-2172

Mahishi, L.H.; Tripathi, G. & Rawal, S.K. (2003). Poly(3-Hydroxybutyrate) (PHB) Synthesis by Recombinant *Escherichia coli* Harbouring *Streptomyces aureofaciens* PHB Biosynthesis Genes: Effect of Various Carbon and Nitrogen Sources. *Microbiological Research*, Vol. 158, No. 1, (January 2003), pp. 19-27, ISSN 0944-5013

Marangoni, C.; Furigo Jr., A. & de Aragão, G.M.F. (2002). Production of Poly(3-Hydroxybutyrate-co-3-Hydroxyvalerate) by *Ralstonia eutropha* in Whey and Inverted Sugar with Propionic Acid Feeding. *Process Biochemistry*, Vol. 38, No. 2, (October 2002), pp. 137-141, ISSN 1359-5113

Making Green Polymers Even Greener: Towards Sustainable Production of Polyhydroxyalkanoates from
Agroindustrial By-Products

97

Martínez, V.; García, P., García, J.L. & Prieto, M.A. (2011). Controlled Autolysis Facilitates the Polyhydroxyalkanoate Recovery in *Pseudomonas putida* KT2440. *Microbial Biotechnology*, Vol. 4, No. 4, (July 2011), pp. 533-547, ISSN 1751-7915

Nath, A.; Dixit, M., Bandiya, A., Chavda, S. & Desai, A.J. (2008). Enhanced PHB Production and Scale up Studies Using Cheese Whey in Fed Batch Culture of *Methylobacterium* sp. ZP24. *Bioresource Technology*, Vol. 99, No. 13, (September 2008), pp. 5749-5755, ISSN 0960-8524

Nelson, K.E.; Weinel, C., Paulsen, I.T., Dodson, R.J., Hilbert, H., Martins dos Santos, V.A.P., Fouts, D.E., Gill, S.R., Pop, M., Holmes, M., Brinkac, L., Beanan, M., DeBoy, R.T., Daugherty, S., Kolonay, J., Madupu, R., Nelson, W., White, O., Peterson, J., Khouri, H., Hance, I., Chris Lee, P., Holtzapple, E., Scanlan, D., Tran, K., Moazzez, A., Utterback, T., Rizzo, M., Lee, K., Kosack, D., Moestl, D., Wedler, H., Lauber, J., Stjepandic, D., Hoheisel, J., Straetz, M., Heim, S., Kiewitz, C., Eisen, J.A., Timmis, K.N., Düsterhöft, A., Tümmler, B. & Fraser, C.M. (2002). Complete Genome Sequence and Comparative Analysis of the Metabolically Versatile *Pseudomonas putida* KT2440. *Environmental Microbiology*, Vol. 4, No. 12, (December 2002), pp. 799-808, ISSN 1462-2920

Nielsen, J. (2001). Metabolic Engineering. *Applied Microbiology and Biotechnology*, Vol. 55, No. 3, (April 2001), pp. 263-283, ISSN 0175-7598

Nielsen, J. & Jewett, M.C. (2008). Impact of Systems Biology on Metabolic Engineering of *Saccharomyces cerevisiae*. *FEMS Yeast Research*, Vol. 8, No. 1, (February 2008), pp. 122-131, ISSN 1567-1364

Nikel, P.I.; de Almeida, A., Melillo, E.C., Galvagno, M.A. & Pettinari, M.J. (2006a). New Recombinant *Escherichia coli* Strain Tailored for the Production of Poly(3-Hydroxybutyrate) from Agroindustrial By-Products. *Applied and Environmental Microbiology*, Vol. 72, No. 6, (June 2006), pp. 3949-3954, ISSN 0099-2240

Nikel, P.I.; Pettinari, M.J., Galvagno, M.A. & Méndez, B.S. (2006b). Poly(3-Hydroxybutyrate) Synthesis by Recombinant *Escherichia coli arcA* Mutants in Microaerobiosis. *Applied and Environmental Microbiology*, Vol. 72, No. 4, (April 2006), pp. 2614-2620, ISSN 0099-2240

Nikel, P.I.; de Almeida, A., Pettinari, M.J. & Méndez, B.S. (2008a). The Legacy of HfrH: Mutations in the Two-Component System CreBC Are Responsible for the Unusual Phenotype of an *Escherichia coli arcA* Mutant. *Journal of Bacteriology*, Vol. 190, No. 9, (May 2008), pp. 3404-3407, ISSN 0021-9193

Nikel, P.I.; Pettinari, M.J., Galvagno, M.A. & Méndez, B.S. (2008b). Poly(3-Hydroxybutyrate) Synthesis from Glycerol by a Recombinant *Escherichia coli arcA* Mutant in Fed-Batch Microaerobic Cultures. *Applied Microbiology and Biotechnology*, Vol. 77, No. 6, (January 2008), pp. 1337-1343, ISSN 0175-7598

Nikel, P.I.; Zhu, J., San, K.Y., Méndez, B.S. & Bennett, G.N. (2009). Metabolic Flux Analysis of *Escherichia coli creB* and *arcA* Mutants Reveals Shared Control of Carbon Catabolism under Microaerobic Growth Conditions. *Journal of Bacteriology*, Vol. 191, No. 17, (September 2009), pp. 5538-5548, ISSN 0021-9193

Nonato, R.V.; Mantelatto, P.E. & Rossell, C.E.V. (2001). Integrated Production of Biodegradable Plastic, Sugar and Ethanol. *Applied Microbiology and Biotechnology*, Vol. 57, No. 1-2, (October 2001), pp. 1-5, ISSN 0175-7598

Oliveira, F.C.; Freire, D.M. & Castilho, L.R. (2004). Production of Poly(3-Hydroxybutyrate) by Solid-State Fermentation with *Ralstonia eutropha*. *Biotechnology Letters*, Vol. 26, No. 24, (December 2004), pp. 1851-1855, ISSN 0141-5492

Olivera, E.R.; Carnicero, D., García, B., Miñambres, B., Moreno, M.A., Cañedo, L., DiRusso, C.C., Naharro, G. & Luengo, J.M. (2001a). Two Different Pathways Are Involved in the β-Oxidation of *n*-Alkanoic and *n*-Phenylalkanoic Acids in *Pseudomonas putida* U: Genetic Studies and Biotechnological Applications. *Molecular Microbiology*, Vol. 39, No. 4, (February 2001), pp. 863-874, ISSN 1365-2958

Olivera, E.R.; Carnicero, D., Jodra, R., Miñambres, B., García, B., Abraham, G.A., Gallardo, A., Román, J.S., García, J.L., Naharro, G. & Luengo, J.M. (2001b). Genetically Engineered *Pseudomonas*: A Factory of New Bioplastics with Broad Applications. *Environmental Microbiology*, Vol. 3, No. 10, (October 2001), pp. 612-618, ISSN 1462-2920

Ouyang, S.P.; Luo, R.C., Chen, S.S., Liu, Q., Chung, A., Wu, Q. & Chen, G.G.Q. (2007). Production of Polyhydroxyalkanoates with High 3-Hydroxydodecanoate Monomer Content by *fadB* and *fadA* Knockout Mutant of *Pseudomonas putida* KT2442. *Biomacromolecules*, Vol. 8, No. 8, (August 2007), pp. 2504-2511, ISSN 1525-7797

Park, S.J.; Park, J.P. & Lee, S.Y. (2002). Production of Poly(3-Hydroxybutyrate) from Whey by Fed-Batch Culture of Recombinant *Escherichia coli* in a Pilot-Scale Fermenter. *Biotechnology Letters*, Vol. 24, No. 3, (February 2002), pp. 185-189, ISSN 0141-5492

Park, W.H.; Lenz, R.W. & Goodwin, S. (1998). Epoxidation of Bacterial Polyesters with Unsaturated Side Chains. I. Production and Epoxidation of Polyesters from 10-Undecenoic Acid. *Macromolecules*, Vol. 31, No. 5, (March 1998), pp. 1480-1486, ISSN 0024-9297

Philip, S.; Keshavarz, T. & Roy, I. (2007). Polyhydroxyalkanoates: Biodegradable Polymers with a Range of Applications. *Journal of Chemical Technology and Biotechnology*, Vol. 82, No. 3, (March 2007), pp. 233-247, ISSN 1097-4660

Pohlmann, A.; Fricke, W.F., Reinecke, F., Kusian, B., Liesegang, H., Cramm, R., Eitinger, T., Ewering, C., Pötter, M., Schwartz, E., Strittmatter, A., Voß, I., Gottschalk, G., Steinbüchel, A., Friedrich, B. & Bowien, B. (2006). Genome Sequence of the Bioplastic-Producing "Knallgas" Bacterium *Ralstonia eutropha* H16. *Nature Biotechnology*, Vol. 24, No. 10, (October 2006), pp. 1257-1262, ISSN 1087-0156

Povolo, S.; Toffano, P., Basaglia, M. & Casella, S. (2010). Polyhydroxyalkanoates Production by Engineered *Cupriavidus necator* from Waste Material Containing Lactose. *Bioresource Technology*, Vol. 101, No. 20, (October 2010), pp. 7902-7907, ISSN 0960-8524

Pries, A.; Steinbüchel, A. & Schlegel, H.G. (1990). Lactose- and Galactose-Utilizing Strains of Poly(Hydroxyalkanoic Acid)-Accumulating *Alcaligenes eutrophus* and *Pseudomonas saccharophila* Obtained by Recombinant DNA Technology. *Applied Microbiology and Biotechnology*, Vol. 33, No. 4, (July 1990), pp. 410-417, ISSN 0175-7598

Queiroz, A.U.B. & Collares-Queiroz, F.P. (2009). Innovation and Industrial Trends in Bioplastics. *Polymer Reviews*, Vol. 49, No. 2, (April 2009), pp. 65-78, ISSN 1558-3724

Reddy, S.V.; Thirumala, M. & Mahmood, S.K. (2009). A Novel *Bacillus* sp. Accumulating Poly(3-Hydroxybutyrate-*co*-3-Hydroxyvalerate) from a Single Carbon Substrate. *Journal of Industrial Microbiology and Biotechnology*, Vol. 36, No. 6, (June 2009), pp. 837-843, ISSN 1367-5435

Rehm, B.H.A. (2010). Bacterial Polymers: Biosynthesis, Modifications and Applications. *Nature Reviews Microbiology*, Vol. 8, No. 8, (August 2010), pp. 578-592, ISSN 1740-1526

Reinecke, F. & Steinbüchel, A. (2009). *Ralstonia eutropha* Strain H16 as Model Organism for PHA Metabolism and for Biotechnological Production of Technically Interesting

Making Green Polymers Even Greener: Towards Sustainable Production of Polyhydroxyalkanoates from
Agroindustrial By-Products

99

Biopolymers. *Journal of Molecular Microbiology and Biotechnology*, Vol. 16, No. 1-2, (October 2009), pp. 91-108, ISSN 1464-1801

Sabirova, J.S.; Ferrer, M., Lünsdorf, H., Wray, V., Kalscheuer, R., Steinbüchel, A., Timmis, K.N. & Golyshin, P.N. (2006). Mutation in a "*tesB*-Like" Hydroxyacyl-Coenzyme A-Specific Thioesterase Gene Causes Hyperproduction of Extracellular Polyhydroxyalkanoates by *Alcanivorax borkumensis* SK2. *Journal of Bacteriology*, Vol. 188, No. 24, (December 2006), pp. 8452-8459, ISSN 0021-9193

Sauer, U. (2006). Metabolic Networks in Motion: ^{13}C-Based Flux Analysis. *Molecular Systems Biology*, Vol. 2, No. 1, (November 2006), pp. 62, ISSN 1744-4292

Sawaya Jank, M. (2011). Company Production Rankings for Sugarcane, Sugar and Ethanol - South-Central Brazil. *In*: UNICA - Sugarcane Industry Association, 07.06.2011, Available from http://english.unica.com.br/dadosCotacao/estatistica/

Scholz, C.; Fuller, R.C. & Lenz, R.W. (1994). Growth and Polymer Incorporation of *Pseudomonas oleovorans* on Alkyl Esters of Heptanoic Acid. *Macromolecules*, Vol. 27, No. 10, (May 1994), pp. 2886-2889, ISSN 0024-9297

Scholz, C. (2010). Perspectives to Produce Positively or Negatively Charged Polyhydroxyalkanoic Acids. *Applied Microbiology and Biotechnology*, Vol. 88, No. 4, (October 2010), pp. 829-837, ISSN 0175-7598

Serafim, L.S.; Lemos, P.C., Albuquerque, M.G.E. & Reis, M.A.M. (2008). Strategies for PHA Production by Mixed Cultures and Renewable Waste Materials. *Applied Microbiology and Biotechnology*, Vol. 81, No. 4, (December 2008), pp. 615-628, ISSN 0175-7598

Silva-Queiroz, S.R.; Silva, L.F., Pradella, J.G.C., Pereira, E.M. & Gomez, J.G.C. (2009). PHA$_{MCL}$ Biosynthesis Systems in *Pseudomonas aeruginosa* and *Pseudomonas putida* Strains Show Differences on Monomer Specificities. *Journal of Biotechnology*, Vol. 143, No. 2, (August 2009), pp. 111-118, ISSN 0168-1656

Solaiman, D.K.Y.; Ashby, R.D., Foglia, T. & Marmer, W.N. (2006). Conversion of Agricultural Feedstock and Coproducts into Poly(Hydroxyalkanoates). *Applied Microbiology and Biotechnology*, Vol. 71, No. 6, (August 2006), pp. 783-789, ISSN 0175-7598

Steinbüchel, A. (2001). Perspectives for Biotechnological Production and Utilization of Biopolymers: Metabolic Engineering of Polyhydroxyalkanoate Biosynthesis as a Successful Example. *Macromolecular Bioscience*, Vol. 1, No. (January 2001), pp. 1-24, ISSN 1616-5195

Steinbüchel, A. & Lütke-Eversloh, T. (2003). Metabolic Engineering and Pathway Construction for Biotechnological Production of Relevant Polyhydroxyalkanoates in Microorganisms. *Biochemical Engineering Journal*, Vol. 16, No. 2, (November 2003), pp. 81-96, ISSN 1369-703X

Steinbüchel, A. (2005). Non-Biodegradable Biopolymers from Renewable Resources: Perspectives and Impacts. *Current Opinion in Biotechnology*, Vol. 16, No. 6, (December 2005), pp. 607-613, ISSN 0958-1669

Stephanopoulos, G. (1999). Metabolic Fluxes and Metabolic Engineering. *Metabolic Engineering*, Vol. 1, No. 1, (January 1999), pp. 1-11, ISSN 1096-7176

Suriyamongkol, P.; Weselake, R., Narine, S., Moloney, M. & Shah, S. (2007). Biotechnological Approaches for the Production of Polyhydroxyalkanoates in Microorganisms and Plants - a Review. *Biotechnology Advances*, Vol. 25, No. 2, (March-April 2007), pp. 148-175, ISSN 0734-9750

Taidi, B.; Anderson, A.J., Dawes, E.A. & Byrom, D. (1994). Effect of Carbon Source and Concentration on the Molecular Mass of Poly(3-Hydroxybutyrate) Produced by *Methylobacterium extorquens* and *Alcaligenes eutrophus*. *Applied Microbiology and Biotechnology*, Vol. 40, No. 6, (February 1994), pp. 786-790, ISSN 0175-7598

Takagi, Y.; Hashii, M., Maehara, A. & Yamane, T. (1999). Biosynthesis of Polyhydroxyalkanoate with a Thiophenoxy Side Group Obtained from *Pseudomonas putida*. *Macromolecules*, Vol. 32, No. 25, (December 1999), pp. 8315-8318, ISSN 0024-9297

Tyo, K.E.; Fischer, C.R., Simeon, F. & Stephanopoulos, G. (2010). Analysis of Polyhydroxybutyrate Flux Limitations by Systematic Genetic and Metabolic Perturbations. *Metabolic Engineering*, Vol. 12, No. 3, (May 2010), pp. 187-195, ISSN 1096-7176

Wang, F. & Lee, S.Y. (1997). Production of Poly(3-Hydroxybutyrate) by Fed-Batch Culture of Filamentation-Supressed Recombinant *Escherichia coli*. *Applied and Environmental Microbiology*, Vol. 63, No. 12, (December 1997), pp. 4765-4769, ISSN 0099-2240

Wei, X.X.; Shi, Z.Y., Yuan, M.Q. & Chen, G.G.Q. (2009). Effect of Anaerobic Promoters on the Microaerobic Production of Polyhydroxybutyrate (PHB) in Recombinant *Escherichia coli*. *Applied Microbiology and Biotechnology*, Vol. 82, No. 4, (March 2009), pp. 703-712, ISSN 0175-7598

Wiechert, W. (2001). [13]C Metabolic Flux Analysis. *Metabolic Engineering*, Vol. 3, No. 3, (July 2001), pp. 195-206, ISSN 1096-7176

Wong, H.H. & Lee, S.Y. (1998). Poly-(3-Hydroxybutyrate) Production from Whey by High-Density Cultivation of Recombinant *Escherichia coli*. *Applied Microbiology and Biotechnology*, Vol. 50, No. 1, (July 1998), pp. 30-33, ISSN 0175-7598

Wu, Q.; Wang, Y. & Chen, G.G.Q. (2009). Medical Application of Microbial Biopolyesters Polyhydroxyalkanoates. *Artificial Cells, Blood Substitutes and Biotechnology*, Vol. 37, No. 1, (January 2009), pp. 1-12, ISSN 1073-1199

Yang, S.T.; Zhu, H., Li, Y. & Hong, G. (1994). Continuous Propionate Production from Whey Permeate Using a Novel Fibrous Bed Bioreactor. *Biotechnology and Bioengineering*, Vol. 43, No. 11, (May 1994), pp. 1124-1130, ISSN 1097-0290

Young, F.K., Kastner, J.R. & May, S.W. (1994). Microbial Production of Poly-β-Hydroxybutyric Acid from D-Xylose and Lactose by *Pseudomonas cepacia*. *Applied and Environmental Microbiology*, Vol. 60, No. 11, (November 1994), pp. 4195-4198, ISSN 0099-2240

Yu, J. & Chen, L.X.L. (2006). Cost-Effective Recovery and Purification of Polyhydroxyalkanoates by Selective Dissolution of Cell Mass. *Biotechnology Progress*, Vol. 22, No. 2, (March 2006), pp. 547-553, ISSN 1520-6063

Zhang, H.; Obias, V., Gonyer, K. & Dennis, D. (1994). Production of Polyhydroxyalkanoates in Sucrose-Utilizing Recombinant *Escherichia coli* and *Klebsiella* Strains. *Applied and Environmental Microbiology*, Vol. 60, No. 4, (April 1994), pp. 1198-1205, ISSN 0099-2240

Zhong, Z.W.; Song, B. & Huang, C.X. (2009). Environmental Impacts of Three Polyhydroxyalkanoate (PHA) Manufacturing Processes. *Materials and Manufacturing Processes*, Vol. 24, No. 5, (March 2009), pp. 519-523, ISSN 1042-6914

5

Biological Activities and Effects of Food Processing on Flavonoids as Phenolic Antioxidants

Ioannou Irina and Ghoul Mohamed
Nancy University – ENSAIA,
France

1. Introduction

Plants produce a great variety of organic compounds as a response to environmental stresses like microbial attack, insect/animal predation and ultraviolet radiations. The role of these metabolites is to increase plants resistance to these stresses. They can be classified into three major groups according to their biosynthetic route and structural features: terpenoids, alkaloids, and phenolic compounds. Phenolic compounds, which are mainly synthesized from phenylalanine produced by the shikimic acid pathway, are the most widely distributed in the plant kingdom. Plant tissues may contain up to several grams per kilogram of polyphenols. External stimuli such as microbial infections, ultraviolet radiation, low temperature, water and nutrition stresses induce their synthesis. The chemical structure of plant phenolics varies from simple to highly polymerized compounds as lignin and tannins. Phenolics have been categorized in different classes according to their basic carbon skeleton: the most relevant phenolic groups for nutritional health value are phenolic acids (C6-C1), phenylpropanoids (C6-C3) from which derivates the lignin polymer (C6-C3)n, coumarins (cyclized C6-C3) and flavonoids (C6-C3-C6). These compounds have a role in the visual appearance (peel and flesh pigmentation, browning), taste (astringency and bitterness), and health-promoting properties (free radical scavengers). This chapter focuses mainly on the biological activities of flavonoïds and on the effect of processes on the evolution of flavonoids and phenolic compounds during food transformations.

2. Flavonoids: Classification and biological activities

Flavonoids except chalcones, aurones and isoflavones share the same basic skeleton, a flavanone nucleus containing two hexacarbonic aromatic rings formed by fifteen atoms of carbon (A and B) interconnected with an heterocyle C composed of three carbon atoms and one oxygen atom (Figure 1).

This nucleus can undergo many modifications such as hydroxylation, alkylation or glycosylation. Depending on these modifications, the flavonoids are classified into 9 groups (chalcones, aurones, flavanones, dihydroflavanols, flavones, isoflavones, anthocyans, flavonols, flavanols). Compounds belonging to the same group differ between them by the degree and the position of hydroxylation, the presence of substitute on the nucleus and the state of their polymerization (Table 1).

Fig. 1. The flavanone nucleus.

Besides conferring coloration to plants, flavonoids can act as enzyme and microbial inhibitors, chelating agents, protection against UV, against free radicals. Moreover, flavonoids believed to have health properties such as analgesic, anti-inflammatory, anti-allergic and protectors against cardiovascular diseases. These properties are attributed mainly to their antioxidant activity and are variable depending to their structure.

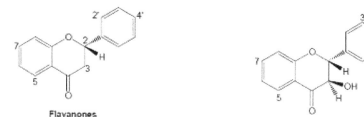

Chalcones

Flavonoid	Substitution					
	2′	3′	4′	5′	6′	4
Davidigeni	OH		OH			OH
Asebogenin	OH		OMe		OH	OH

Aurones

Flavonoid	Substitution					
	4	6	7	3′	4′	5
Leptosodin		OH	OMe	OH	OH	
Maritimetin		OH	OH	OH	OH	

Flavanones

Flavonoid	Substitution					
	5	6	7	3′	4′	5′
Eriodictyol	OH		OH	OH	OH	
Hesperitin	OH		OH	OH	OMe	
Naringenin	OH		OH		OH	

Dihydroflavonols

Flavonoid	Substitution					
	5	6	7	3′	4′	5′
Taxifolin	OH		OH	OH	OH	
Fusecin			OH	OH	OH	

Table 1. Different classes of flavonoids and their structures.

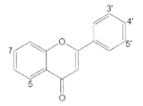

Flavones

Flavonols

Flavonoid	Substitution					
	5	6	7	3'	4'	5'
Apigenin	OH		OH		OH	
Chyrisin			OH			
Luteolin	OH		OH	OH	OH	

Flavonoid	Substitution					
	5	6	7	3'	4'	5'
Kampherol	OH		OH		OH	
Myricetin	OH		OH	OH	OH	OH
Quercetin	OH		OH	OH	OH	

Isoflavones

Anthocyanes

Flavonoid	Substitution					
	5	6	7	3'	4'	5'
Daidzein			OH		OH	
Genistein	OH		OH		OH	

Flavonoid	Substitution					
	3	5	7	3'	4'	5'
Pelargonidin	OH	OH	OH		OH	
Cyanidin	OH	OH	OH	OH	OH	
Delphinidin	OH	OH	OH	OH	OH	OH

Flavanols

Flavonoid	Substitution					
	5	6	7	3'	4'	5'
Catechin	OH		OH	OH	OH	
Gallocatechin	OH		OH	OH	OH	OH

Table 1. Different classes of flavonoids and their structures. (Continuation)

2.1 Anti-oxidant and anti-free radical activities of flavonoids

The most described property of flavonoids is their capacity to protect the organism against free radicals and oxygenated reactive species (ORS) produced during the metabolism of oxygen (Grace, 1994). The cellular damage by the free radicals causes a change of the net charge of cells, thus modifying their osmotic pressure and inducing their swelling and their death. The free radicals act also on the mediators of the inflammatory diseases, and accelerate the tissue damage. Moreover, cells lesions lead to an increase in the production of the ORS which induces the consumption and the depletion of the endogenous chelating agents. To protect against oxygenated reactive species, the organism and living cells have developed several mechanisms (Halliwell, 1995) including enzymes like the superoxyde dismutase, the catalase and the glutathion peroxidase, and also non-enzymatic homologues such as the glutathion, the ascorbic acid and l'α-tocopherol.

The protective effect of flavonoids is due to several mechanisms such as free radicals trapping, enzymes inhibition and metallic ions chelation. These properties depend on the structure of the flavonoids and the degree of substitution and saturation (Table 1).

2.1.1 Free radicals trapping

The flavonoids can prevent the damage caused by the free radicals according to various ways; one of them is the direct trapping of the radicals. In this case, the flavonoids are oxidized by the radicals (R•) leading to less reactive and more stable species according to the following mechanisms (Halliwell, 1995):

$$\text{Flavonoid (OH)} + R\bullet \rightarrow \text{Flavonoid (O}\bullet) + RH$$

The formed flavonoxy radical (flavonoid (O•)) is stabilized by resonance. The non-paired electron can be delocalized on the whole of the aromatic cycle. But, it can continue to evolve according to several processes (dimerisation, dismutation, recombination with other radicals, oxidation in quinon) either while reacting with radicals or other antioxidants, or with biomolecules. The flavonoxy (FL-O•) radical can react with another radical to form stable quinine as follows:

The flavonoxy radical can interact with oxygen to give a quinone and a superoxide anion. This reaction is responsible for an undesirable prooxidant effect of flavonoids. So the capacity of flavonoids to act as antioxidant depends not only on the redox potential of the couple Flavonoid (O•)/ Flavonoid (OH), but also on the reactivity of generated flavonoxy radical (Van Acker et al., 1995).

2.1.2 Effect on the mediator of nitric oxide synthesis

Several flavonoids reduce the cellular lesions related to ischaemia, by interfering with the activity of nitric oxide synthase. The nitric oxide is produced by various types of cells such as the endothelium cells and the macrophages. The nitric oxide is produced through the constitutive activity of nitric oxide synthase. It plays a role for the maintenance of the dilation of the blood-vessels, the relaxation of the smooth muscles, the signal of transduction and the inflammation (Parihar et al., 2008; Valko et al., 2007). However, at high concentrations it induces an irreversible oxidative damage on cellular walls; because the activated macrophages increase their simultaneous productions of nitric oxide and the superoxide anions. The nitric oxide reacts with the free radicals producing peroxynitrite anion (ONOO-), a more reactive species:

$$H^+ + O_2^- + NO^\cdot \rightarrow ONOO^- + H^+ \leftrightarrow ONOOH$$

When the flavonoids are used as antioxidants, the free radicals are trapped thus reducing the conversion of nitric oxide into peroxynitrite (Shutenko et al., 1999). Flavonoids can also react with nitric oxide directly (Van Acker et al., 1995). Therefore, it was speculated that the trapping of nitric oxide by the flavonoids is in the origin of their protective effect of the cardiovascular system.

2.1.3 Inhibition of the enzymes activities

It is well known that flavonoids are able to inhibit the activities of several enzymes implicated in radical's generation. Among these enzymes, the xanthine oxidase, lipoxygenase, cyclo-oxygenase, peroxidase and tyrosin kinase are the most studied.

The xanthine dehydrogenase and the xanthine oxydase are implied in the metabolism of the xanthine into uric acid. The xanthine deshydrogenase is the configuration available under the normal physiological conditions, but it changes into xanthine oxydase during cells reperfusion (reoxygenation) and reacts with molecular oxygen to release the superoxyde radical (O_2^-). The flavonoids act as a strong inhibitor of the xanthine oxydase and as a trapper of the superoxide radical (Sanhueza et al., 1992).

The flavonoids have also the capacity in one hand, to inhibit the metabolism of the acid arachidonic (Ferrandiz & Alcaraz, 1991) by inhibiting the lipooxygenase and thus preventing the production of the chimiotactic compounds from this acid. This characteristic gives to the flavonoids the anti-inflammatory and anti-thrombogenic properties. In the other hand, flavonoids have also the capacity to reduce the release of peroxidases and proteolytic enzymes and thus the production of the ROS (Middleton & Kandaswami, 1992).

The activity of the tyrosin kinase is affected by the presence of the flavonoids (Nijveldt et al., 2001). This enzyme is implied in several cellular functions such as the enzymatic catalysis, the transport through the membrane, the transduction of the signals for hormones or growth factors and the transfer of energy in the synthesis of ATP. So, the inhibition of this enzyme by the flavonoids interferes with the way of transduction of the signals controlling the cellular proliferation.

2.1.4 Chelation of the metal ions

The ions of iron (Fe^{2+}) and copper (Cu^+), are essential for certain physiological functions of living cells (Van Acker et al., 1995). They can be, either as components of hemoproteins, or of cofactors of various enzymes implicated in antioxidant defense system of cells. Besides their beneficial role, they are also responsible for the production of the hydroxyl radical by the reduction of hydrogen peroxide ($\cdot$OH) according to the following reaction:

$$H_2O_2 + Fe^{2+}(Cu^+) \rightarrow \cdot OH + -OH + Fe^{3+}(Cu^+)$$

The flavonoids form a stable complex with transition metals (Fe^{3+}, Al^{3+}, Cu^{2+}, Zn^{2+}); the stoichiometry of the complex and the site of chelation depend on the nature of the flavonoid mainly the presence of the catechol part (Le Nest et al., 2004) and the pH (Cornard & Merlin, 2002[a,b]). Moreover, this phenomenon of chelation is accompanied sometimes by the oxidation of the flavonoid (Cu^{2+}, Fe^{3+}). The chelation is occurred generally on the hydroxyl groups in position 3' and 4' of the B cycle, on the position 3 of hydroxyl group of A cycle and on the positions 3 and 4 of carbonyl group of C cycle (Figure 2).

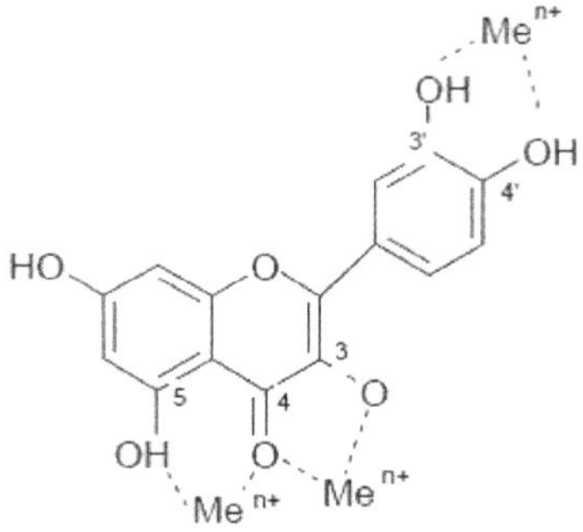

Fig. 2. Chelation of flavonoids.

When the flavonoids have several chelating metal sites, they can be polymerized. The copolymerization of the flavonoids and iron is responsible for anemia disease observed in large consumers of tea (Damas et al. 1985). The capacity of the flavonoids to complex metals is probably at the origin of the inhibition of many enzymes whose active site contains metals.

2.2 Pharmacologic effects of flavonoids

A general presentation of the hypothetical links between the mechanism of action and clinical effects of the consumption of flavonoids is summarized in Table 2 and Figure 3. The various clinical effects of the flavonoids are described in details in the following paragraphs.

Biological activities of flavonoids (anti-cancer, anti-inflammatory, antioxidant) depend on the presence of several functional groups on the backbone of these compounds (Limem et al., 2008). The most important are the hydroxyl groups and their positions (3 and 4' on C and B cycles respectively or 3' and 4' on the cycle B), the double bond between carbon 2 and 3 and 4 oxo function of the cycle C. The absence or the substitution of these groups leads to a significant reduction of biological activities of the flavonoids. The relation between flavonoids structure and biological activities is well summarized (Figure 4).

Consumption	Effect	References
Quercetin –rich fruits and vegetables (during 15 years)	Decrease of apoplexy incidence	Imai et al., 1997
Green tea (during 10-11 years)	Decrease of cancers risk (lung, liver, colon) Retard cancers progression Decrease of breast cancer risk	Nakachi et al., 1996/1998
Dadzein, genistein and coumestrol -rich food	Decrease of prostate cancer risk	Strom et al., 1999
375mg of curcumin (3 times a day for 12 weeks)	Effective treatment of anterior uveitis Increase of blood levels of glutathione peroxidase	Lal et al., 1999
55mg/day of isoflavonoids for 8 weeks	No effect on lipid peroxidation	Hodgson et al., 1999
Extract of red vine leaf (360 or 720 mg/day for 12 weeks)	Small effect on chronic venous insufficiency	Kieswetter et al., 2000
Green tea extract + linoleic acid (3g/day for 4weeks)	Decrease of blood levels of malondialdehyde No effect on other markers of oxidative stress or production of nitric oxide	Freese et al., 1999
Red vine phenolic compounds (3time 660mg/day for 2 weeks)	Increase in serum antioxidant capacity	Carbonneau et al., 1997
Blackcurrant and apple juices (750 up 1500ml/day for one week)	Decrease of blood levels of malondialdehyde Increase in glutathione peroxidase (no other changes in antioxidant status)	Young et al., 1999
Green or red tea (2g/day for 2 days)	Transient increase in blood antioxidant parameters (radical scavenging)	Serafini et al., 1996
113ml of alcohol-free red or white wine for 3 weeks	Transient increase in blood antioxidant parameters (radical scavenging)	Muzes et al., 1990
1L/day of soy milk for 4 weeks	No effect on blood cholesterol Decrease in oxidative damage of DNA bases	Mitchell & Collins, 1999
Flavonones, vitamin C	Decrease of prostate cancer risk	Rossi et al., 2007
Lyophilized grape powder (flavans, anthocyanins, qercetin, myricetin, kaempferol, resveratrol)	Reduction in plasma triglyceride concentration, cholesterol (LDL), apolipoproteins B, E and TNF α	Zern et al., 2005
Quercetin	Reduced risk of mortality due to ischemic heart disease Reduced risk of lung cancer	Knekt et al., 2002
Quercetin, naringenin, hesperitin	Reduced risk of breast cancer	
Quercetin, myricetin	Reduced risk of asthma	
Kaempferol, naringin, hesperitin	Reduced risk of type 2 diabetes	
Myricetin	Reduced risk of cerebrovascular diseases	

Table 2. Effects of flavonoids consumption on biological activities.

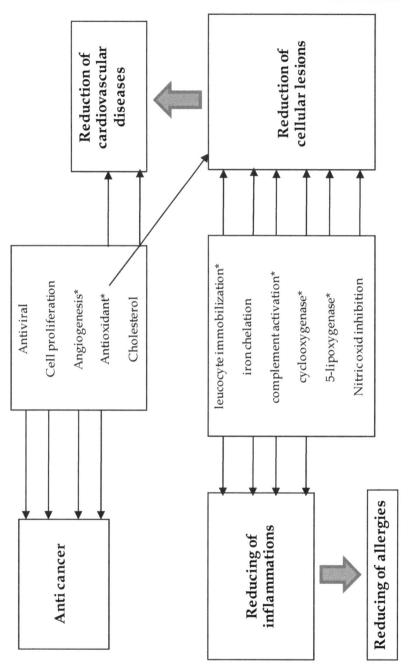

Fig. 3. Assumption of the links between the mechanisms of action of flavonoids and their effects on diseases (Nijveldt et al., 2001). * indicates the reduction of the enzyme activity or the function indicated.

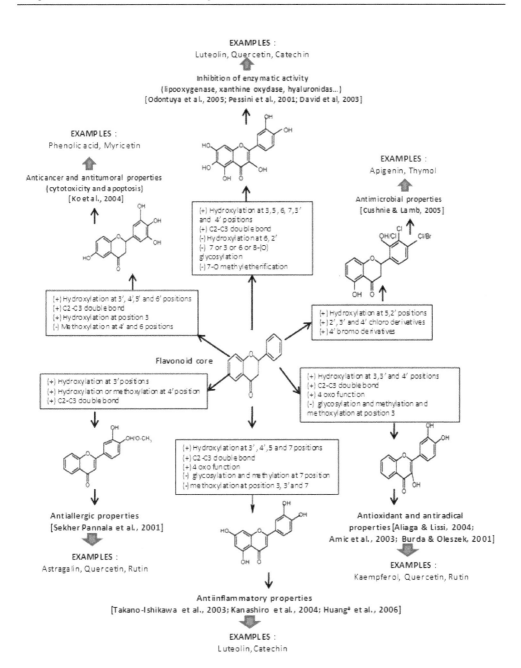

Fig. 4. Flavonoid structural elements necessary for biological activity: (+), the presence of structural elements promotes the cited activity; (-), the presence of structural elements reduces the cited activity.

2.3 Extraction of flavonoids

Extraction is the most important step in the development of analytical methods for plant extracts analysis. A summary of experimental conditions of the extraction methods is reported in Table 3. Basis unit operations of extraction is often the plant drying and its milling to obtain an homogenous powder and improve the extraction kinetic of the molecules. Methods as sonication, heating under reflux, extraction with Soxhlet apparatus are the most used (Ong, 2004). However, these methods are often long and need large volumes of organic solvents, with low extraction rates. Molecules we want to extract can be polar, non-polar or heat sensitive; thus the extraction method must take all these parameters into account.

To reduce the use of organic solvents and to improve the extraction rate, other methods such as extraction assisted by microwave, supercritical extraction, accelerated extraction by solvents, the pressurized liquid extraction, the pressurized extraction by hot water and the pressurized extraction by hot water associated to surfactants were introduced to the phenol extraction from plants. These different techniques were summarized in Table 3.

Extraction method	Solvents	Temperature (°C)	Pressure	Time
Sonication	Methanol, Ethanol, Mix alcohol/water	Can be heated		1 h
Soxhlet extraction	Methanol, Ethanol, Mix alcohol/water	Depending on the solvent used		3-18 h
Microwave extraction	Methanol, Ethanol, Mix alcohol/water	80-150	Depending on the extraction container	10-40
Extraction by supercritical fluid	Carbon dioxid, Mix carbon dioxid/Methanol	40-100	250-450 bar	30-100 min
Extraction by accelerated solvent	Methanol	80-200	100 bar	20-40 min
Extraction by pressurized liquid	Methanol	80-200	10-20 bar	20-40 min
Pressurized extraction by hot water	Water, water with 10-30% ethanol	80-300	10-50 bar	40-50 min
Pressurized extraction by hot water with surfactant	Water with surfactant (triton X100 ou SDS)	80-200	10-20 bar	40-50 min

Table 3. Experimental conditions for the phenol extraction.

2.4 Flavonoid occurrence in foods

Since several decades, many studies dealt with the analysis of foods to determine its composition in flavonoids. Many reviews were published, where the main flavonoids in foods are gathered. Tomás-Barberan et al. (2000) focused on fruits and vegetables. In 2009, INRA

(French National Institute Of Agricultural Research) developed a database on flavonoids in foods (http://www.phenol-explorer.eu). Table 4 was built according to data collected on the database of INRA; it presents some examples of foods containing flavonoids cited. Flavonoids chosen are the main found in foods, their quantity is specified into brackets.

Flavonoids	Foods (flavonoid content in mg/100g or 100ml)
Flavanons:	
- Naringenin	Red wine (0.05), Grapefruit (1.56), Mexican oregano (372), Almond (0.02)
- Hesperidin	Grape fruit juice from concentrate (1.55), Lemon juice from concentrate (24.99), Orange juice from concentrate (51.68), Peppermint dried (480.65)
Flavons:	
- Luteolin	Olive oil extra virgin (0.36), Thyme fresh (39.50), Olive black (3.43), Artichoke heads (42.10)
- Apigenin	Olive oil extra virgin (1.17), Italian oregano (3.50), Marjoram dried (4.40).
Flavonols:	
- Kaempferol	Red wine (0.23), Red raspberry pure juice (0.04), Tea black bottled (0.13), Capers (104.29), Cumin (38.60).
- Quercetin	Red wine (0.83), Buckwheat whole grain flour (0.11), Chocolate dark (25), Black elderberry (42), Orange pure juice (1.06), Mexican oregano (42), Onions red raw (1.29), almond (0.02)
Flavan-3-ols:	
- Catechin	Beer regular (0.11), Wine red (6.81), Barley whole grain flour (1.23), Cocoa powder (107.75), Grape black (5.46), Strawberry (6.36), Plum (4.60), Pistachio (3.50), Broad bean pod (16.23)
- Epicatechin	Red wine (3.78), Chocolate dark (70.36), Blackberry (11.48), European cranberry (4.20), Apricot (4.19), Custard apple (5.63), Tea green infusion (7.93)
Anthocyanins:	
- Petunidin 3-O-glucoside	Red wine (1.40), Highbush blueberry (6.09), Black grape (2.76), Black common bean (0.80)
- Malvidin 3-O-glucoside	Red wine (9.97), White wine (0.04), Black grape (39.23), Red raspberry (0.62)

Table 4. Examples of composition in flavonoids of certain foods.

According to Table 4, foods containing great quantity of flavonoids are fruit and vegetables; the processing of these raw foods modify the flavonoid content according to the process conditions. For example, in olive oil extra virgin, there is 1.17 mg of apigenin for 100g, but if this oil is refined the apigenin content decrease to 0.03 mg/100 g. Thus processes induce some consequences on flavonoid composition in foods.

3. Effect of food processing

Processes used in food engineering are numerous. We focus on the effect of unit operations on the degradation of the phenolic compounds as flavonoids and their antioxidant activity.

Among unit operations, we distinguish different categories: (i) the thermal processes such as pasteurization, baking, cooling, freezing, (ii) the non-thermal processes such as high pressure, pulsed electric fields, filtration, (iii) the mechanical processes such as peeling, cutting or mixing and (iv) the domestic processes that is to say processes by means of preparation of the convenience foods at consumers home.

3.1 Thermal processes

Thermal processes have a large influence in flavonoid availability in foods which depends on their magnitude and duration. Different heating methods (drying, microwaving, heating by an autoclave, roasting, water immersion, pasteurization, pressured-steam heating, blanching) were used and their effects were analyzed (Table 5). On this table, are gathered examples of significant studies to show the effect of thermal processes on the degradation of phenolic compounds.

As shown in Table 5, most of thermal processes lead to a degradation of phenolic compounds except in some cases as the apple juice processing where an increase of temperature from 40°C to 70°C allows increasing flavonoid content (50%) (Gerard & Roberts, 2004). A roasting of 130°C, 33 min increases the phenol content of cashew nuts (Chandrasekara & Shahidi, 2011); same results were noticed for peanuts (Yu et al., 2005). In these cases, an increase of temperature improves the extraction of phenolic compounds from foods; others results showed losses of phenolic compounds in different quantities. A loss of about 22% in total flavonoids has been observed in boiled products at a temperature of 50°C during 90s (Viña & Chaves, 2008). For the roasting process at 120°C, 20 min provokes a decrease of 12% of total flavonoid content (Zhang et al., 2010) and 15.9% for 160°C, 30min (Zielinski et al., 2009). Sharma & Gujral (2011) noticed for a roasting at 280°C during 20s, a loss of 8% in phenolic content. Steam heating at 0.2 MPa during 40 min induces a decrease of 25% in flavonoid content (Huang[b] et al., 2006; Zhang, et al., 2010). Similar findings were reported with microwaving at 700W during 10 min (Zhang, et al., 2010), 900 W during 120 s (Sharma & Gujral, 2011) and autoclaving at 100°C, 15 min (Choi et al., 2006). However, one blanching per immersion in water at 100°C during 4 min does not deteriorate flavonoids (Viña et al., 2007). Drying processes lead also to flavonoids degradation. The proportion lost depends on the drying method. Freeze-drying is the less aggressive method whereas hot air drying leads to major losses. As intermediate solutions microwave and vacuum drying can be used (Dong et al., 2011; Viña & Chaves, 2008; Zainol et al., 2009; Zhang et al., 2009). Pasteurization induces losses in phenolic compounds, significant losses are noticed for tomatoes' sauce pasteurized at 115°C during 5 min (Valverdú-Queralt et al., 2011), likewise a loose of 40% for a temperature of 85 °C during 5 min is measured by Hartman et al. (2008) for strawberries.

A few studies identified phenolic compounds in foods and followed their degradation during heat treatment. They noticed that individual phenolic compounds are also subject to heat degradation. The identification and quantification of these compounds were performed with high performance liquid chromatography. Rutin in buckwheat groats is reported to be more stable to heat then vitexin, isovitexin , homoorientin and orientin during roasting at 160°C for 30 min (Zielinski, et al., 2009). However, an increase of the dehulling time (10 to 130 min) leads to greater losses of rutin in the same product grains (Dietrych-Szostak & Oleszek, 1999). Boiling including soaking (100°C/121°C) with/or without draining stages induces 1-90% losses of quercetin and kaempferol in Brazilien beans (Ranilla et al., 2009).

Thermal pasteurization treatments (90°C, 60s) for strawberry juices have no effect on quercetin and kaempferol contents (Odriozola-Serrano et al., 2008), whereas it reduces naringin, rutin, quercetin and naringenin content for grapefruit juices (Igual et al, 2011). For Fuleki & Ricardo-Da-Silva (2003), pasteurization of grape juice increased the concentration of catechins in cold-pressed juices, but it decreased concentrations in hot-pressed juices. The concentration of most procyanidins was also increased by pasteurization.

However, the above results may not be comparable, because on the one hand, the food matrix is different from one assay to another and on the other hand, the food matrix can act as a barrier to heat effect or induce the degradation. It is not easy then to dissociate the thermal processing effect from the food matrix effects. Thus, some authors studied the effects of thermal processes on model solutions of phenolic compounds; these studies are led especially on flavonoids. The data indicated that flavonoids in aqueous solutions show different sensitivity to heat treatment depending on their structures. However, whatever their structure a significant degradation is observed for temperature above 100°C. For rutin, a higher stability compared to its aglycon form (quercitin) is observed (Buchner et al., 2006; Friedman, 1997; Makris & Rossiter, 2000; Takahama, 1986). These findings are attributed to the prevention of carbanion formation because of the glycosylation of the 3-hydroxyl group in the C-ring (Buchner, et al., 2006; Friedman, 1997; Takahama, 1986). Authors reported also that Luteolin was more stable to heat than rutin and luteolin-7-glucoside when heated at 180°C for 180min (Murakami et al., 2004). The degradation of flavonoids is not only a function of temperature and magnitude of heating; it may depend also on other parameters such as pH, phytochemicals, structure and even the presence or absence of oxygen. Indeed, original flavonol concentration has no effect on the degradation of rutin and quercetin. It is suggested that the reaction pathways are not influenced by the different flavonol solutions molarities (Buchner, et al., 2006). Moreover, under weak basic (Buchner, et al., 2006; Friedman, 1997; Takahama, 1986) and neutral (Friedman, 1997; Takahama, 1986) reaction conditions, more degradation of rutin and quercetin is observed (Buchner, et al., 2006). The absence of oxygen highly reduces quercetin degradation and prevents rutin breaking up during heating. The presence of oxygen is shown to accelerate quercetin and rutin degradation due to the presence of the reactive oxygen species (Buchner, et al., 2006; Makris & Rossiter, 2000). Chlorogenic acid is observed to protect rutin against degradation when a mixture of the two substances is heated at 180°C (Murakami, et al., 2004).

Sometimes, authors dealt with the antioxidant activity of foods or solutions studied. It is difficult to summarize the evolution of the antioxidant activity according to conditions heat processes. Too numerous factors are implied in its evolution. Decreases in phenol content do not lead systematically to a decrease of the antioxidant activity. Indeed, the degradation products of phenolic compounds can also have an antioxidant activity sometimes higher than the initial phenolic compounds (Buchner, et al., 2006; Murakami, et al., 2004); for high temperatures, these products can be Maillard products. Thus, an increase of antioxidant activity is noticed in many studies using thermal processes (Chandrasekara & Shahidi, 2011; Hartman et al., 2008; Sharma & Gujral., 2011). However interactions are important phenomena which act on the antioxidant activity of molecules. Depending on this environment, synergies between antioxidant compounds and the food matrix can occur (Wang et al., 2011). In some cases, the antioxidant capacity of flavonoids in a food matrix is enhanced (Freeman et al., 2010) ; while in other cases, the antioxidant capacity is reduced (Hidalgo et al., 2010). Thus, in other studies, antioxidant activity remains constant (Leitao et al., 2011) or can be decreased (Davidov-Pardo et al., 2011).

			Food product/Flavonoid	Processing conditions	Impact on flavonoid content	References
Heat processes	Food products	Total phenol content	Nuts	Roasting (130°C, 33 min)	Increase of phenol content	Chandrasekara & Shahidi, 2011
			Eucommia ulmoides flower tea	Microwave drying (Power : 140, 240, 480, 560 and 700 W; time durations: 1, 2, 3, and 4 min)	Stability of total flavonoid content	Dong et al., 2011
			Barley	Roasting (280°C, 20s) Microwave cooking (900 W, 120s)	A 8% loss in phenol content A 49.6 % loss in phenol content	Sharma & Gujral, 2011,
			Buckwheat	Roasting 20min and 40min at 80°C and 120°C Pressurized steam-heating (0.1 MPa, 20 min ; 0.2 MPa, 40 min) Microwaving (700W, 10 min)	20-30% increases depending on the conditions 18-30% increases depending on the conditions 20 % increase in flavonoid content	Zhang et al., 2010
			Tomatoes	Pasteurization (115°, 5 min)	Losses in phenol content	Valverdú-Queralt et al., 2010
			C. asiatica leaf, root and petiole	Air-oven drying Vacuum oven drying Freeze drying	A 97% loss in flavonoid content A 87.6% loss in flavonoid content A 73% loss in flavonoid content	Zainol et al., 2009
			Buckwheat seeds Buckwheat groats	Heating at 160°C for 30 min	A 15.9% loss in flavonoid content A 12.2% loss in flavonoid content	Zielinski et al., 2009
			Strawberry	Pasteurization (85°C, 5 min)	A 40% loss in phenol content	Hartman et al, 2008
			Celery	Dry air (48°C,1h) Water immersion (50°C, 90s)	A 60% loss in flavonoid content A 22% loss in flavonoid content	Viña et Chaves, 2008
			Brussels sprouts	Blanching (50°C)	Stability of total flavonoid content	Viña et al., 2007
			Mushroom (Shiitake)	Autoclave : (100, 121°C, 10 or 30 min)	Increase of free flavonoids (64%) Decrease of bound flavonoids: 50% (100°C, 30min), 75% (121°C, 10 min), 90% (121°C, 30 min) Stability under (100°C, 10 min)	Choi et al., 2006
			Sweet potato	Steaming (40 min)	14 % increase in flavonoid content	Huang[b] et al., 2006
			Peanut	Roasting (175°C, 5min)	40% increase in total phenol content	Yu et al., 2005
			Apple juice	Heating at 40_C, 50_C, 60_C and 70_C in a	50% increase between 40°C and 70°C	Gerard & Roberts, 2004

Table 5. Effects of heat processes on phenolic content.

			Food product/Flavonoid	Processing conditions	Impact on flavonoid content	References
Heat Processes	Food products	Individual phenolic compound	Grapefruit juices	Pasteurization (95°C, 80s)	Decrease of naringin, rutin, quercetin and naringenin content	Igual et al., 2011
			Bean (Quercetin , kaempferol)	Atmospheric (100°C) and pressure boiling (121°C) with and without soaking and draining	Increases of 1-90% of quercetin and kaempferol derivatives with soaking and drainning	Ranilla et al., 2009
			Buckwheat (Vitexin, isovitexin, rutin)	Roasting at 160°C for 30 min.	Losses of 80% of vitexin, isovitexin and rutin. Disappearance of homoorientin and orientin.	Zielinski et al., 2009
			Strawberry juices (kaempferol, quercetin, myricetin, anthocyanins)	High-intensity pulsed electric fields Pasteurization (90°C, 60s ; 90°C, 30s)	Stability of kaempferol, quercetin and myricetin. 10% increase of anthocyanins content (90°C, 60s)	Odriozola-Serrano et al., 2008
			Grape juice (Catechin, procyanidin)	Flash pasteurization (85°C)	Increase of Catechins in cold-pressed juice Decrease of Catechins in hot-pressed juice Increase of Procyanidins	Fuleki & Ricardo-Silva, 2003
			Buckwheat (Rutin, isovitexin)	Heating for (10,70, 130 min) to 150°C then steaming (0.35 MPa, 20 min)	Increase of rutin and isovitexin Steaming induces more losses	Dietrych-Szostak & Oleszek, 1999
	Model solutions		Aqueous flavonol solutions (quercetin and rutin)	Heating at 100°C for 300 min under pH 5 and 8 with air or nitrogen perfusion	Quercetin is more sensitive to heat under weak basic pH The presence of oxygen accelerates the degradation of quercetin and rutin	Buchner et al, 2006
			Aqueous flavonol solutions (quercetin and rutin)	Heating at 97°C for 240min under pH 8	Quercetin is more sensitive to heat than rutin The presence of oxygen accelerates the degradation of quercetin and rutin	Makris &Rossiter, 2000
			Rutin, luteolin, luteolin-7-glucoside	Heating at 100°C for 300 or 360min Heating at 180°C for 120 or 180min	Flavonoids are generally stable at 100°C Luteolin is more stable to heat than rutin and luteolin-7-glucoside (180°C,180min)	Murakami et al., 2004
			Aqueous flavonol solutions (quercetin and rutin)	Heating at 97°C for 240min under pH 8	Quercetin is more sensitive to heat than rutin The presence of oxygen accelerates the degradation of quercetin and rutin	Makris & Rossiter, 2000

Table 5. Effects of heat processes on phenolic content. (Continuation)

3.2 Non thermal processes

Certain authors showed the capacity of innovative processes (microwave, infra-red, high-pressure processing) to less degrade the phenolic antioxidants in food as regard to thermal processes. Odriozola-Serrano et al. (2008) studied the effect of high-intensity pulsed electric fields (HIPEF) process on quercetin and kaempferol contents of strawberry juices and

Food product	Processing conditions	Impact on flavonoids content	References
Onions	Cutting	Induction of flavonol biosynthesis	Pérez-Gregorio et al., 2011
Tomatoes	Peeling, Dicing	Great losses in phenol content	Valverdú-Queralt et al., 2011
Potatoes	Cutting	Induction of flavonol biosynthesis	Tudela et al., 2002
Asparagus	Chopping	18.5% decrease of rutin content	Makris and Rossiter, 2001
Onions	Peeling, trimming	Losses of 39%	Ewald et al., 1999

Table 6. Mechanical processing effects on phenol content.

	Food product	Processing conditions	Impact on flavonoid content	References
Domestic processes	Onion bulbs Asparagus spears	Boiling (60min)	A 20.5% decrease in total flavonoid content in onion bulbs A 43.9% decrease in total flavonoid content in Asparagus spears	Makris and Rossiter, 2001
	Onions	Sautéing (5min)	Increase of quercetin conjugates and total flavonoid contents	Lombard et al., 2005
		Baking (15min, 176°C)		
		Boiling (5min)	A 18.8% decrease of total flavonoid content	
	Onions	Boiling (3min)	Boiling gave limited reduction in flavonoids content	Ewald et al., 1999
		Microwaving (650w)		
		Warm-holding (60°C, 1h, 2h)		
	Brown - skinned Onions Red skinned- Onions	Boiling (20min)	A 14.3% loss of quercetin conjugates A 2 1.9% loss of quercetin conjugates	Price et al., 1997
		Frying (5min, 15min)	23-29% Losses of quercetin conjugates	
	Onions	Boiling (5min)	A 20% loss of total flavonoids	Lee et al., 2008
		Microwaving (1min, High heat)	No significant effect on total flavonoid content	
		Sautéing (3min)	No significant effect on total flavonoid content	

Table 7. Effects of domestic treatment on phenol content.

reported that such a process has no damage on these compounds. In 2009, the same study was led on tomatoes' juice; pulsed electric field has no effect on phenol content and led to a better conservation during the storage (Odriozola-Serrano et al, 2009). The use of high pressure, instead pasteurisation, on fruit smoothies is better to keep phenolic content constant (Keenan et al., 2011). Suarez-Jacobo et al. (2011) found the same results for an apple juice, phenolic content and antioxidant activity remain constant.

Few studies deal with filtration, Pap et al. (2010) recommended for blackcurrant juice filtration an enzymatic pre-treatment instead a reverse osmosis process, since it results in a juice concentrates highest in anthocyanins and flavonols. Hartman et al. (2008) also used an enzymatic treatment for strawberry mash; no loss of phenolic compounds was noticed.

3.3 Mechanical processes

Processes studied in literature concern essentially peeling, trimming, chopping, slicing, crushing, pressing and sieving of flavonoid-rich foods (Table 6). Processing is expected to affect content, activity and availability of bioactive compounds (Nicoli et al., 1999). According to authors, major losses of flavonoids took place during the pre-processing step when parts of product was removed: onions peeling and trimming resulted in 39% flavonoids losses (Ewald, et al., 1999) and asparagus chopping yielded a 18.5% decrease of rutin content (Makris & Rossiter, 2001). Great losses are also noticed for the peeling and the dicing of tomatoes (Valverdú-Queralt et al., 2011).

Slicing significantly affected the rutin content of asparagus (Makris & Rossiter, 2001). However, cutting increased flavonol content in fresh cut-potatoes (Tudela, et al., 2002) and fresh-cut onions (Pérez-Gregorio et al., 2011). In fact, wounding enhances flavonol biosynthesis through the induction of phenylalanine ammonia-lyase enzyme which is related to the wound-healing process in order to fight pathogen attack after tissue wounding (Tudela, et al., 2002).

3.4 Domestic processes

Several studies simulated food home preparation conditions in order to investigate their effects on flavonoid degradation (Table 7). Common domestic processes such as boiling, frying, baking, sautéing, steam-cooking and microwaving were studied.

Boiling resulted in flavonoids losses which are leached in cooking water, 43.9% for asparagus spears and 20.5% for onions (Makris & Rossiter, 2001). Similar losses in onions were reported (Lee et al., 2008; Lombard et al., 2005; Price et al., 1997). Microwaving does not markedly affect flavonoid content in onions (Ewald et al., 1999; Lee, et al., 2008; Lombard, et al., 2005; Price, et al., 1997; Tudela et al., 2002). As regards sautéing operations, contradictory findings were reported. Lee et al. (2008) reported a decrease of flavonoid content at almost of 21% whereas Lombard et al. (2005) showed an increase of the total flavonoid of 25% in onions (Lombard, et al., 2005).

Frying is reported to decrease onion flavonoid content between 25 and 33% (Lee, et al., 2008; Price, et al., 1997).Steaming and baking do not significantly affect the flavonoid content of onions (Lee, et al., 2008). Conversely, baking is found to increase quercetin conjugate and total flavonol content (7%) in onions as these compounds were concentrated in the tissues, as water and other volatiles were lost during cooking (Lombard, et al., 2005).

These contradictory results can be attributed easily to the diversity of food products used and the lack of the standardization of domestic processes.

Table 8 summarizes the possible evolution of phenolic antioxidants and their antioxidant activities according to the data collected in this chapter.

Phenolic Antioxidants		Antioxidant activity	
Evolution	Possible Cause	Evolution	Possible Cause
Increase	- Better extraction of phenolic compounds. - A stress inducing phenol synthesis as mechanical processes.	Increase	- Degradation products have an antioxidant activity. - Increase of the total phenol content. - Positive Synergies occur between phenolic antioxidants.
Decrease	- Degradation of phenolic compounds.	Decrease	- Degradation of the phenolic antioxidants. - Negative synergies occur between phenolic antioxidants
No change	- No degradation. - Compensation of an increase and a decrease.	No change	- No degradation of the phenol antioxidants. - Compensation of an increase and a decrease.

Table 8. Possible evolutions of phenolic antioxidants content and their antioxidant activity during food transformations.

4. Conclusion

Phenolic antioxidants have a great importance in human food diet: (i) they are widely widespread in raw foods as fruit and vegetables, tea, coffee, cocoa, (ii) they gather numerous properties beneficial for human health as anti-oxidant, anti-inflammatory, anti-allergic, antimicrobial and anticancer properties and (iii) they can be preserved during food transformation by using adapted process conditions and also nonaggressive processes. However, provide to consumers enriched food products in antioxidants is not so easy; indeed, despite the number of studies on the effect of food processes on the degradation of phenolic antioxidants and their antioxidant activities, it is difficult to generalize results. Many factors influence the evolution of these parameters: (i) the kind of raw food (genotype, cultivation method), (ii) the lack of standardization of measurement methods: phenolic content, antioxidant activity by ABTS, DPPH, ORAC, (iii) the influence of the food matrix: existence of interactions between molecules and iv) the lack of standardization of processes applied (conditions, materials).

5. References

Aliaga, C. & Lissi, E. (2004). Comparison of the free radical scavenger activities of quercitin and rutin: an experimental and theoretical study. *Canadian Journal of Chemistry*, Vol.82, pp.1668-73

Amic, D.; Davidovic-Amic, D.; Beslo, D. & Trinajstic, N. (2003). Structure–radical scavenging activity relationships of flavonoids. *Croatian Chemistry Acta*, Vol.76, pp.55–61

Buchner, N.; Krumbein, A.; Rhon, S. & Kroh, L. W. (2006). Effect of thermal processing on the flavonols rutin and quercetin. *Rapid Communications in Mass Spectrometry*, Vol.20, pp. 3229-3235

Burda, S. & Oleszek, W. (2001). Antioxidant and antiradical activities of flavonoids. *Journal of Agricultural and Food Chemistry*, Vol.49, pp.2774–2779

Carbonneau, M.A. ; Leger, C.L. ; Monnier, L. ; Bonnet, C. ; Michel, F. ; Fouret, G. ; Dedieu, F. & Descomps, B. (1997). Supplementation with wine phenolic compounds increases the antioxidant capacity of plasma and vitamin E of low-density lipoprotein without changing the lipoprotein Cu(2C)-oxidizability: possible explanation by phenolic location. *European Journal of Clinical Nutrition*, Vol.51, pp.682–690

Chandrasekara, N. & Shahidi, F. (2011). Effect of Roasting on Phenolic Content and Antioxidant Activities of Whole Cashew Nuts, Kernels, and Testa. *Journal of Agricultural and Food Chemistry*, Vol.59, pp.5006-5014

Choi, Y.; Lee, S. M.; Chun, J.; Lee, H. B., & Lee, J. (2006). Influence of heat treatment on antioxidant activities and polyphenolic compounds of Shiitake (Lentinus edodes) mushroom. *Food Chemistry*, Vol.99, pp.381-387

Cornard[a], J.P. & Merlin, J.C. (2002). Complexes of aluminium(III) with isoquercitrin: spectroscopic characterization and quantum chemical calculations. *Polyhedron*, Vol.21, pp.27-28

Cornard[b], J.P. & Merlin, J.C. (2002). Spectroscopic and structural study of complexes of quercetin with Al(III). *Journal of Inorganic Biochemistry*, Vol.92, pp.1-19

Cushnie, T.P.T. & Lamb, A.J. (2005). Antimicrobial activity of flavonoids. *International Journal of Antimicrobial Agents*, Vol.26, pp.343-56

Damas, J.; Bourdon, V.; Remacle-Volon, G. & Lecomte, J. (1985). Pro-inflammatory flavonoids which are inhibitors of prostaglandin biosynthesis. *Prostaglandins Leukotrienes and Medecine*, Vol.19, pp.11–24

Davidov-Pardo, G.; Arozarena, I. & Marín-Arroyo, M.R. (2011). Stability of polyphenolic extracts from grape seeds after thermal treatments. *European Food Research Technology*, Vol.232, pp.211-220

Dietrych-Szostak, D. & Oleszek, W. (1999). Effect of Processing on the Flavonoid Content in Buckwheat (Fagopyrum esculentum Moench) Grain. *Journal of Agricultural and Food Chemistry*, Vol.47, No.10, pp.4384-4387

David, S.C.; Sies, H. & Scewe, T. (2003). Inhibition of 15-Lipoxygenase by flavonoids: structure-activity relations and mode of action. *Biochemistry Pharmacology*, Vol.65, pp.773-81

Dong, J.; Ma, X.; Fu, Z. & Guo, Y. (2011). Effects of microwave drying on the contents of functional constituents of Eucommia ulmoides flower tea. *Industrial Crops and Products* (in press).

Ewald, C.; Fjelkner-Moding, S.; Johansson, K.; Sjoholm, I. & Akesson, B. (1999). Effect of processing on major flavonoids processed onions, green beans, and peas. *Food Chemistry*, Vol.64, pp.231-235

Ferrandiz, M.L. & Alcaraz, M.J. (1991). Anti-inflammatory activity and inhibition of arachidonic acid metabolism by flavonoids. *Agents Actions*, Vol.32, pp.283–8

Freeman, B. L.; Eggett, D. L. & Parker, T. L. (2010). Synergistic and antagonistic interactions of phenolic compounds found in navel oranges. *Journal of Food Science*, Vol.75, No.6, pp.C570-C576

Freese, R.; Basu, S.; Hietanen, E.; Nair, J.; Nakachi, K.; Bartsch, H. & Mutanen, M. (1999). Green tea extract decreases plasma malondialdehyde concentration, but does not affect other indicators of oxidative stress, nitric oxide production, or haemostatic factors during a high-linoleic acid diet in healthy females. *European Journal of Nutrition*, Vol.38, pp.149–157

Friedman, M. (1997). Chemistry, Biochemistry, and Dietary Role of Potato Polyphenols. A Review. *Journal of Agricultural and Food Chemistry*, Vol.45, No.5, pp.1523-1540

Fuleki, T. & Ricardo-Da-Silva, J.M. (2003). Effects of Cultivar and Processing Method on the Contents of Catechins and Procyanidins in Grape Juice. *Journal of Agricultural and Food Chemistry*, Vol.51, pp.640-646

Gerard, K.A. & Roberts, J.S. (2004). Microwave heating of apple mash to improve juice yield and quality. *Food Science and Technology*. Vol.37, pp.551-557

Grace, P.A. (1994). Ischaemia-reperfusion injury. *British Journal of Surgery*, Vol.81, pp.637–47

Halliwell, B. (1995). How to characterize an antioxidant: an update. *Biochemical Society Symposium*. Vol.61, pp.73–101

Hartmann, A.; Patz, C.-D.; Andlauer, W.; Dietrich, H. & Ludwig, M. (2008). Influence of processing on quality parameters of strawberries. *Journal of Agricultural and Food chemistry*, Vol.56, No.20,pp.9484-9489

Hidalgo, M. ; Sanchez-Moreno, C. & De Pascual-Teresa, S. (2010). Flavonoid-flavonoid interaction and its effect on their antioxidant activity. *Food Chemistry*, Vol.121, pp.691-696

Hodgson, J.M.; Puddey, I.B.; Croft, K.D.; Mori, T.A.; Rivera, J. & Beilin, J.L. (1999). Isoflavonoids do not inhibit in vivo lipid peroxidation in subjects with high-normal blood pressure. *Atherosclerosis*, Vol. 145, pp.167–172

Huang[a], W.H.; Lee, A.R. & Yang, C.H. (2006). Antioxidative and anti-inflammatory activities of polyhydroxyflavonoids of Scutellaria baicalensis GEORGI. *Bioscience Biotechnology and Biochemistry*, Vol.70, pp.2371–80

Huang[b], Y.; Chang, Y.; & Shao, Y. (2006). Effects of genotype and treatment on the antioxidant activity of sweet potato in Taiwan. *Food Chemistry*, Vol.98, pp.529-538

Igual, M.; García-Martínez, E.; Camacho, M.M. & Martínez-Navarrete, N.(2011). Changes in flavonoid content of grapefruit juice caused by thermal treatment and storage. *Innovative Food Science and Emerging Technologies*, Vol.12, pp.153-162

Imai, K.; Suga, K. & Nakachi, K. (1997). Cancer-preventive effects of drinking green tea among a Japanese Population. *Preventive Medicine*, Vol. 26, pp.769-775

Kanashiro, A.; Kabeya, L.M.; Polizello, A.C.; Lopes, N.P.; Lopes, J.L. & Lucisano-Valim Y.M. (2004). Inhibitory activity of flavonoids from Lychnophora sp. On generation of reactive oxygen species by neutrophils upon stimulation by immune complexes. *Phytotherapy Research*, Vol.18, pp.61–65

Keenan, D.F.; Brunton, N.; Gormley, R. & Butler, F. (2011). Effects of thermal and high hydrostatic pressure processing and storage on the content of polyphenols and some quality attributes of fruit smoothies. *Journal of Agricultural and Food chemistry*, Vol.59, No.2, pp.601-607

Kiesewetter, H.; Koscielny, J.; Kalus, U.; Vix, J.M.; Peil, H.; Petrini, O.; Van Toor, B.S. & De Mey, C. (2000) Efficacy of orally administered extract of red vine leaf AS 195 (folia Vitis viniferae) in chronic venous insufficiency (stages I-II). A randomized, double-blind, lacebo-controlled trial. *Arzneimittel-Forschung*, Vol.50, No.2, pp.109-111

Knekt, P.; Kumpulainen, J.; Järvinen, R.; Rissanen, H.; Heliövaara, M.; Reunanen, A.; Hakulinen, T. & Aromaa, A. (2002). Flavonoid intake and risk of chronic diseases. *American Journal Clinical Nutrition*. Vol.76, pp.560–568

Ko, C.H.; Shen, S.C. & Chen, Y.C. (2004). Hydroxylation at C40 or C6 is essential for apoptosis-inducing activity of flavanones through activation of the caspase-3 cascade and production of reactive oxygen species. *Free Radical Biology & Medecine*, Vol.36, pp.897–910

Leitao, C.; Marchioni, E.;Bergaentzlé, M.; Zhao, M.; Didierjean, L. ; Taidi, B. & Ennahar, S. (2011). Effects of Processing Steps on the Phenolic content and antioxidant Activity of Beer. *Journal of Agricultural and Food Chemistry*, Vol.59, pp.1249-1255

Lal, B.; Kapoor, A.K.; Asthana, O.P.; Agrawal, P.K.; Prasad, R.; Kumar, P. & Srimal, R.C. (1999). Efficacy of curcumin in the management of chronic anterior uveitis. *Phytotherapy Research*, Vol.13, pp.318–322

Lee, S. U.; Lee, J. H.; Choi, S. H.; Lee, J. S.; Ohnisi-Kameyama, M.; Kozukue, N.; Levin, C. E. & Friedman, M. (2008). Flavonoid Content in Fresh, Home-processed, and Light-Exposed Onions and in Dehydrated Commercial Onion Products. *Journal of Agricultural and Food Chemistry*, Vol.56, pp.8541-8548

Le Nest, G.; Caille, O.; Woudstra, M.; Roche, S.; Guerlesquin, F. & Lexa D. (2004). Zn–polyphenol chelation: complexes with quercetin, (+)-catechin, and derivatives: I optical and NMR studies. *Inorganica Chimica Acta*, Vol. 357, No.3, pp. 775-784

Limem, I.; Guedon, E.; Hehn, A.; Bourgaud, F.; Chekir Ghedira, L.; Engasser, J.M. & Ghoul, M. (2008). Production of phenylpropanoid compounds by recombinant microorganisms expressing plant-specific biosynthesis genes. *Process Biochemistry*, Vol.43, pp.463–479

Lombard, K.; Peffley, E.; Geoffriau, E.; Thompson, L. & Herring, A. (2005). Quercetin in onion (Allium cepa L.) after heat-treatment simulating home preparation (Allium cepa L.) after heat-treatment simulating home preparation. *Journal of Food Composition and Analysis*, Vol.18, pp.571-581

Makris, D. P. & Rossiter, J. T. (2000). Heat-induced, Metal-Catalyzed Oxidative Degradation of Quercetin and Rutin (Qercetin 3-O-Rhamnosylglucoside) in Aqueous Model. *Journal of Agricultural Food Chemistry*, Vol.48, pp.3830-3838

Makris, D. P. & Rossiter, J. T. (2001). Domestic Processing of Onion Bulbs (Allium cepa) and Asparagus Spears (Asparagus officinalis): Effect of Flavonol Content and Antioxidant Status. *Journal of Agricultural and Food Chemistry*, Vol.49, pp.3216-3222

Middleton, E.J. & Kandaswami, C. (1992). Effects of flavonoids on immune and inflammatory cell functions. *Biochemistry and Pharmacology*, Vol.43, pp.1167–79

Mitchell, J.H. & Collins, A.R. (1999). Effects of a soy milk supplement on plasma cholesterol levels and oxidative DNA damage in men — a pilot study. *European Journal of Nutrition*. Vol.38, pp.143–148

Murakami, M.; Yamaguchi, T.; Takamura, H. & Matoba, T. (2004). Effects of Thermal Treatment on Radical-scavenging Activity of Single and Mixed Polyphenolic Compounds. *Food Chemistry and Toxicology*, Vol.69, pp.FCT7-FCT10

Muzes, G.; Deak, G.; Lang, I.; Nekam, K.; Niederland, V. & Feher, J. (1990). Effect of silimarin (Legalon) therapy on the antioxidant defence mechanism and lipid peroxidation in alcoholic liver disease (double blind protocol). *Orvosi Hetilap*, Vol.131, pp.863–866

Nakachi, K.; Suemasu, K.; Suga, K.; Takeo, T.; Imai, K. & Higashi, Y. (1998). Influence of drinking green tea on breast cancer malignancy among Japanese patients. *Japanese Journal of Cancer Research*, Vol.89, pp. 254–261

Nakachi, K.; Imai, K. & Suga, K. (1996). Epidemiological evidence for prevention of cancer and cardiovascular disease by drinking green tea. In: H. Ohigashi, T. Osawa, J.Watanabe, T. Yoshikawa (Eds.), *Food Factors for Cancer Prevention*, pp. 105–108, Springer, Tokyo

Nicoli, M. C. ; Anese, M. & Parpinel, M. (1999). Influence of processing on the antioxidant properties of fruit and vegetables. *Trends in Food Science and Technology*, Vol.10, No.3, pp.94-100

Nijveldt, R.J.; Van Nood, E.; Van Hoorn, D.E.G.; Boelens, P.; Van Norren, K. & Van Leeuwen P.A. (2001). Flavonoids: a review of probable mechanisms of action and potential applications. *American Journal of Clinical Nutrition*, Vol.74, pp.418-425

Odontuya, G .; Hoult J.R.S. & Houghton P.J. (2005). Structure-activity relationship for antiinflammatory effect of luteolin and its derived glucosides. *Phytotherapy Research*, Vol.19, pp.782-786

Odriozola-Serrano, I.; Soliva-Fortuny, R.; Hernández-Jover, T. & Martín-Belloso, O. (2009). Carotenoid and phenolic profile of tomato juices processed by high intensity pulsed electric fields compared with conventional thermal treatments. *Food Chemistry*, Vol.112, No.1, pp.258-266

Odriozola-Serrano, I.; Soliva-Fortuny, R. & Martín-Belloso, O. (2008). Phenolic acids, flavonoids, vitamin C and antioxidant capacity of strawberry juices processed by high-intensity pulsed electric fields or heat treatments. *European Food Research Technology*, Vol.228, pp.239-248

Ong, E.S. (2004). Extraction methods and chemical standardization of botanicals and herbal preparations. *Journal of Chromatography B: Analytical Technologies in the Biomedical and Life Sciences*, Vol.812, pp.23-33

Pap, N.; Pongrácz, E.; Jaakkola, M.; Tolonen, T.; Virtanen, V.; Turkki, A.; Horváth-Hovorka, Z.; Vatai, G. & Keiski, R.L. (2010). The effect of pre-treatment on the anthocyanin and flavonol content of black currant juice (Ribes nigrum L.) in concentration by reverse osmosis. *Journal of Food Engineering*, Vol.98, pp.429-436

Parihar, A.; Parihar, M.S.; Milner, S. & Bhat, S. (2008). Oxidative stress and anti-oxidative mobilization in burn injury. *Burns*, Vol.34, pp.6-17

Pérez-Gregorio, M. R.; Garcia-Falcon, M. S. & Simal-Gandara, J. (2011). Flavonoids changes in fresh-cut onions during storage in different packaging systems. *Food Chemistry*, Vol.124, pp.652-658

Pessini, A.C.; Tako, T.T.; Cavalheiro, E.C.; Vichnewski, W.; Sampaio, S.V. & Giglio, J.R. (2001). A hyaluronidase from Tityus serrulatus scorpion venom: isolation, characterization and inhibition by flavonoids. *Toxicon*, Vol.39, pp.1495-1504

Price, K. R.; Bacon, J. R. & Rhodes, M. J. C. (1997). Effect of Storage ans Domestic Processing on the Content and Composition of Flavonol Glucosides in Onion (Allium cepa). *Journal of Agricultural and Food Chemistry*, Vol.45, pp.938-942

Rossi, M.; Garavello, W.; Talamini, R.; Negri, E.; Bosetti, C.; Maso, L.D.; Lagiou, P.; Tavani, A. ; Polesel, J. ; Barzan,

L.; Ramazzotti, V.; Franceschi, S. & La Vecchia, C. (2007). Flavonoids and the risk of oral and pharyngeal

cancer: A case-control study from Italy. *Cancer Epidemiology Biomarkers and Prevention*, Vol.16, pp.1621-1625

Ranilla, L. G.; Genovese, M. I. & Lajolo, F. M. (2009). Effect of Different Cooking Conditions on Phenolic Compounds and Antioxidant Capacity of Some Selected Brazilian Bean (Phaselous vulgaris L.) Cultivars *Journal of Agricultural and Food Chemistry*, Vol.57, No.13, pp.5734-5742

Sanhueza, J.; Valdes, J.; Campos, R.; Garrido, A. & Valenzuela, A. (1992). Changes in the xanthine dehydrogenase/xanthine oxidase ratio in the rat kidney subjected to ischemia-reperfusion stress: preventive effect of some flavonoids. *Research Communications in Chemical Pathology and Pharmacology*, Vol.78, pp.211–218

Sekher Pannala A.; Chan, T.S.; O'Brien, P.J. & Rice-Evans, C.A. (2001). Flavonoid Bring chemistry and antioxidant activity: fast reaction kinetics. *Biochemical and Biophysical Research Communications*, Vol.282, pp.1161–1168

Serafini, M.; Ghiselli, A. & Ferro-Luzzi, A. (1996). In vivo antioxidant effect of green and black tea in man, *European Journal of Clinical Nutrition*, Vol.50, pp.28–32

Sharma, P. & Gujral, H.S. (2011). Effect of sand roasting and microwave cooking on antioxidant activity of barley. *Food Research International*, Vol.44, pp.235-240

Strom, S.; Yamamura, Y.; Duphorne, C.M.; Spitz, M.R.; Babaian, R.J.; Pillow, P.C. & Hursting, S.D. (1999). Phytoestrogen intake and prostate cancer: a case-control study using a new database. *Nutrition and Cancer*, Vol.33, pp.20–25

Shutenko, Z.; Henry, Y.; Pinard, E.; Seylaz, J.; Potier, P.; Berthet, F.; Girard, P. & Secombe, R. (1999). Influence of the antioxidant quercetin in vivo on the level of nitric oxide determined by electron paramagnetic resonance in rat brain during global ischemia and reperfusion. *Biochemistry and Pharmacology*, Vol.57, pp.199–208

Suárez-Jacobo, A.; Rüfer, C.E.; Gervilla, R.; Guamis, B.; Roig-Sagués, A.X. & Saldo, J. (2011). Influence of ultra-high pressure homogenization on antioxidant capacity, polyphenol and vitamin content of clear apple juice. *Food Chemistry*, Vol. 127, pp. 447-454

Takahama, U. (1986). Spectrophotometric study on the oxidation of rutin by horseradish peroxidase and characteristics of the oxidized products. *BBA - General Subjects*, Vol.882, No.3, pp.445-451

Takano-Ishikawa, Y.; Goto, M. & Yamaki, K. (2003). Inhibitory effects of several flavonoids on E-selectin expression on human umbilical vein endothelial cells stimulated by tumor necrosis factor-a. *Phytotherapy Research*,Vol.17, pp.1224–1227

Tomás-Barberán, F. A., Ferreres, F., & Gil, M. I. (2000). Antioxidant phenolic metabolites from fruit and vegetables and changes during postharvest storage and processing. In Atta-ur-Rahman. *Studies in Natural Products Chemistry*, pp.739-795, Elsevier Science

Tudela, J. A.; Cantos, E.; Espin, J. C.; Tomás-Barberán, F. A. & Gil, M. I. (2002). Induction of Antioxidant Flavonol Biosynthesis in Fresh-Cut Potatoes. Effect of Domestic Cooking. *Journal of Agricultural and Food Chemistry*, Vol.50, pp.5925-5931

Valko, M.; Leibfritz, D.; Moncol, J.; Cronin, M.T.D.; Mazur, M. & Tesler, J. (2007). Free radicals and antioxidant in physiological functions and human disease. *The International Journal of Biochemistry and Cell Biology*, Vol.39, pp.44-84

Valverdú-Queralt, A.; Medina-Remón, A.; Andres-Lacueva, C. & Lamuela-Raventos, R.M. (2011). Changes in phenolic profile and antioxidant activity during production of diced tomatoes. *Food Chemistry*, Vol.126, pp. 1700-1707

Van Acker, S.A.; Tromp, M.N.; Haenen, G.R.; Van der Vijgh, W.J. & Bast, A. (1995). Flavonoids as scavengers of nitric oxide radical. *Biochemical and Biophysical Research Communications*, Vol.214, No.3, pp.755-759

Viña, S. Z. & Chaves, A. R. (2008). Effect of heat treatment and refrigerated storage on antioxidant properties of pre-cut celery (Apium graveolens L.). *International Journal of Food Science and Technology*, Vol.43, pp.44-51

Viña, S. Z. ; Olivera, D. F. ; Marani, C. M. ; Ferreyra, R. M.; Mugridge, A.; Chaves, A. R. & Mascheroni, R. H. (2007). Quality of Brussels sprouts (Brassica oleracea L. gemmifera DC) as affected by blanching method. *Journal of Food Engineering*, Vol.80, pp.218-225

Young, J.F.; Nielsen, S.E.; Haraldsdottir, J.; Daneshvar, B.; Lauridsen, S.T.; Knuthsen, P.; Crozier, A.; Sandstrom, B. &Dragsted, L.O. (1999). Effect of fruit juice intake on urinary quercetin excretion and biomarkers of antioxidative status. *American Journal of Clinical Nutrition*. Vol.69, pp.87–94

Yu, J.; Ahmedna, M. & Goktepe, I. (2005). Effects of processing methods and extraction solvents on concentration and antioxidant activity of peanut skin phenolics. *Food Chemistry*, Vol. 90, pp.199–206

Wang, S.; Meckling, K.A.; Marcone, M.F.; Kakuda, Y & Tsao, R. (2011). Synergistic, Additive, and Antagonistic Effects of Food Mixtures on Total antioxidant Capacities. *Journal of Agricultural and Food Chemistry*, Vol.59, pp.960-968

Zainol, M. M.; Abdul-Hamid, A.; Bakar, F. A. & Dek, S. P. (2009). Effect of different drying methods on the degradation of selected flavonoids in Centella asiatica. *International Food Research Journal*, Vol.16,No.4, pp.531-537

Zern, T.L.; Wood, R.J.; Greene, C.; West, K.L.; Liu, Y.; Aggarwal, D.; Shachter, N.S. & Fernandez, M.L. (2005). Grape polyphenols exert a cardioprotective effect in pre-and postmenopausal women by lowering plasma lipids and reducing oxidative stress. *Journal of Nutrition*, Vol.135, No.8, pp.1911-1917

Zhang, M.; Chen, H.; Li, J.; Pei, Y. & Liang, Y. (2010). Antioxidant properties of tartary buckwheat extracts as affected by different thermal processing methods. *LWT- Food Science and Technology*, Vol.43, pp.181-185

Zhang, M.; Hettiarachchy, N. S.; Horax, R.; Chen, P. & Over, K. F. (2009). Effect of maturity stages and drying methods on the retention of selected nutrients and phytochemicals in bittermelon (momordica charantia) leaf. *Journal of Food Science*, Vol.74, No.6, pp.C441-C446

Zielinski, H.; Mishalska, A.; Amigo-Benavent, M.; Del Castillo, M. D. & Piskula, M. K. (2009). Changes in Protein Quality and Antioxidant Properties of Buckwheat Seeds and Groats Induced by roasting. *Journal of Agricultural and Food Chemistry*, Vol.57, pp.4771-4777

Part 2

Microbial Biotechnology as an Effective Tool in Biopharmaceutical Production

6

Improvement of Heterologous Protein Secretion by *Bacillus subtilis*

Hiroshi Kakeshita[1,2], Yasushi Kageyama[1],
Katsuya Ozaki[1], Kouji Nakamura[2] and Katsutoshi Ara[1]
[1]Biological Science Laboratories,
Kao Corporation,
[2]Graduate School of Life and Environmental Sciences,
University of Tsukuba,
Japan

1. Introduction

The Gram-positive bacterium, *Bacillus subtilis* and related species are widely used as hosts for the extracellular production of industrially worthy enzymes, such as amylases, proteases, xylanase, and lipases (Braun et al., 1999; Tjalsma et al., 2000; Westers et al., 2004). These species possess a very high capacity for secreting a variety of exoenzymes into the growth medium, thereby reducing downstream purification processes. In addition, many of these are generally regarded as safe (GRAS) microorganisms, and do not produce endotoxins. Therefore, the secretion system of these species presents many advantages in terms of production capacity, structural authenticity, product purification, and safety. Nevertheless, the secretion of heterologous proteins from eukaryotes by these species is frequently inefficient (Table1). Hence, these species are never selected as the best cell factory for pharmaceutical proteins (Westers et al., 2004).

In pharmaceutical industry, the production of recombinant proteins in *Escherichia coli* is well established. In many cases, proteins are produced in cytoplasm of *E. coli*, and therefore, the production of recombinant proteins involves refolding and purification from inclusion bodies. However, the production of soluble recombinant proteins is relatively more cost-effective and less time-consuming. In fact, many studies have been performed regarding methods to overcome the problem of inclusion bodies and to improve protein solubility for the expression of heterologous proteins (Kapust & Waugh, 1999; Baneyx & Mujacic, 2004; Sørensen & Mortensen, 2005; Rabhi-Essafi et al., 2007). Therefore, developing human protein producing hosts is a major challenge in the field of biotechnology and protein production in *Bacilli*.

In *B. subtilis*, one long-standing major problem is the presence of high levels of extracellular protease for the production of heterologous proteins. In recent years, many proteases have been identified via the completed genome sequence of *B. subtilis* (Kunst et al., 1997; Westers et al., 2004), thereby allowing the construction of many protease-depleted strains for the production of heterologous proteins.

In addition, considerable efforts have been targeted at developing *B. subtilis* as a host for the production of heterologous proteins (Wu & Wong, 2002; Li et al., 2004; Westers et al., 2004; Kodama et al., 2007a, 2007b). However, many problems still remain for the secretion of human proteins, and these should be analyzed from the complementary perspectives of both the target protein and the secretion pathway, in order to improve human protein secretion.

We have used human interferon-α and interferon-β as heterologous model proteins to investigate the effects of *B. subtilis* secretion.

In this report, the knowledge which has become available in recent years aimed at improving heterologous protein secretion is discussed, and co-production of a Tat system is shown to provide a useful tool to enhance the secretion of heterologous proteins.

Product	Yield	Reference
α-amylase (AmyQ)	1 – 3 g/L	Palva, 1982
Avid-stable α-amylase	3.1 g/L	Heng et al., 2005
Cutinase	20 mg/L	Brockmeier et al., 2006
Proinsulin (PI)	1 g/L	Olmos-Soto and Contreras-Flores, 2003
LipaseA	600 mg/L	Lesuisse et al., 1993
Streptavidin	35 -50 mg/L	Wu et al., 2002a
scFv	10 -15 mg/L	Wu et al., 2002b
Interleukin (IL) -3	100 mg/L	Westers at al., 2006
hEGF	7.0 mg/L	Lam et al., 1998
human Interferon (IFN)-α2b	0.5-1.0 mg/L	Palva et al., 1983
human Interferon (IFN)-γ	20 mg/L	Rojas Contreras et al., 2010

Table 1. Protein products from *B. subtilis*

2. Signal peptide and propeptide

The major of Bacterial secreted proteins are translocated across the cytoplasmic membrane via the Sec pathway (Antelmann et. al. 2004). Secretory proteins are identified by a signal peptide at the protein's N-terminus. A signal peptide consists of a positively charged N-domain, a hydrophobic H domain, and a C domain containing a specific cleavage site. Most signal peptides are Sec dependent signal peptides, which are cleaved by a type I signal peptidase at the AXA cleavage site (Tjalsma et al., 2000), as an example, *B. subtilis* α-amlyase (AmyE) (Fig. 1).

2.1 Signal peptide

For the production of a heterologous protein in the culture medium of *B. subtilis*, it is necessary to use a signal peptide that directs the protein very efficiently to the translocase. However, heterologous protein secretion often results in inefficient and unsatisfyingly low yields. The relationship between signal peptides and target proteins remains unknown. Accordingly, previous studies have indicated the need for individually optimal signal peptides for every heterologous secretion target.

Recently, Brockmeier et al. (2006) established a new strategy for the optimization of heterologous protein secretion in *B. subtilis*, by screening a library of all natural signal peptides of the strain. Accordingly, the best signal peptide for the secretion of one target protein is not automatically the best, or even sufficient, for the secretion of a different target protein (Brockmeier et al., 2006).

In our study, human interferon-α (hIFN-α) was used as a heterologous model protein, to investigate the secretion of the *B. subtilis* several major signal peptides. (Fig. 2). We found that for the secretion production of hIFN-α, the AmyE signal peptide is one of the best signal peptides (unpublished data).

Fig. 1. The amino acid sequence of N-terminus-pre-pro AmyE. The putative SPase cleavage site is indicated by a closed arrowhead, and the post-secretory processing site is indicated by an open arrowhead, as described in the references (Takase et al., 1988; Sasamoto et al., 1989). Numbers above the AmyE amino acid sequence refer to the locations of the encoded amino acid residues of AmyE (adapted from Kakeshita et al., 2011a).

2.2 Propeptide

Some secreted bacterial proteins have cleavage propeptides located between their signal peptide and the mature protein. The propeptide is processed after translocation. Long propeptides (60 to 200 residues) are present for most bacterial extracellular proteases, which are auto-catalytically cleaved and possess intramolecular chaperon activities, for example, *B. subtilis* AprE (Braun et al., 1996; Ikemura & Inouye, 1998; Wang et al., 1998; Yabuta et al., 2001; Yabuta et al., 2002; Zhu et al., 1989). On the other hand, short propeptides (with fewer than 60 residues) are present for a few secreted proteins, including *B. subtilis* α-amylase (AmyE) (Davis et al., 1977; Mezes et al., 1983; Takase et al., 1998) (Fig. 1). In *B. subtilis*, the AmyE propeptide is cleaved by unknown proteins, and is dispensable for secretion, folding, and stability (Takase et al., 1998; Sasamoto et al., 1989).

However, the secretion efficiency of the *Staphylococcus aureus* nuclease (Nuc) was found to be enhanced by a propeptide in *E. coli* (Suciu & Inouye 1996) and *Lactococcus lactis* (Le Loir et al., 1998). In addition, in *L. lactis*, the nine-residue synthetic propeptide, LEISSTCDA, which is fused immediately after the signal peptide cleavage site, is known to enhance heterologous protein secretion (Le Loir et al., 1998; Le Loir et al., 2005; Zhuang et al., 2008; Zhang et al., 2010). Therefore, we evaluated whether the fusion of the AmyE signal peptide and the propeptide could improve the secretion of hIFNα-2b, compared to that with only AmyE signal peptide.

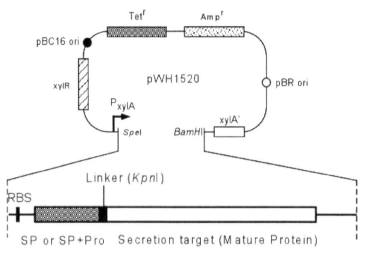

Fig. 2. Construction of plasmids for production and secretion of heterologous proteins. The restriction sites used for each construction are indicated. P_{xylA}, promoter of $xylA$; RBS, ribosome binding site; SP, signal peptide; Pro, propeptide.

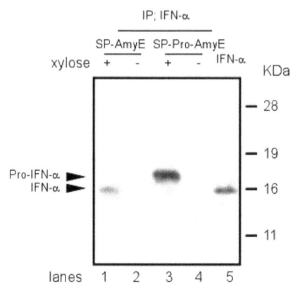

Fig. 3. Western blot analysis of hIFN-α production by $B.$ $subtilis$ Dpr8 with pHKK3101 (AmyE SP-hIFN-α) or pHKK3201 (AmyE SP-Pro-hIFN-α). Samples were collected at 20 h after xylose induction, separated by 15% SDS-PAGE, and stained with Western blotting using anti hIFN-α2b polyclonal antibodies. Dpr8 with pHKK3101 (lanes 1 and 2); Dpr8 with pHKK3201 (lanes 3 and 4); 0.6% xylose induced (lanes 1 and 3), none induced (lanes 2 and 4), and commercially purified hIFN-α 10 ng (lane 5). Arrowheads indicate the positions of the Pro-hIFN-α2b and hIFN-α2b. (adapted from Kakeshita et al., 2011a)

We showed that the secretion production and activity of hIFN-α2b with propeptide increased by more than 3-fold, compared to that without propeptide. The amount of secreted hIFN-α2b with propeptide was 15mg /L. This result indicated that the propeptide of AmyE enhanced the secretion of hIFNα-2b (Fig. 3, Kakeshita et al., 2011a).

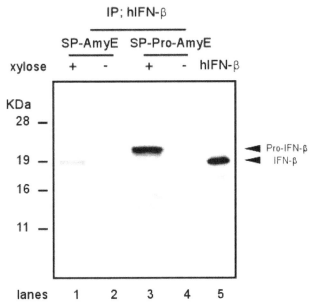

Fig. 4. Western blot analysis of hIFN-β production by *B. subtilis* Dpr8 with pHKK3111 (AmyE SP-hIFN-β) or pHKK3211 (AmyE SP-Pro hIFN-β). Samples were collected at 20 h after xylose induction, separated by 15% SDS-PAGE, and stained with Western blotting using anti hIFN-β polyclonal antibodies. Dpr8 with pHKK3111 (lanes 1 and 2); Dpr8 with pHKK3211 (lanes 3 and 4); 0.6% xylose induced (lanes 1 and 3), none induced (lanes 2 and 4), and commercially purified hIFN-β 50 ng (lane 5). Arrowheads indicate the positions of the Pro-hIFN-β and hIFN-β. (adapted from Kakeshita et al., 2011b)

In *L. lactis*, directed mutagenesis experiments demonstrated that the positive effect of LEISSTCDA on protein secretion was due to the insertion of negatively charged residues in the N-terminus of the mature moiety (Le Loir et al., 2001). In hIFN-α2b with AmyE propeptide, the first 10 amino acid residues of this mature protein have a net charge of -1. On the other hand, hIFN-α2b without propeptide has a net charge of 0. In addition, we demonstrated that propeptide mutants of neutral or positive charge resulted in a reduction in the amount of secreted hIFN-α2b, compared with propeptides of negative charge. This result suggested that negative charges in the mature protein can enhance the secretion of hIFN-α2b (Kakeshita et al., 2011a).

We then indicated that the AmyE propeptide enhanced the secretion of the hIFN-β protein from *B. subtilis*, as well. The secretion production and activity of hIFN-β with propeptide increased by more than 4-fold (Fig. 4, Kakeshita et al., 2011b). The amount of secreted hIFN-β with propeptide was 3.7mg /L. These results indicated that the propeptide of AmyE enhanced the secretion and extracellular production of a heterologous protein in *B. subtilis*.

2.3 Deletion of the C-terminus of SecA

In *B. subtilis*, most secreted proteins are translocated across the cytoplasmic membrane via the Sec system (Tjalsma et al., 2000; Tjalsma et al., 2004; Yamane et al., 2004). Nearly all of the components of the Sec system identified in *E. coli* have also been identified in *B. subtilis* and are biochemically well-characterized (van Wely et al., 2001; Harwood et al., 2008). Among these components, the peripheral membrane protein, SecA is considered to play a pivotal role in secretion. The SecYEG complex acts as a receptor for SecA, and functions as a preprotein conducting channel (Hartl et al., 1990; Fekkes et al., 1997). In *E. coli*, SecB is a molecular chaperone that functions in the post-translational protein translocation pathway, and binds to the C-terminal SecB binding site of *E. coli* SecA. In *B. subtilis*, this region of SecA is highly conserved. However, genome sequencing revealed that SecB is absent in *B. subtilis* (Kunst et al., 1997). *B. subtilis* Ffh interacts directly with SecA, and promotes the formation of soluble SecA-preprotein complexes (Bunai et al., 1999). These results suggest that the signal recognition particle (SRP) of *B. subtilis* not only acts as a targeting factor in co-translational translocation, but also stimulates the process of post-translocation across the membrane (Harwood & Cranenburgh, 2008; Ling et al., 2007; Tjalsma et al., 2000; Yamane et al., 2004). In additon, it has been shown that SecB binding site of *B. subtilis* SecA is not essential for viability and protein secretion (van Wely et al., 2000). The SecB binding site is connected by a C-terminal Linker (CTL) with the α-helical scaffold domain (HSD) in SecA. A cross-species comparison of the amino acid sequence of SecA revealed that the CTL is not well-conserved between *B. subtilis* and other species, including *E. coli*. We examined the effects of modifying the C-terminal region of SecA on growth and the extracellular production of heterologous proteins in *B. subtilis*, and demonstrated that the C-terminal domain (CTD) of SecA is not essential for viability or protein secretion. Furthermore, we showed that the productivity of hINF-α2b increased by 2.2-fold, compared to wild type SecA (Kakeshita et al., 2010). The crystal structure of *B. subtils* SecA indicated that CTL binds to the surface of NBF-I. The CTL-binding grove is a highly conserved and hydrophobic surface, and this grove is predicted to be one of the mature preprotein binding sites in SecA (Hunt et al., 2002). Therefore, deletion of the CTL of SecA is likely to affect SecA - preprotein interaction, and likely caused an increase in the secretion of heterologous proteins.

2.4 Co-expression of PrsA

PrsA is essential for viability and protein secretion. In protein secretion, PrsA is suggested to mediate protein folding at the late stage of secretion (Konitinen et al., 1991; Kontinen & Sarvas, 1993; Vitikainen et al., 2001). We examined the effect of co-expression of an extra-cytoplasmic molecular chaperone, PrsA. It is known that co-expression of an extra-cytoplasmic molecular chaperone, PrsA enhances the secretion of several model proteins: α -amylase, Single-chain antibody (SCA), and recombinant Protective antigen (rPA) (Kontinen & Sarvas, 1993; Vitikainen et al., 2001; Wu et al., 1998; Williams et al., 2003).

We demonstrated that co-expression of PrsA can act in concert with the AmyE propeptide to enhance the secretion production of hIFN-β. The amount of secreted hIFN-β with propeptide was 5.5mg /L. (Fig. 5, Kakeshita et al., 2011b).

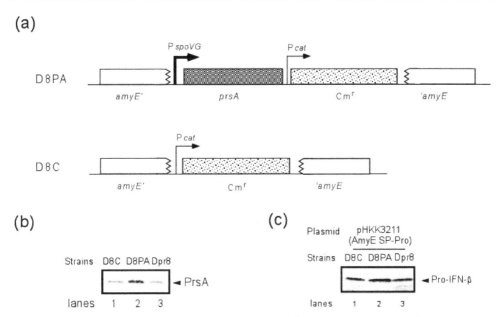

Fig. 5. Comparison of the amounts of secreted hIFN-β from *B. subtilis* D8C and D8PA, PrsA co-expressing strains. (a) Schematic representation of the gene structure around the amyE locus in the *B. subtilis* mutant strains D8PA and D8C. P_{spoVG} and *prsA* represent the *B. subtilis* spoVG promoter and *B. subtilis* PrsA, respectively. P_{cat} and Cmr represent the chloramphenicol-resistant gene promoter and coding region, respectively. (b) Western blot analysis of PrsA protein from *B. subtilis* D8C, D8PA, and Dpr8. (c) Western blot analysis of hIFN-β production by *B. subtilis* D8C, D8PA, and Dpr8. D8C with pHKK3211 (lane 1); D8PA with pHKK3211 (lane 2); Dpr8 with pHKK3211 (lane 3). Arrowheads indicate the positions of Pro-IFN-β. (Adapted from Kakeshita et al., 2011b).

3. Tat pathway

The majority of bacterial secreted proteins are translocated across the cytoplasmic membrane via the Sec pathway, which acts on unfolded proteins using the energy provided by ATP hydrolysis (Tajalsma et al., 2000; Antelman et al., 2000). Recently, a novel and different secretion protein translocation pathway, the twin-arginine translocation (Tat) pathway was discovered (Santini et al., 1998; Berks et al., 2000; van Dijl et al., 2002). The bacterial twin-arginine translocation (Tat) machinery is able to transport folded proteins across the cytoplasmic membrane (Robinson et al., 2001). The Tat pathway might have advantages over the Sec pathway for the production of heterologous proteins, because many proteins fold tightly before they reach the Sec machinery, and thus cannot engage with it for translocation across the cytoplasmic membrane.

B. subtilis contains two substrate specific Tat systems, TatAyCy and TatAdCd. The TatAyCy translocase is required for translocation of YwbN. On the other hand, a TatAdCd translocase translocates the phosphodiesterase PhoD (Jongbloed JD et al., 2002; Pop et al., 2002).

3.1 Twin-arginine signal peptide

Proteins are targeted to the Tat pathway by tripartite N-terminal signal peptides, the amino-terminal portion (n region) of which contain a conserved twin-arginine (RR) motif (R-R-X-#-#, where # is a hydrophobic residue).

In a previous study by Jongbloed et al., a database search for the presence of this motif in amino-terminal protein sequences identified a total number of 27 putative RR-signal peptides.

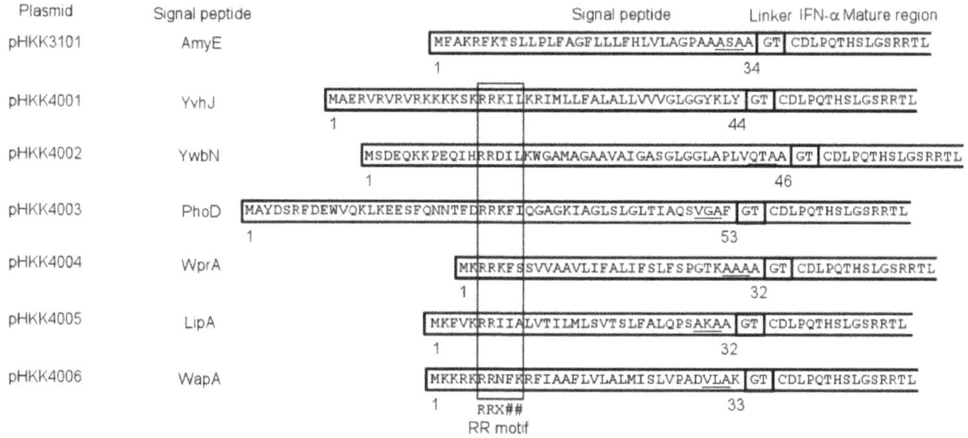

| Plasmid | Signal peptide | Signal peptide | Linker | IFN-α Mature region |

pHKK3101 AmyE MFAKRFKTSLLPLFAGFLLLFHLVLAGPAAASAA GT CDLPQTHSLGSRRTL
          1   34

pHKK4001 YvhJ MAERVRVRVRKKKKSKRRKIIKRIMLLFALALLVVVGLGGYKLY GT CDLPQTHSLGSRRTL
          1   44

pHKK4002 YwbN MSDEQKKPEQIHRRDILKWGAMAGAAVAIGASGLGGLAPLVQTAA GT CDLPQTHSLGSRRTL
          1   46

pHKK4003 PhoD MAYDSRFDEWVQKLKEESFQNNTFDRRKFIQGAGKIAGLSLGLTIAQSVGAF GT CDLPQTHSLGSRRTL
          1   53

pHKK4004 WprA MKRRKFSSVVAAVLIFALIFSLFSPGTKAAAA GT CDLPQTHSLGSRRTL
          1   32

pHKK4005 LipA MKFVKRRIIALVTILMLSVTSLFALQPSAKAA GT CDLPQTHSLGSRRTL
          1   32

pHKK4006 WapA MKKRKRRNFKRFIAAFLVLALMISLVPADVLAK GT CDLPQTHSLGSRRTL
          1   RRX##   33
                RR motif

Fig. 6. Schematic representation of the signal sequences used for secretion of human Interferon-α in *B. subtilis*. Schematic structure of the proteins encoded by each indicated plasmid. The twin-arginine motif is boxed, and the residues at positions -3 to -1 relative to the predicted SPase I cleavage site are underlined. The six base pairs of the KpnI site add the amino acids Gly–Thr to the end of each signal peptide coding sequence; therefore, in the table, each sequence ends with GT. Numbers under the signal peptides refer to the respective locations of the encoded amino acid residues.

We therefore selected six candidate Tat signal peptides, shown in Fig. 6, from the list generated by Jongbloed et al. for testing in the hIFN-α secreted assay. To determine the secretion ability for hINF-α2b, the six signal peptide genes considered to belong to the Tat pathway of *B. subtilis* were PCR-amplified. The PCR-amplified signal peptide genes were inserted upstream of the hIFN-α mature peptide gene in pHKK3101, yielding six secretion expression vectors. pHKK3101 expressing hIFN-α with the AmyE signal peptide, as the Sec-type signal peptide, was used as the control plasmid. The resultant recombinants were transformed into *B. subtilis* Dpr8, respectively, and the secretion expression of hIFN-α mediated by these signal peptides was detected by immunoblotting analysis. The hIFN-α was expressed in these strains and hIFN-α production was induced with the addition of 0.6% of xylose to the exponentially growing cultures (OD660 = 0.3), and both culture supernatants and intracellular lysates were analyzed as described in Kakeshita et al. (2010). As shown in Fig. 7a, in the extracellular fraction, only one band corresponding to mature protein (16 kDa) was detected for the samples of *B. subtilis* Dpr8 cells harboring pHKK3101 (AmyE signal), pHKK4004 (WprA), pHKK4005 (LipA), and pHKK4006 (WapA) by Western blot and immunoblot. This result suggested that the obtained three signal peptides (WprA, LipA, WapA) directed efficient secretion expression.

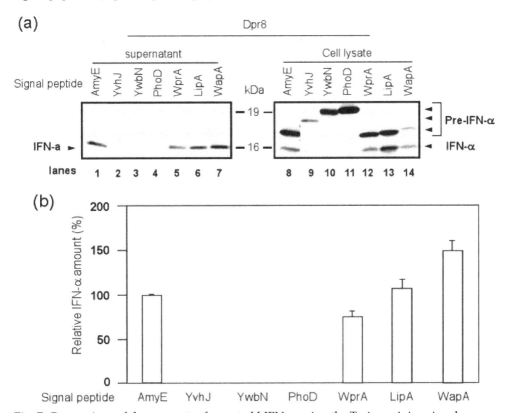

Fig. 7. Comparison of the amounts of secreted hIFN-α using the Twin arginine signal peptides from B. subtilis Dpr8. (a) Western blot analysis of hIFN-α production in B. subtilis Dpr8 harboring seven recombinants. Cells were grown at 30 °C in 2xL medium. Samples were collected at 20 h after xylose induction, separated by 15% SDS-PAGE, and subjected to Western blotting using anti hIFN-β polyclonal antibodies. Protein samples present in the supernatant (lanes 1, 2, 3, 4, 5, and 6) and cell fractions (lanes 7, 8, 9, 10, 11, and 12) of stationary-phase cultures were prepared by centrifugation, analyzed by SDS-PAGE, and immunodetected with anti-hIFN-α antibodies. Dpr8/pHKK3101 (lanes 1 and 8); Dpr8/pHKK4001 (lanes 2 and 9); Dpr8/pHKK4002 (lanes 3 and 10); Dpr8/pHKK4003 (lanes 4 and 11); Dpr8/pHKK4004 (lanes 5 and 12); Dpr8/pHKK4005 (lanes 6 and 13); Dpr8/pHKK4006 (lanes 7 and 14); precursor, pre hIFN-α; mature, hIFN-α. S, supernatant; C, cell fractions. (b) Quantification of secreted hIFN-α mature form in the culture medium and cell fraction. The hIFN-α production corresponding to the supernatant of B. subtilis Dpr8 carrying pHKK3101 (AmyE signal peptide) was set as 100%. Data represent the mean of three experiments, and error bars represent standard error.

Especially, WapA demonstrated the highest efficiency of hIFN-α secretion expression, which was 1.5-fold as high as the Sec dependent signal peptide, AmyE (Fig. 7b).

However, No hIFN-α was detected in the supernatants of Dpr8/pHKK4001 (YvhJ), Dpr8/pHKK4002 (YwbN), or Dpr8/pHKK4003 (PhoD). In the intracellular lysates of Dpr8/pHKK3101, Dpr8/pHKK4004, Dpr8/pHKK4005, and Dpr8/pHKK4006, two bands were detected. As deduced from the molecular mass of each band, these bands ware assigned to the unprocessed precursor (17 kDa) and the mature protein (16 kDa), respectively. On the other hand, only one band corresponding to the unprocessed protein was detected for the samples of Dpr8/pHKK4001 (YvhJ), Dpr8/pHKK4002 (YwbN), and Dpr8/pHKK4003 (PhoD).

These results suggested that the three obtained signal peptides, YvhJ, YwbN, and PhoD cannot be secreted hIFN-α2b into the supernatant.

3.2 Co-expression of the tat system

We examined the effect of co-expression of the Tat-machinary, TatAd/Cd or TatAy/Cy. To examine the effects of the co-expression of *B. subtilis* tat genes on hIFN-α secretion, we constructed TatAd/TatCd and TatAy/TatCy under the control of the *spoVG* promoter in plasmids. It is known that the *spoVG* promoter is a powerful promoter (Zuber & Losick 1983). The resulting constructs were subsequently integrated into the chromosome of *B. subtilis* strain Dpr8 via a double crossover event at the *amyE* locus, leaving the native tat genes intact (Fig. 8a).

The resultant strains, D8tatD and D8tatY were transformed with pHKK3101, pHKK4001, pHKK4002, pHKK4003, pHKK4004, pHKK4005, and pHKK4006 for expression of hIFN-α.

As shown in Fig. 8b and c, when the LipA signal peptide was fused to hIFN-α, a densitometric analysis of the western blotting demonstrated that the amounts of hIFN-α secreted by D8tatD and D8tatY were increased by roughly 2-fold, compared with that in strain Dpr8 (Fig. 8c). When the WprA signal peptide was fused to hIFN-α, in D8tatD, the amount of secreted hIFN-α was increased by 71% compared with that in the parental strain, Dpr8, whereas the enhanced production of hIFN-α increased by 29%. On the other hand, When the WapA signal peptide was fused to hIFN-α, the amounts of hIFN-α secreted by D8tatD and D8tatY were increased by only 10-20%, compared with that in strain Dpr8 (Fig. 8c). Then, when the AmyE signal peptide was fused to hIFN-α, the amounts of hIFN-α secreted by D8tatD and D8tatY were increased by 37% and 25%, respectively compared with that in strain Dpr8 (Fig. 8c). Therefore, WapA signal peptide and AmyE signal peptide are not able to enhance of secretion by co–expression of Tat system. In addition, when the YvhJ, YwbN, and PhoD signal peptides, respectively were fused to hIFN-α, the bands of hIFN-α secreted by D8tatD and D8tatY could not be detected in the resulting supernatants (data not shown).

We demonstrated that co-expression of TatAd/Cd or TatAy/Cy with LipA signal peptide can act in concert to enhance the secretion production of hIFN-α. In addition, WprA signal peptide was enhanced the secretion production of hIFNα by co-expression of TatAd/Cd, not TatAy/Cy. On the other hands, AmyE signal peptide and WapA peptide are Tat pathway independent.

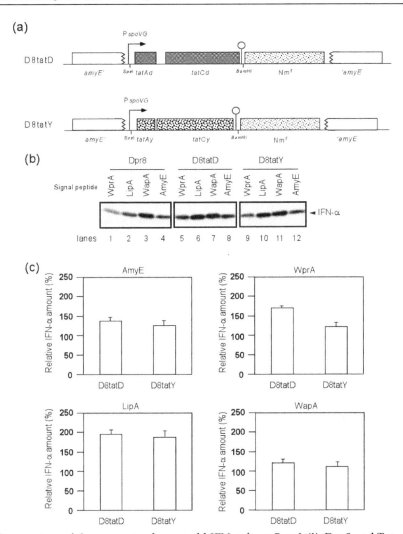

Fig. 8. Comparison of the amounts of secreted hIFN-α from *B. subtilis* Dpr8 and Tat overexpressing strains. (a) Schematic representation of the gene structure around the *amyE* locus in the *B. subtilis* D8tatD and D8tatY mutant strain genomes. Construction of strains D8tatD and D8tatY was by double crossover integration of plasmids pHKK2001 (*tatAd-Cd*) and pHKK2002 (*tatAy-Cy*) into the *amyE* locus of *B. subtilis* Dpr8. The resulting strain contains the native *phoD-tatAd-tatCd* locus, as well as one copy of *tatAd-Cd* and *tatAy-Cy* under the control of the P$_{spoVG}$ promoter. The stem-loop structures and the bent arrows indicate the putative Rho-independent terminators and promoters, respectively. (b) Western blot analysis of hIFN-α production by *B. subtilis* Dpr8, D8tatD, and D8tatY (carrying pHKK3101, pHKK4004, pHKK4005, or pHKK4006) was performed in the same manner as for hIFN-α. (c)   Quantification of secreted hIFN-α in mature form in the culture medium. The hIFN-α production corresponding to the *B. subtilis* Dpr8 strain was set as 100%. Data represent the mean of three experiments, and error bars represent standard error.

4. Conclusions

In recent years, considerable efforts have been targeted at developing *B. subtilis* as a host for the production of heterologous proteins. However, the secretion of heterologous proteins from eukaryotes by these species produces small yields and is frequently inefficient. Initially, we considered the major problem to be the presence of high levels of extracellular protease in *B. subtilis*. Nevertheless, even after obtaining many depleted protease strains, the problem of inefficient secretion was not resolved. Currently, it is considered that the largest problem is the detection of the pre-mature form of human protein in cell lysate, when human proteins with signal peptide are over expressed in *B. subtilis* (Fig. 7a). Normally, the pre-mature forms of target secretion proteins are not detected in cell lysates. If the pre-mature form of target a secretion protein is detected, it indicates a problem in the secretion pathway, for example, non-functional or depleted SecA, SecY, Ffh, or FtsY (Sadaie et al. 1991; Takamatsu et al., 1992; Honda et al., 1993; Oguro et al., 1995; Tjalsma et al., 2000; Tjalsma et al., 2004; Yamane et al., 2004). Therefore, we must solve this primary problem, which is the accumulation of the precursor of human proteins in *B. subtilis* cells.

We indicated that the propeptide of AmyE enhanced the secretion of the extracellular production of a heterologous protein in *B. subtilis*. In *L. lactis*, the nine-residue synthetic propeptide, LEISSTCDA, which is fused immediately after the signal peptide cleavage site, is known to enhance heterologous protein secretion (Le Loir et al., 1998). In addition, LEISSTCDA enhances secretion efficiency (Le Loir et al., 2001). Therefore, it is considered that a short type propeptide may be one answer to improve the accumulation of precursor.

On the other hand, we indicated that the deletion of the C-terminal domain of SecA enhanced the secretion of heterologous proteins. *secA* is an essential gene, and SecA is considered to play a pivotal role in secretion (Sadaie et al. 1991; Takamatsu et al., 1992; Tjalsma et al., 2000; Tjalsma et al., 2004; Yamane et al., 2004). In addition, we exhibited that the co-expression of PrsA or the Tat system can be able to enhance the secretion production. In the future, it may be necessary to modify the components of the secretion machinery for higher secretion efficiency.

5. Acknowledgments

We are grateful to Naotake Ogasawara, Junichi Sekiguchi, Fujio Kawamura, Kunio Yamane and members of MGP group in Kao Corporation for valuable discussions.

This work is the subproject, 'Development of a Technology for Creation of a Host Cell' included within the industrial technology project, 'Development of a Generic Technology for Production Process Starting Productive Function' of the Ministry of Economy, Trade and Industry, entrusted by the New Energy and Industrial Technology Development Organization (NEDO), Japan.

6. References

Antelmann H, Tjalsma H, Voigt B, Ohlmeier S, Bron S, van Dijl JM, Hecker M (2001) A proteomic view on genome-based signal peptide predictions. Genome Research Vol.11, pp.1484-1502, ISSN 1088-9051 (Print), 1549-5469 (Electronic).

Baneyx F, Mujacic M (2004) Recombinant protein folding and misfolding in *Escherichia coli*. Nature Biotechnology, Vol.11, pp.1399-1408, ISSN 1087-0156, EISSN 1546-1696.

Berks BC, Sargent F, Palmer T (2000) The Tat protein export pathway. Molecular Microbiology, Vol.35, pp.260-274, ISSN 0950-382X(Print), 1365-2958 (Electronic).

Braun P, Tommassen J, Filloux A (1996) Role of the propeptide in folding and secretion of elastase of *Pseudomonas aeruginosa*. Molecular Microbiology Vol.19, pp.297-306, ISSN 0950-382X, EISSN: 1365-2958.

Braun P, Gerritse G, van Dijl JM, Quax WJ (1999) Improving protein secretion by engineering components of the bacterial translocation machinery. Current Opinion in Biotechnology, Vol.10, pp.376–381, ISSN 0958-1669.

Brockmeier U, Caspers M, Freudl R, Jockwer A, Noll T, Eggert T (2006) Systematic screening of all signal peptides from *Bacillus subtilis*: a powerful strategy in optimizing heterologous protein secretion in Gram-positive bacteria. Journal of Molecular Biology, Vol.362, pp.393-402, ISSN 0022-2836.

Bunai K, Yamada K, Hayashi K, Nakamura K, Yamane K (1999) Enhancing effect of *Bacillus subtilis* Ffh, a homologue of the SRP54 subunit of the mammalian signal recognition particle, on the binding of SecA to precursors of secretory proteins in vitro. Journal of Biochemistry, Vol.125, pp151-159, ISSN 0021-924X (Print), 1756-2651 (Electronic).

Davis A, Moore IB, Parker DS, Taniuchi H (1977) Nuclease B: a possible precursor of nuclease A, an extracellular nuclease of *Staphylococcus aureus*. The Journal of Biological Chemistry, Vol.252, pp.6544-6553, ISSN 0021-9258 (Print), 1083-351X (Electronic).

Fekkes P, van der Does C, Driessen AJ (1997) The molecular chaperone SecB is released from the carboxy-terminus of SecA during initiation of precursor protein translocation. The EMBO Journal, Vol.16, pp.6105-6113, ISSN 0261-4189.

Hartl FU, Lecker S, Schiebel E, Hendrick JP, Wickner W (1990) The binding cascade of SecB to SecA to SecY/E mediates preprotein targeting to the *E. coli* plasma membrane. Cell Vol. 63, pp. 269-279, ISSN 0092-8674.

Harwood, CR, Cranenburgh R (2008) *Bacillus* protein secretion: an unfolding story. Trends in Microbiology, Vol.16, pp.73-79, ISSN 0966-842X.

Heng C, Chen Z, Du L, Lu F (2005) Expression and secretion of an acid-stable α-amylase gene in *Bacillus subtilis* by SacB promoter and signal peptide, Biotechnol ogy Letters, Vol.27, pp.1731-1737, ISSN 0141-5492 (Print), 1573-6776 (Electronic).

Honda K, Nakamura K, Nishiguchi M, Yamane K (1993) Cloning and characterization of a *Bacillus subtilis* gene encoding a homolog of the 54-kilodalton subunit of mammalian signal recognition particle and *Escherichia coli* Ffh. Journal of Bacteriology, Vol.175, pp.4885-4894, ISSN 0021-9193 (Print), 1098-5530 (Electronic).

Hunt JF, Weinkauf S, Henry L, Fak JJ, McNicholas P, Oliver DB, Deisenhofer J (2002) Nucleotide control of interdomain interactions in the conformational reaction cycle of SecA. Science, Vol.297, pp.2018-2026, ISSN 0036-8075.

Ikemura H, Inouye M (1988) In vitro processing of prosubtilisin produced in *Escherichia coli*. The Journal of Biological Chemistry, Vol.263, pp.12959-12963, ISSN 0021-9258 (Print), 1083-351X (Electronic).

Jongbloed JD, Antelmann H, Hecker M, Nijland R, Bron S, Airaksinen U, Pries F, Quax WJ, van Dijl JM, Braun PG (2002) Selective contribution of the twin-arginine translocation pathway to protein secretion in *Bacillus subtilis*. The Journal of

Biological Chemistry, Vol.277,pp.44068-44078, ISSN 0021-9258 (Print), 1083-351X (Electronic).

Kakeshita H, Kageyama Y, Ara K, Ozaki K, Nakamura K (2010) Enhanced extracellular production of heterologous proteins in *Bacillus subtilis* by deleting the C-terminal region of the SecA secretory machinery. Molecular Biotechnology Vol.46, pp.250-257, ISSN 1073-6085 (Print), 1559-0305 (Electronic).

Kakeshita H, Kageyama Y, Ara K, Ozaki K, Nakamura K (2011a) Propeptide of *Bacillus subtilis* Amylase Enhances Extracellular Production of Human Interferon-α in *Bacillus subtilis*. Applied Microbiology and Biotechnology, Vol.89, pp.1509-1517, ISSN 0175-7598 (Print), 1432-0614 (Electronic). .

Kakeshita H, Kageyama Y, Endo K, Tohata M, Ara K, Ozaki K, Nakamura K (2011b) Secretion of biologically-active human interferon-β by *Bacillus subtilis*. Biotechnology Letters, Vol.33, pp.1847-1852, ISSN 0141-5492 (Print), 1573-6776 (Electronic)

Kapust RB, Waugh DS (1999) *Escherichia coli* maltose-binding protein is uncommonly effective at promoting the solubility of polypeptides to which it is fused. Protein Science, Vol.8, pp.1668-1674, ISSN (Print) 0961-8368, ISSN (Electronic) 1469-896x.

Kodama T, Endo K, Ara K, Ozaki K, Kakeshita H, Yamane K, Sekiguchi J (2007a) Effect of *Bacillus subtilis* spo0A mutation on cell wall lytic enzymes and extracellular proteases, and prevention of cell lysis. Journal of Bioscience and Bioengineering, Vol.103, pp.13-21, ISSN 1389-1723 (Print), 1347-4421 (Electronic) .

Kodama T, Endo K, Sawada K, Ara K, Ozaki K, Kakeshita H, Yamane K, Sekiguchi J (2007b) *Bacillus subtilis* AprX involved in degradation of a heterologous protein during the late stationary growth phase. Journal of Bioscience and Bioengineering, Vol.104, pp.135-143, ISSN 1389-1723 (Print), 1347-4421 (Electronic).

Kontinen VP, Saris P, Sarvas M (1991) A gene (*prsA*) of *Bacillus subtilis* involved in a novel, late stage of protein export. Molecular Microbiology, Vol.5, pp.1273-1283, ISSN 0950-382X (Print), 1365-2958 (Electronic).

Kontinen V, Sarvas M (1993) The PrsA lipoprotein is essential for protein secretion in *Bacillus subtilis* and sets a limit for high-level secretion. Molecular Microbiology, Vol.8, pp.727–737, ISSN 0950-382X (Print), 1365-2958(Electronic).

Kunst F, Ogasawara N, Moszer I, Albertini AM, Alloni G, Azevedo V, Bertero MG, Bessieres P, Bolotin A, Borchert S, Borriss R, Boursier L, Brans A, Braun M, Brignell SC, Bron S, Brouillet S, Bruschi CV, Caldwell B, Capuano V, Carter NM, Choi SK, Codani JJ, Connerton IF, Cummings NJ, Daniel RA, Denizot F, Devine KM, Dusterhoft A, Ehrlich SD, Emmerson PT, Entian KD, Errington J, Fabret C, Ferrari E, Foulger D, Fritz C, Fujita M, Fujita Y, Fuma S, Galizzi A, Galleron N, Ghim S Y, Glaser P, Goffeau A, Golightly EJ, Grandi G, Guiseppi G, Guy BJ, Haga K, Haiech J, Harwood CR, Henaut A, Hilbert H, Holsappel S, Hosono S, Hullo MF, Itaya M, Jones L, Joris B, Karamata D, Kasahara Y, Klaerr-Blanchard M, Klein C, Kobayashi Y, Koetter P, Koningstein G, Krogh S, Kumano M, Kurita K, Lapidus A, Lardinois S, Lauber J, Lazarevic V, Lee SM, Levine A, Liu H, Masuda S, Mauel C, Medigue C, Medina N, Mellado RP, Mizuno M, Moestl D, Nakai S, Noback M, Noone D, O'Reilly M, Ogawa K, Ogiwara A, Oudega B, Park SH, Parro V, Pohl TM, Portetelle D, Porwollik S, Prescott AM, Presecan E, Pujic P, Purnelle B, Rapoport G, Rey M, Reynolds S, Rieger M, Rivolta C, Rocha E, Roche B, Rose M, Sadaie Y, Sato T,

Scanlan E, Schleich S, Schroeter R, Scoffone F, Sekiguchi J, Sekowska A, Seror SJ, Serror P, Shin BS, Soldo B, Sorokin A, Tacconi E, Takagi T, Takahashi H, Takemaru K, Takeuchi M, Tamakoshi A, Tanaka T, Terpstra P, Tognoni A, Tosato V, Uchiyama S, Vandenbol M, Vannier F, Vassarotti A, Viari A, Wambutt R, Wedler E, Wedler H, Weitzenegger T, Winters P, Wipat A, Yamamoto H, Yamane K, Yasumoto K, Yata K, Yoshida K, Yoshikawa HF, Zumstein E, Yoshikawa H, Danchin A (1997) The complete genome sequence of the Gram-positive bacterium *Bacillus subtilis*. Nature, Vol.390, pp.249-256, ISSN 0028-0836, EISSN 1476-4687.

Lam KH, Chow KC, Wong WK, (1998) Construction of an efficient *Bacillus subtilis* system for extracellular production of heterologous protein. Journal of Biotechnology, Vol.63, pp.167– 177, ISSN 0168-1656 (Print), 1873-4863 (Electronic).

Le Loir Y, Gruss A, Ehrlich SD, Langella P (1998) A nine-residue synthetic propeptide enhances secretion efficiency of heterologous proteins in *Lactococcus lactis*. Journal of Bacteriology, Vol.180, pp.1895-1903, ISSN 0021-9193 (Print), 1098-5530 (Electronic).

Le Loir Y, Nouaille S, Commissaire J, Brétigny L, Gruss A, Langella P (2001) Signal peptide and propeptide optimization for heterologous protein secretion in *Lactococcus lactis*. Applied and Environmental Microbiology, Vol.67, pp.4119-4127, ISSN 0099-2240 (Print), 1098-5336 (Electronic).

Le Loir Y, Azevedo V, Oliveira SC, Freitas DA, Miyoshi A, Bermúdez-Humarán LG, Nouaille S, Ribeiro LA, Leclercq S, Gabriel JE, Guimaraes VD, Oliveira MN, Charlier C, Gautier M, Langella P (2005) Protein secretion in *Lactococcus lactis* : an efficient way to increase the overall heterologous protein production. Microbial Cell Factories, Vol.40, pp.44-49, doi:10.1186/1475-2859-4-2, ISSN: 1475-2859.

Lesuisse E, Schanck K, Colson C, (1993) Purification and preliminary characterization of the extracellular lipase of *Bacillus subtilis* 168, an extremely basic pH-tolerant enzyme, European Journal of Biochemistry, Vol.216, pp.155–160, ISSN 0014-2956 (Print), 1432-1033 (Electronic).

Ling L, Xu Z, Li W, Shuai J, Lu P, Hu C (2007) Protein secretion pathways in *Bacillus subtilis*: implication for optimization of heterologous protein secretion. Biotechnology Advances, Vol.25, pp.1-12, ISSN 0014-2956 (Print), 1432-1033 (Electronic)

Li W, Zhou X, Lu P (2004) Bottlenecks in the expression and secretion of heterologous proteins in *Bacillus subtilis*. Research in Microbiology, pp.155, pp.605-610, ISSN 0923-2508 (Print), 1769-7123 (Electronic).

Mézes PSF, Yang YQ, Hussain M, Lampen, JO (1983) *Bacillus cereus* 569/H β-lactamase I: cloning in *Escherichia coli* and signal sequence determination. FEBS Letters, Vol.161, pp.195-200, ISSN 0014-5793 (Print), 1873-3468 (Electronic).

Oguro A, Kakeshita H, Honda K, Takamatsu H, Nakamura K, Yamane K (1995) *srb*: a *Bacillus subtilis* gene encoding a homologue of the α-subunit of the mammalian signal recognition particle receptor. DNA Research Vol.2, pp.95-100, ISSN 1340-2838 (Print), 1756-1663 (Electronic).

Olmos-Soto J and Contreras-Flores R, (2003) Genetic system constructed to overproduce and secrete proinsulin in *Bacillus subtilis*, Applied and Environmental Microbiology, Vol.62, pp.369– 373, ISSN 0099-2240 (Print), 1098-5336 (Electronic).

Palva I, Lehtovaara P, Kaariainen L, Sibakov M, Cantell K, Schein CH, Kashiwagi K, Weissmann C (1983) Secretion of interferon by *Bacillus subtilis*. Gene, Vol.22, pp.229–235, ISSN 0378-1119.

Palva I (1982) Molecular cloning of □-amylase gene from *Bacillus amyloliquefaciens* and its expression in *B. subtilis*. Gene, Vol. 19, pp81-87. ISSN 0378-1119.

Pop O, Martin U, Abel C, Müller JP (2002) The twin-arginine signal peptide of PhoD and the TatAd/Cd proteins of *Bacillus subtilis* form an autonomous Tat translocation system. The Journal of Biological Chemistry, Vol. 277, pp. 3268-3273, ISSN 0021-9258 (Print), 1083-351X (Electronic).

Randall RE, Goodbourn S (2008) Interferons and viruses: an interplay between induction, signalling, antiviral responses and virus countermeasures. Journal of General Virology, Vol.89, pp.1-47, ISSN: 0022-1317 (Print), 1465-2099 (Electronic).

Rabhi-Essafi I, Sadok A, Khalaf N, Fathallah DM (2007) A strategy for high-level expression of soluble and functional human interferon alpha as a GST-fusion protein in *E. coli*. Protein Engineering Design and Selection, Vol.5, pp.201-209, ISSN (Print): 1741-0126. ISSN (Electronic): 1741-0134.

Robinson C, Bolhuis A (2001) Protein targeting by the twin-arginine translocation pathway. Nature Reviews Molecular Cell Biology, Vol.2, pp.350–356, ISSN 1471-0072 (Print), 1471-0080 (Electronic).

Rojas Contreras JA, Pedraza-Reyes M, Ordoñez LG, Estrada NU, Barba de la Rosa AP, De León-Rodríguez A. (2010) Replicative and integrative plasmids for production of human interferon γ in *Bacillus subtilis*. Plasmid, Vol.64, pp.170-176, ISSN 0147-619X (Print) 1095-9890 (Electronic).

Sadaie Y, Takamatsu H, Nakamura K, Yamane K (1991) Sequencing reveals similarity of the wild-type div+ gene of *Bacillus subtilis* to the *Escherichia coli secA* gene. Gene, Vol.98, pp.101-105, ISSN 0378-1119 (Print), 1879-0038 (Electronic).

Santini, C. L., B. Ize, A. Chanal, M. Muller, G. Giordano, and L.-F. Wu. 1998. A novel sec-independent periplasmic protein translocation pathway in *Escherichia coli*. The EMBO Journal, Vol.17, pp.101-112, ISSN 0261-4189.

Sasamoto H, Nakazawa K, Tsutsumi K, Takase K, Yamane K (1989) Signal peptide of *Bacillus subtilis* α-amylase. Journal of Biochemistry, Vol.106, pp.376-382, ISSN 0021-924X (Print), 1756-2651 (Electronic)

Sørensen HP, Mortensen KK (2005) Soluble expression of recombinant proteins in the cytoplasm of *Escherichia coli*. Microbial Cell Factories, Vol.4, doi:10.1186/1475-2859-4-1, ISSN: 1475-2859.

Srivastava P, Bhattacharaya P, Pandey G, Mukherjee KJ (2005) Overexpression and purification of recombinant human interferon alpha2b in *Escherichia coli*. Protein Expression and Purification, Vol.41, pp.313-322, ISSN 1046-5928 (Print), 1096-0279 (Electronic).

Suciu D, Inouye M (1996) The 19-residue pro-peptide of staphylococcal nuclease has a profound secretion-enhancing ability in *Escherichia coli*. Molecular Microbiology, Vol.21, pp.181-195, ISSN 0950-382X (Print), 1365-2958 (Electronic)..

Takase T, Mizuno H, Yamane K (1988) NH$_2$-terminal processing of *Bacillus subtilis* α-amylase, The Journal of Biological Chemistry, Vol.263, pp.11548-11553, ISSN 0021-9258 (Print).

Takamatsu H, Fuma S, Nakamura K, Sadaie Y, Shinkai A, Matsuyama S, Mizushima S, Yamane K. (1992) In vivo and in vitro characterization of the *secA* gene product of *Bacillus subtilis*. Journal of Bacteriology, Vol.174, pp.4308-4316, ISSN 0021-9193 (Print), 1098-5530 (Electronic).

Tjalsma H, Bolhuis A, Jongbloed JD, Bron S, van Dijl JM (2000) Signal peptide-dependent protein transport in *Bacillus subtilis*: A genome-based survey of the secretome. Microbiology and Molecular Biology Reviews, Vol.64, pp.515–547, ISSN 1092-2172 (Print), 1098-5557 (Electronic).

Tjalsma H, Antelmann H, Jongbloed JD, Braun PG, Darmon E, Dorenbos R, Dubois JY, Westers H, Zanen G, Quax WJ, Kuipers OP, Bron S, Hecker M, van Dijl JM (2004) Proteomics of protein secretion by *Bacillus subtilis*: separating the "secrets" of the secretome. Microbiology and Molecular Biology Reviews, Vol.68, pp.207–233, , ISSN 1092-2172 (Print), 1098-5557 (Electronic).

van Dijl JM, Braun, PG, Robinson C, Quax WJ, Antelmann H, Hecker M, Muller J, Tjalsma H, Bron S, Jongbloed JD (2002) Functional genomic analysis of the *Bacillus subtilis* Tat pathway for protein secretion. Journal of Biotechnology, Vol.98, pp.243–254 ISSN: 0168-1656.

van Wely KH, Swaving J, Klein M, Freudl R, Driessen AJ (2000) The carboxyl terminus of the *Bacillus subtilis* SecA is dispensable for protein secretion and viability. Microbiology, Vol.146, pp.2573-2581, ISSN: 1350-0872 (Print), 1465-2080(Electronic).

van Wely KH, Swaving J, Freudl R, Driessen AJ (2001) Translocation of proteins across the cell envelope of Gram-positive bacteria. FEMS Microbiology Reviews, Vol.25, pp.437-454, ISSN 0168-6445 (Print), 1574-6976 (Electronic).

Vitikainen M, Pummi T, Airaksinen U, Wu H, Sarvas M, Kontinen VP (2001) Quantitation of the capacity of the secretion apparatus and requirement for PrsA in growth and secretion of α-amylase in *Bacillus subtilis*. Journal of Bacteriology, Vol.183, pp.1881–1890, ISSN 0021-9193 (Print), 1098-5530 (Electronic).

Wang L, Ruan B, Ruvinov S, Bryan PN (1998) Engineering the independent folding of the subtilisin BPN' pro-domain: correlation of pro-domain stability with the rate of subtilisin folding. Biochemistry, Vol.37, pp.3165-3171, ISSN 0006-2960 (Print), 1520-4995 (Electronic).

Westers L, Westers H, Quax WJ (2004). *Bacillus subtilis* as cell factory for pharmaceutical proteins: A biotechnological approach to optimize the host organism. Biochimica et Biophysica Acta, Vol.1694, pp.299–310, ISSN 0006-3002 (Print).

Westers L, Dijkstra DS, Westers H, van Dijl JM, Quax WJ (2006) Secretion of functional human interleukin-3 from *Bacillus subtilis*. Journal of Biotechnolgy, Vol.123, pp.211-224, ISSN 0168-1656 (Print), 1873-4863 (Electronic).

Williams RC, Rees ML, Jacobs MF, Pragai Z, Thwaite JE, Baillie LW, Emmerson PT, Harwood CR (2003) Production of *Bacillus anthracis* protective antigen is dependent on the extracellular chaperone, PrsA. The Journal of Biological Chemistry, Vol.278, pp.18056–18062, ISSN 0021-9258 (Print), 1083-351X (Electronic).

Wu SC, Ye R, Wu XC, Ng SC, Wong SL (1998) Enhanced secretory production of a single-chain antibody fragment from *Bacillus subtilis* by coproduction of molecular chaperones. Journal of Bacteriology, Vol.180, pp.2830–2835, ISSN 0021-9193 (Print), 1098-5530 (Electronic).

Wu SC, Wong SL (2002a) Engineering of a *Bacillus subtilis* strain with adjustable levels of intracellular biotin for secretory production of functional streptavidin. Applied and Environmental Microbiology, Vol.68, pp.1102-1108, ISSN 0099-2240 (Print), 1098-5336 (Electronic).

Wu SC, Yeung JC, Duan Y, Ye R, Szarka SJ, Habibi HR, Wong SL (2002b) Functional production and characterization of a fibrin-specific single-chain antibody fragment from *Bacillus subtilis*: effects of molecular chaperones and a wall-bound protease on antibody fragment production. Applied and Environmental Microbiology, Vol.68, pp.3261-3269, ISSN 0099-2240 (Print), 1098-5336 (Electronic).

Yabuta Y, Takagi H, Inouye M, Shinde U (2001) Folding pathway mediated by an intramolecular chaperone: propeptide release modulates activation precision of pro-subtilisin. The Journal of Biological Chemistry. Vol.276, pp. 44427-44434, ISSN 0021-9258 (Print), 1083-351X (Electronic).

Yabuta Y, Subbian E, Takagi H, Shinde U, Inouye M (2002) Folding pathway mediated by an intramolecular chaperone: dissecting conformational changes coincident with autoprocessing and the role of Ca^{2+} in subtilisin maturation. Journal of Biochemistry, Vol.131, pp.31-37, ISSN 0021-924X (Print), 1756-2651 (Electronic).

Yamane K, Bunai K, Kakeshita H (2004) Protein traffic for secretion and related machinery of *Bacillus subtilis*. Bioscience, Biotechnology, and Biochemistry Vol.68, pp.2007-2023, ISSN 0916-8451 (Print), 1347-6947 (Electronic).

Zhang Q, Zhong J, Liang X, Liu W, Huan L (2010) Improvement of human interferon α secretion by *Lactococcus lactis*. Biotechnology Letters, Vol.32, pp.1271-1277, ISSN 0141-5492 (Print), 1573-6776 (Electronic).

Zhu X, Ohta Y, Jordan F, Inouye M (1989) Pro-sequence of subtilisin can guide the refolding of denatured subtilisin in an intermolecular process. Nature, Vol.339, pp.483-484, ISSN 0028-0836 (Print).

Zhuang Z, Wu ZG, Chen M, Wang PG (2008). Secretion of human interferon-β 1b by recombinant *Lactococcus lactis*. Biotechnology Letters, Vol.30, pp.1819-1823, ISSN 0141-5492 (Print), 1573-6776 (Electronic).

Zuber P, Losick R (1983). Use of a *lacZ* fusion to study the role of the *spo0* genes of *Bacillus subtilis* in developmental regulation. Cell, Vol.35, pp.275-283, ISSN 0092-8674 (Print).

7

Increasing Recombinant Protein Production in *E. coli* by an Alternative Method to Reduce Acetate

Hendrik Waegeman and Marjan De Mey

Ghent University, Centre of Expertise-Industrial Biotechnology and Biocatalysis,
Belgium

1. Introduction

Since the development of recombinant DNA technology (Cohen et al., 1973), it became possible to express heterologous genes in pro- or eukaryotic hosts, *i.e.* genes which they naturally not express. This development enabled the production of all kinds of products of which the high-added value recombinant proteins, became increasingly important and as such boosted biopharmaceutical and industrial enzyme applications. Up to now, the FDA (Food and Drug Administration) and EMEA (European Medicines Agency) have licensed the application of more than 150 recombinant proteins to be used as a pharmaceutical (Ferrer-Miralles et al., 2009). Global sales of biopharmaceuticals are estimated to account for US$70–80 Billion today (Walsh, 2010). Industrial enzymes (e.g. proteases, amylases, lipases, cellulases, pullulanases, pectinases) are used in various industrial segments and the industrial enzyme market is still expanding, estimated to reach US$ 3.74 Billion by the year 2015 (Global Industry Analysts, 2011). To date, the majority of this industrial enzyme market value is generated by recombinant processes (Hodgson, 1994; Demain & Vaishnav, 2009).

It is clear that recombinant protein production has evolved to one of the most important branches in modern biotechnology, representing a billion-dollar business, both in the production of biopharmaceuticals and industrial enzymes.

A pivotal choice in the design of a recombinant protein bioprocess is the selection of a suitable host strain. This selection is influenced by different factors: (i) ease of cultivation and growth characteristics, (ii) ease of genetic manipulation and availability of molecular tools, (iii) ability of post-translational modifications (e.g. glycosylation patterns, disulfide bond formation), (iv) downstream processing, and (v) regulatory aspects (generally regarded as safe, SAFE (Lotti et al., 2004; Sahdev et al., 2008; Durocher & Butler, 2009).

These aspects will determine whether the designed recombinant protein bioprocess will end up in an economical viable bioprocess which can compete with the present process.

In contrast to biopharmaceuticals, industrial enzyme bioprocesses are only economical viable as a low production cost is assured. This implies that higher yields, titres and production rates are necessary which can only be obtained by fast growing organisms. This is reflected by the distribution of the most commonly used organisms in these two industries. Whereas slow growing organisms as plants and animals are used as host in half of the biopharmaceutical processes, they count only for 12% of the processes in the industrial enzyme market (Demain & Vaishnav, 2009; Ferrer-Miralles et al., 2009). Bacteria on the other hand, have a market share of 30% in both industries. However, yeasts and molds, which grow much faster in comparison with higher eukaryotes, are used in 58 % of the cases in the industrial enzyme market and only in 18% of the cases in the in the biopharmaceutical market.

Several bacteria have been explored as host for recombinant protein production. Recently, much interest is raised in the use of *Bacillus* strains as host for recombinant protein production because of their advantageous features as gram-positive (Terpe, 2006). However, till today *Escherichia coli* remains a very popular and predominantly used bacterium for recombinant protein production. This is primarily because this well-characterised organism can easily and rapidly grow on cheap substrates and can be simply modified through a broad variety of molecular tools. But even more, the further exploration of other potential microbial hosts are often restricted due to limited information about genetics and metabolism and/or the availability of molecular tools.

2. *Escherichia coli* for recombinant protein production

Besides the advantage of many available molecular tools, the easily cultivable and genetically and metabolically well-known *Escherichia coli* can be grown to high biomass concentrations in high cell density cultures allowing the production of high amounts of heterologous protein (Makrides, 1996). Nonetheless, *E. coli* suffers from some major drawbacks as well.

i.   The production of heterologous proteins to high titres concurs mostly with the initiation of a stress response and/or metabolic burden, both associated with the use of multi-copy plasmids, resulting in misfolding and degradation of the heterologous protein and formation of inclusion bodies (Noack et al., 1981; Parsell & Sauer, 1989; Bentley et al., 1990; Gill et al., 2000; Hoffmann & Rinas, 2004; Ventura & Villaverde, 2006).

ii.   As prokaryotic, *Escherichia coli* lacks the ability to perform enhanced post-translational modifications making the production of more complex eukaryotic proteins in *E. coli* challenging. This inability to form disulfide bonds or to execute glycosylation results in the production of instable and non-functional proteins.

iii. Secretory production of recombinant proteins into the culture medium includes several advantages, especially in cases of toxic recombinant proteins. However, compared to other hosts, *E. coli* does not naturally secrete proteins in high amounts. Nonetheless, *E. coli* possesses different secretions systems for the transport of proteins from the cytoplasmic to the perisplasmic or extracellular environment (Tseng et al., 2009). Crucial hereby is the signal peptide which is linked to the protein allowing recognition and transport by the secretion system.

iv.   The main difficulty when using *E. coli* as host is the production of acetate as by-product during fermentations as a result of overflow metabolism occurring when cells grow rapidly and cannot metabolise the delivered carbon source fast enough (Andersen & von Meyenburg, 1980; Holms, 1986). It is generally observed that even low concentrations of acetate can hamper growth and obstruct the production of recombinant proteins (Jensen & Carlsen, 1990; Nakano et al., 1997).

Many efforts have been made to overcome these hurdles and hence to increase recombinant protein production in *E. coli* or to express more complex proteins in this host. These engineering attempts are summarized in Fig. 1.

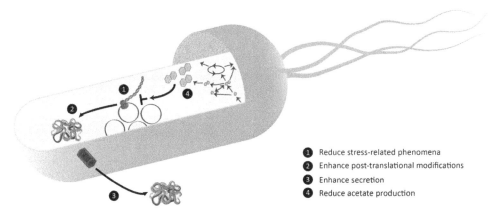

① Reduce stress-related phenomena
② Enhance post-translational modifications
③ Enhance secretion
④ Reduce acetate production

Fig. 1. Overview of different engineering approaches to increase recombinant protein production in *Escherichia coli*

The primarily used approach to produce recombinant proteins is to clone the gene of interest on a multi-copy plasmid under the control of a strong promoter in order to achieve high transcription rates and hence high recombinant protein concentrations. However, problems such as metabolic burden, segregational instability, misfolding and proteolytic breakdown or aggregation in inclusion bodies, and difficulties in controlling gene expression are usually associated with multi-copy plasmids and the use of strong promoters (Noack et al., 1981; Parsell & Sauer, 1989; Bentley et al., 1990; Dong et al., 1995; Kurland & Dong, 1996; Gill et al., 2000; Hoffmann & Rinas, 2004; Ventura & Villaverde, 2006). Most engineering strategies to tackle these problems focus on prevention of misfolding, neutralisation of increased protease activity or stress response (Chou, 2007). An elaborated review of these efforts is given in (Waegeman & Soetaert, 2011).

Two post-translational modifications which are pivotal for the stability and activity of many more complex eukaryotic proteins are disulfide bonds and glycosylation. The former is being facilitated in *E. coli* by secreting the recombinant protein into the more oxidizing periplasmic space using the Sec or Tat secretion system, by altering the redox state of the cytoplasm through modifications in the thioredoxin reductase gene (*trxB*) and gluthatione reductase genes (*gor*) or by cytoplasmic overexpression of periplasmic disulfide oxidoreductases (such as DsbC) which enhance the rate of disulfide isomerisation. An excellent review of these engineering strategies can be found in (de Marco, 2009).

Besides the proper formation of disulfide bonds, *E. coli* also lacks the ability of glycosylation. In order to make *E. coli* produce N-linked glycoproteins the gene cluster *pgl*, responsible for glycosylation in *Campylobacter jejuni* (Szymanski et al., 1999; Abu-Qarn et al., 2008) was successfully transferred (Wacker et al., 2002). Moreover, combination of the *pgl* system with a simple, genetically encoded glycosylation tag, expands the glycosylation possibilities of *E. coli* (Fisher et al., 2011).

The secretory production of recombinant proteins into the fermentation broth includes several advantages compared to cytoplasmic production. Although *E. coli* has different secretion systems for transport of proteins, secretion of recombinant proteins is rather complex. Many research efforts focus on the utilisation of these existing transport routes for the secretion of heterologous proteins (Choi & Lee, 2004; Jong et al., 2010) including selection and modification of the signal peptide, coexpression of proteins that assist in translocation and folding, improvement of periplasmic release when transport occurs in two steps or protection of the target protein from degradation and contamination (Abdallah et al., 2007).

3. An alternative approach to reduce acetate production and improve recombinant protein production in *Escherichia coli*

Throughout the years, various *Escherichia coli* strains with different genotypes have been examined for their potential to produce recombinant proteins in high titres. A comprehensive overview of all *E. coli* strains used in recombinant protein production processes and their characteristics is given in (Waegeman & Soetaert, 2011). Although *E. coli* B and *E. coli* K12 strains are equally used as host for recombinant protein production (47% and 53%, respectively), *E. coli* BL21 is by far the most commonly used strain (35%) in academia. In industry, this number is probably even much higher.

Escherichia coli BL21 displays higher biomass yields compared to *E. coli* K12 resulting in substantially lower acetate amounts which in return has a positive effect on the recombinant protein production (El-Mansi & Holms, 1989; Shiloach et al., 1996). The second reason of the extensive use of *E. coli* BL21 as microbial host for recombinant protein production is that this strain is deficient in the proteases Lon and OmpT, which decreases the breakdown of recombinant protein and result in higher yields (Gottesman, 1989; Gottesman, 1996). However, until recently the genome sequence of *E. coli* BL21 was not available making genetically modifications not always straightforward and therefore challenging. Consequently, still a lot of attention and effort is going towards *E. coli* K12-derived strains as most favourable *E. coli* strain for recombinant protein production (Ko et al., 2010; Ryu et al., 2010; Striedner et al., 2010).

Many different strategies have been applied to increase recombinant protein formation and decrease acetate formation in *E. coli* K12 strains including optimisation of the bioprocess conditions as metabolic engineering of the production host (De Mey et al., 2007b). These approaches comprise attempts which can be categorised in 3 classes: (i) deletion of acetate pathway genes, (ii) avoiding overflow metabolism by limiting the glucose uptake system through alteration of the carbon source, applying elaborate feeding strategies, or engineering the glucose uptake system, and (iii) avoiding overflow metabolism by redirecting central metabolic fluxes and preserving sufficient precursors of the amino acids, the building blocks of proteins (Fig. 2).

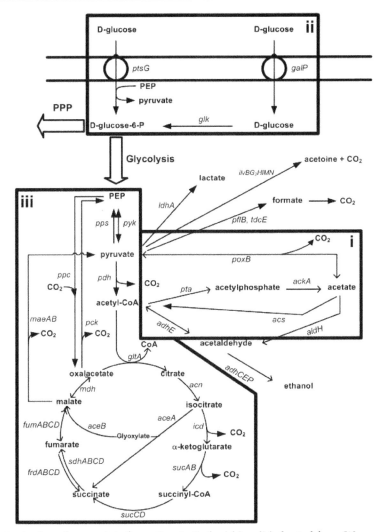

Fig. 2. Strategies to reduce acetate formation in *Escherichia coli* (adapted from Waegeman & Soetaert, 2011): (i) blocking the acetate pathway by knocking out genes that encode for acetate pathway enzymes, (ii) reducing the glucose uptake rate, and (iii) redirecting central metabolic fluxes. PPP, pentose phosphate pathway; *aceA*, isocitrate lyase; *aceB*, malate synthase; *ackA*, acetate kinase; *acn*, aconitase; *acs*, acetyl-CoA synthase; *adhCEP*, ethanol dehydrogenase; *adhE*, aldehyde dehydrogenase; *fumABCD*, fumarase; *galP*, galactose permease; *glk*, glucokinase; *icd*, isocitrate dehydrogenase; *ilvBG₂HIMN*, acetolacetate decarboxylase; *ldhA*, lactate dehydrogenase; *maeAB*, malic enzyme; *mdh*, malate dehydrogenase; *pck*, phosphoenolpyruvate carboxykinase; *pdh*, pyruvate dehydrogenase; *pflB, tdcE*, pyruvate formate lyase; *poxB*, pyruvate oxidase; *ppc*, phosphoenolpyruvate carboxylase; *pps*, phosphoenolpyruvate synthase; *pta*, acetylphosphotransferase; *pyk*, pyruvate kinase; *sdhABCD*, succinate dehydrogenase; *sucAB*, a-ketoglutarate dehydrogenase; *sucCD*, succinate thiokinase.

The first, rational effort to decrease acetate production is to block the acetate pathway by knocking out genes that encode for acetate pathway enzymes, e.g. *ackA* (acetate kinase), *pta* (phosphate acetyltransferase) and *poxB* (pyruvate oxidase) (Diaz-Ricci et al., 1991; Yang et al., 1999; Contiero et al., 2000; Dittrich et al., 2005; De Mey et al., 2007a). These attempts resulted in a considerably decrease of acetate production but in return pyruvate, lactate or formate formation, which are also undesired by-products, increased to a large extent.

A second widely followed approach to minimise acetate formation during high cell density fermentations is to limit rapid uptake of glucose causing overflow metabolism. Overflow metabolism occurs when high glycolytic fluxes, due to rapid glucose uptake, are not further processed in the TCA cycle developing a bottleneck at the pyruvate node and consequently pyruvate is converted to acetate.

Strategies based on optimising the bioprocess conditions to reduce the glucose uptake rate comprise applying specific glucose feeding patterns, the application of alternative substrates, the addition of supplements to the medium, the control of a range of fermentation parameters and the application of systems to remove acetate from the fermentation broth (Farmer & Liao, 1997; Nakano et al., 1997; Akesson et al., 1999; Akesson et al., 2001b; Akesson et al., 2001a; Fuchs et al., 2002 ; Chen et al., 2005; Eiteman & Altman, 2006). Although all these attempts were in many cases successful to reduce acetate production, they imply a severe lower growth rate and they do not utilise the full potential of the microbial host.

Engineering of the glucose uptake system is being successfully applied as well to overcome overflow metabolism. By deleting one of the phosphotransferase system genes, e.g. *ptsG*, *ptsH* or *ptsI*, the uptake through the major glucose transporter is several impeded, resulting in a reduced glycolytic flux and reduced acetate pathway (Chou et al., 1994; Siguenza et al., 1999; De Anda et al., 2006; Wong et al., 2008). To restore the strong reduction in growth rate as consequence of hampering the main glucose transporter De Anda et al. (2006) overexpressed the alternative glucose transporter gene *galP* (coding for a galactose permease) and exploited the native glucose kinase (Glk) transporter. The resulting strain *E. coli* W3110 Δ*ptsH galP*+ displayed a very low acetate yield and a significantly increased recombinant protein yield compared to the *E. coli* W3110 wild-type, without reduction in growth rate. Wong et al (2008) restored glucose transport by co-expressing the gene *glf*, encoding for a passive glucose transporter of *Zymomonas mobilis*. However, this only resulted in a decreased acetate formation in M9 minimal media, not in LB media.

A third approach to overcome overflow metabolism is to redirect the fluxes around the bottleneck, the phosphoenolpyruvate-pyruvate-oxaloacetate node, instead of restricting the glucose uptake. Farmer & Liao (1997) increased anaplerotic and glycolate fluxes by overexpressing phosphenolpyruvate carboxylase (encoded by *ppc*) and by deleting the FadR regulator. This notable strategy resulted in a more than 75% decrease in acetate yield compared to its wild type. Alternatively, another important success was achieved by the overexpression of a heterologous anaplerotic pyruvate carboxylase from *Rhizobium etli* resulting in a 57% reduction in acetate formation and a 68% increase in recombinant protein production (March et al., 2002). Similarly, De Mey et al. (2010) achieved an increase in recombinant protein production by deleting the genes coding for acetate pathway enzymes combined with the overexpression of *ppc*.

An alternative approach to enhance recombinant protein production is mimicking the *E. coli* BL21 phenotype in *E. coli* K12 by interfering on the regulatory level of gene expression instead of targeting genes directly involved in the conversion of metabolites in the acetate pathway our around the phosphoenolpyruvate-pyruvate-oxaloacetate node. ^{13}C metabolic flux analysis showed that the low acetate production in *E. coli* BL21(DE3) is caused by activation of the glyoxylate pathway (Noronha et al., 2000), a pathway which is normally not activated under glucose excess in *E. coli* K12 strains. Furthermore, acetate assimilation pathways are more active in *E. coli* BL21 compared to in *E. coli* K12 (Phue et al., 2005).

3.1 Influence of transcriptional regulators ArcA and IclR on *Escherichia coli* phenotypes

Regulation of gene expression is very complex and transcriptional regulators can be subdivided in global and local regulators depending on the number of operons they control. Global regulators control a vast number of genes, which must be physically separated on the genome and belong to different metabolic pathways (Gottesman, 1984). According to EcoCyc (Keseler et al., 2011) *E. coli* K12 MG1655 contains 40 master regulators and sigma factors. Nonetheless, only seven global regulators control the expression of 51% of all genes: ArcA, Crp, Fis, Fnr, Ihf, Lrp and NarL. In contrast to global regulators, local regulators control only a few genes, e.g. 20% of all transcriptional regulators control the expression of only one or two genes (Martinez-Antonio & Collado-Vides, 2003).

The global regulator ArcA (anaerobic redox control) was first discovered in 1988 by Iuchi and Lin and the regulator seemed to have an inhibitory effect on expression of aerobic TCA cycles genes under anaerobic conditions (Iuchi & Lin, 1988). Later on, it was unravelled that ArcA is a component of the dual-component regulator ArcAB, in which ArcA is the regulatory protein and ArcB acts as sensory protein (Iuchi et al., 1990).

Acording to EcoCyc (Keseler et al., 2011) ArcA is involved in the regulation of 168 genes and itself is regulated by 2 regulators (FnrR, RpoD). Statistical analysis of gene expression data (Salmon et al., 2005) showed that ArcA regulates the expression of a wide variety of genes involved in the biosynthesis of small macromolecules, transport, carbon and energy metabolism, cell structure, etc. The regulatory activity of ArcA is dependent on the oxygen concentration in the environment. The most profound effects of ArcA are noticed under microaerobic conditions (Alexeeva et al., 2003) but recently it was reported that also under aerobic conditions ArcA has an effect on central metabolic fluxes (Perrenoud & Sauer, 2005).

Similarly to the global transcriptional regulator ArcA, the local transcriptional regulator isocitrate lyase regulator IclR has a reductive effect on the flux through the TCA cycle (Rittinger et al., 1996). IclR represses the expression of the *aceBAK* operon, which codes for the glyoxylate pathway enzymes isocitrate lyase (encoded by *aceA*), malate synthase (encoded by *aceB*), and isocitrate dehydrogenase kinase/phosphatise (encoded by *aceK*) (Yamamoto & Ishihama, 2003). The last enzyme phosphorylates the TCA cycle enzyme isocitrate dehydrogenase (Icd) controlling the switch between the flux through the TCA cycle and the glyoxylate pathway. It is reported that when IclR levels are low or when IclR is inactivated, i.e. for cells growing on acetate (Cortay et al., 1991; Cozzone, 1998; El-Mansi et al., 2006), or in slow growing glucose utilising cultures (Fischer & Sauer, 2003; Maharjan et al., 2005), repression on glyoxylate genes is released and the glyoxylate pathway is activated.

As both transcriptional regulators, ArcA and IclR, are involved in controlling the flux through the TCA cycle and glyoxylate pathway, they are interesting targets for metabolic engineering for mimicking the *E. coli* BL21 phenotype in *E. coli* K12.

To investigate their effect, single knockouts as a knockout combination were made in *E. coli* MG1655 (K12-strain). The different mutants and wild type were cultivated in a 2L stirred tank bioreactor under glucose abundant (batch cultivation) conditions in order to precisely determine extracellular fluxes and growth rates and consequently to evaluate the physiological and metabolic consequences of *arcA* and *iclR* deletions on *E. coli* MG1655. In order to evaluate if these effects are corresponding with the characteristics of *E. coli* BL21, this *E. coli* strain was also tested. The growth rates and the average carbon and redox balances of the different strains are shown in Table 1.

E. coli strain	μ_{max} (h^{-1})	Carbon (%)	Redox (%)
MG1655	0.66 ± 0.02	97	101
MG1655 ΔarcA	0.60 ± 0.01	96	94
MG1655 ΔiclR	0.61 ± 0.02	95	95
MG1655 ΔarcA ΔiclR	0.44 ± 0.03	99	101
BL21(DE3)	0.59 ± 0.02	93	99

Table 1. Average maximum growth rate, carbon balance and redox balance for batch cultures of the investigated strains

The *arcA* and *iclR* single knockouts strains have a slightly lower maximum growth rate. In contrary the combined *arcA-iclR* double knockout strain in *E. coli* MG1655 exhibits a substantial reduction of 38% in μ_{max}. Fig. 3 shows the effect of these mutations on various product yields under abundant glucose conditions. The corresponding average redox and carbon balances close very well (Table 1).

Product yields in c-mole/c-mole glucose for *E. coli* MG1655, MG1655 Δ*arcA*, MG1655 Δ*iclR*, MG1655 Δ*arcA* Δ*iclR* and BL21 under glucose abundant conditions. Oxygen yield is shown as a positive number for a clear representation, but O_2 is actually consumed during experiments. The values represented in the graph are the average of at least two separate experiments and the errors are standard deviations calculated on the yields.

Both the *arcA* and *iclR* knockout strains show an increased biomass yield in *E. coli* MG1655. When combining these deletions in *E. coli* MG1655 the yield is further increased to 0.063 ± 0.01 c-mole/c-mole glucose, which approximates the theoretical biomass yield of 0.65 c-mole/c-mole glucose (assuming a P/O-ratio of 1.4) (Varma et al., 1993a; Varma et al., 1993b) and slightly higher compared to the *E. coli* BL21(DE3) wild-type. The higher biomass yield in *E. coli* MG1655 Δ*arcA* Δ*iclR* is accompanied by a 70% and 16% reduction in acetate and CO_2 yields, respectively. This reduction in CO_2 yield could indicate that the glyoxylate pathway is more active in the double knock-out mutant as is observed in *E. coli* BL21 (Noronha et al., 2000).

The deletion of local transcriptional regulator *iclR* reduces the acetate formation with 50% in *E. coli* MG1655. When the global transcriptional regulator *arcA* is additionally deleted, the acetate yield is even further decreased to a comparable value of *E. coli* BL21(DE3).

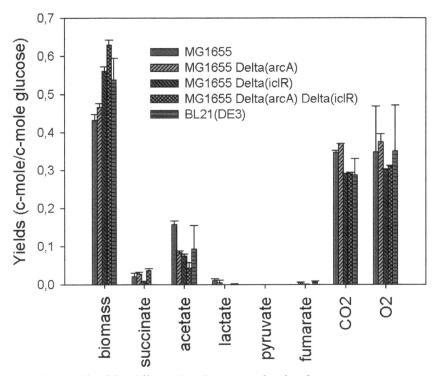

Fig. 3. Product yields of the different *E. coli* strains in batch cultures.

[13]C-metabolic flux analysis confirmed our hypothesis that the deletion of both *arcA* and *iclR* in *E. coli* MG1655 alters central metabolism fluxes profoundly (Fig. 4). A higher flux at the entrance of the TCA cycle was observed due to *arcA* deletion resulting in a reduced production of acetate and less carbon loss. Due to the *iclR* deletion, the glyoxylate pathway is activated resulting in a redirection of 30% of the isocitrate molecules directly to succinate and malate without CO_2 production.

Moreover, similar central metabolic fluxes were observed in the combined *arcA-iclR* double knockout in *E. coli* MG1655 as in *E. coli* BL21(DE3). These results suggest that the expression levels of *arcA* and *iclR* are low in *E. coli* BL21. We could confirm that deletion of both *arcA* and *iclR* in *E. coli* BL21 had no severe implications on the phenotype (Waegeman et al., 2011c). Only a slight decrease in growth rate was observed. Thus, this proves that ArcA and IclR are poorly active in *E. coli* BL21 whereas in *E. coli* K12 both regulators play an important role. This can be explained by mutations in the promoter region of *iclR* and a less efficient codon usage of *arcA* in *E. coli* BL21 (Waegeman et al., 2011a).

Thus, by deletion of a local and global transcriptional regulator, ArcA and IclR respectively, we could mimic the physiological and metabolic properties of *E. coli* BL21 in an *E. coli* K12 strain. Furthermore, only a small part of the tremendously elevated biomass yield was attributed to increased glycogen content (Waegeman et al., 2011a) making this strain an attractive candidate for recombinant protein production.

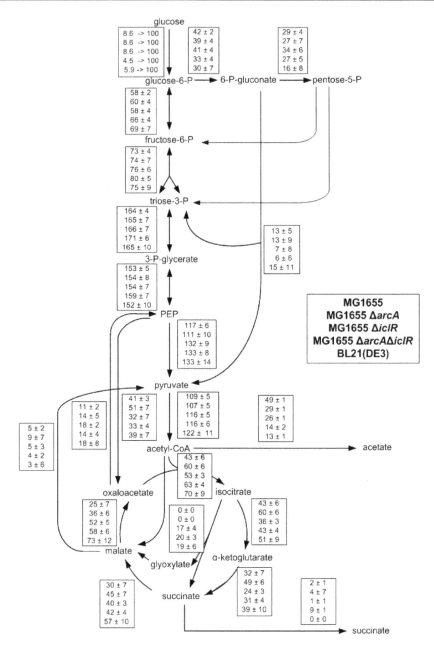

Fig. 4. Metabolic Flux distribution in *E. coli* MG1655, its derivate single knockout strains
Δ*arcA* and Δ*iclR*, and the double knockout strain Δ*arcA* Δ*iclR*, and *E. coli* BL21 cultivated in
glucose abundant conditions. More specific details about the metabolic flux calculations can
be found in (Waegeman et al., 2011a)

3.2 *Escherichia coli* MG1655 Δ*arcA* Δ*iclR* as potential candidate for recombinant protein production

Our previous research has shown that similar metabolic and physiological characteristics as *E. coli* BL21 can be achieved in *E. coli* K12 by combined deletion of the global transcriptional regulator ArcA and the local transcriptional regulator IclR.

To investigate whether these metabolic alterations in *E. coli* MG1655 also beneficially influence recombinant protein production, we compared the recombinant protein production of the metabolically engineered strain to *E. coli* BL21(DE3) using GFP (Green Fluorescent Protein) as a biomarker (Fig. 5).

Batch cultures were performed in 2L stirred tank bioreactors. Yields are calculated by dividing GFP and biomass concentrations during the cultivation phase when biomass concentrations are higher than 2 gL^{-1}. The values represented in the graph are the average of at least two separate experiments and the errors are the standard deviations calculated on the yields.

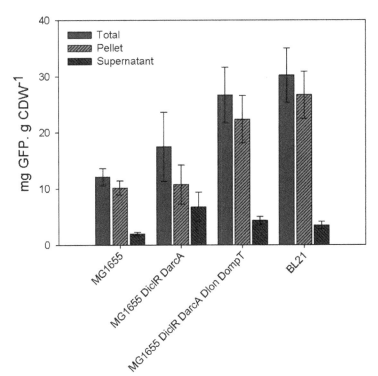

Fig. 5. Overall GFP yields of the different strains in batch cultures.

To our regret, the combined *arcA-iclR* double knockout mutant did not perform as we anticipated. Although the increased biomass yield and decreased acetate yield in the double knockout beneficially influence recombinant protein production as a higher GFP yield was observed to its wild type *E. coli* MG1655 (30% increase in the double knockout strain), still a striking difference of more than 40% was detected compared to *E. coli* Bl21. Additionally, we observed that at higher cell densities (>2gL⁻¹ CDW) the GFP concentrations decreases again suggesting proteases activity (Waegeman et al., 2011b).

Proteases play an important role in the degradation of foreign proteins (Gottesman & Maurizi, 1992) and it is generally believed that recombinant proteins are better produced in *E. coli* BL21 and his derivates because these strains lack the cytoplasmic ATP-dependent protease Lon (Gottesman, 1989) and the periplasmic OmpT (Gottesman, 1996).

Although also other proteases are known for the degradation of proteins, but in a lesser extent towards recombinant proteins (Jürgen et al., 2010) and since *E. coli* BL21 lacks the proteases Lon and OmpT, these proteases were also deleted in the *E. coli* MG1655 ΔarcA ΔiclR strain.

The additional deletion of the proteases Lon and OmpT, resulting in the quadruple knockout strain *E. coli* MG1655 ΔarcA ΔiclR Δlon ΔompT, could impede the breakdown of GFP at higher cell densities. The GFP yield obtained at the end of the glucose growth phase in bioreactor experiments approximates the GFP yield of *E. coli* BL21 (DE3).

4. Conclusion

To date, recombinant protein production has evolved to one of the most important branches in modern biotechnology, representing a billion-dollar business, both in the production of biopharmaceuticals and industrial enzymes. Although many organisms have been used as host, *Escherichia coli* is predominantly utilised as microbial host, representing 30% of the bioprocesses in both industries.

Although, *E. coli* strains are popular because they are fast growers, metabolically and genetically well characterised, and many molecular tools are available, these strains display several drawbacks. Besides problems related to stress response, post-transcriptional modification and secretion of recombinant proteins, a major drawback is the formation of acetate in aerobic cultures which retards growth and impedes protein production.

Logically, many endeavours have been reported to decrease acetate formation and increase recombinant protein production in this host. However, among the different *E. coli* strains, *E. coli* BL21 and his derivates show a significant low acetate formation compared to *E. coli* K12 strains, making BL21 a standard host in industrial recombinant protein production bioprocesses. Though, *E. coli* BL21 is not the optimal host due to plasmid instability and, until recently, unknown genome sequence making genetic modifications challenging.

Traditionally, acetate formation in *E. coli* K12 strains is tackled by blocking the acetate pathways or avoiding overflow metabolism through limiting the glucose uptake rate or redirecting the fluxes around the bottleneck, the phosphoenolpyruvate-pyruvate-oxaloacetate node. Alternatively, we propose to copy similar physiological and metabolic properties of *E. coli* BL21 in *E. coli* K12. This was achieved by combined deletion of the global transcriptional regulator ArcA and the local regulator IclR. Albeit this metabolically engineered *E. coli* K12 derivate displayed higher biomass yield and lower acetate yield resulting in a substantially increase in recombinant protein yield, the protein yield was still considerably lower than the yield observed in *E. coli* BL21. This difference in recombinant protein production is caused by proteolytic activity in *E. coli* K12, which does not occur in *E. coli* BL21 due to absence of the proteases Lon and OmpT. Additional deletion of these proteases in our combined *arcA-iclR* double knockout strain, hampered this proteolytic activity yielding recombinant protein levels similar to *E. coli* BL21.

In conclusion, by deleting only four genes, i.e. *arcA*, *iclR*, *lon*, and *ompT* it was possible to mimic the phenotype of *E. coli* BL21 in *E. coli* K12. The metabolically engineered quadruple knockout strain exhibited not only a tremendous increase in biomass yield and severe decrease in acetate yield, the recombinant protein production increased by a factor 2, resulting in a strain that can compete with *E. coli* BL21 for the industrial production of recombinant proteins. These results are incentive to further optimization of *E. coli* as microbial host making *E. coli* an often-chosen host in industrial bioprocesses.

5. Acknowledgment

The research of Hendrik Waegeman was financially supported by the Special Research Fund (BOF) of Ghent University

6. References

Abdallah, A.M., van Pittius, N.C.G., Champion, P.A.D., Cox, J., Luirink, J., Vandenbroucke-Grauls, C.M.J.E., Appelmelk, B.J.& Bitter, W. (2007). Type VII secretion-mycobacteria show the way. *Nat Rev Microbiol*, Vol. 5, pp. 883-891

Abu-Qarn, M., Eichler, J.& Sharon, N. (2008). Not just for Eukarya anymore: protein glycosylation in Bacteria and Archaea. *Curr Opin Struct Biol*, Vol. 18, pp. 544-550

Akesson, M., Hagander, P.& Axelsson, J.P. (2001a). Avoiding acetate accumulation in *Escherichia coli* cultures using feedback control of glucose feeding. *Biotechnology and Bioengineering*, Vol. 73, pp. 223-230

Akesson, M., Hagander, P.& Axelsson, J.P. (2001b). Probing control of fed-batch cultivations: analysis and tuning. *Control Engineering Practice*, Vol. 9, pp. 709-723

Akesson, M., Karlsson, E.N., Hagander, P., Axelsson, J.P.& Tocaj, A. (1999). On-line detection of acetate formation in *Escherichia coli* cultures using dissolved oxygen responses to feed transients. *Biotechnology and Bioengineering*, Vol. 64, pp. 590-598

Alexeeva, S., Hellingwerf, K.J.& de Mattos, M.J.T. (2003). Requirement of ArcA for redox regulation in *Escherichia coli* under microaerobic but not anaerobic or aerobic conditions. *J Bacteriol*, Vol. 185, pp. 204-209

Andersen, K.B.& von Meyenburg, K. (1980). Are growth rates of *Escherichia coli* in batch cultures limited by respiration? *J Bacteriol*, Vol. 144, pp. 114-123

Bentley, W.E., Mirjalili, N., Andersen, D.C., Davis, R.H.& Kompala, D.S. (1990). Plasmid-encoded protein: the principal factor in the "metabolic burden" associated with recombinant bacteria. *Biotechnol Bioeng*, Vol. 35, pp. 668-681

Chen, X., Cen, P.& Chen, J. (2005). Enhanced production of human epidermal growth factor by a recombinant *Escherichia coli* integrated with in situ exchange of acetic acid by macroporous ion-exchange resin. *Journal of Bioscience and Bioengineering*, Vol. 100, pp. 579-581

Choi, J.H.& Lee, S.Y. (2004). Secretory and extracellular production of recombinant proteins using *Escherichia coli*. *Appl Microbiol Biotechnol*, Vol. 64, pp. 625-635

Chou, C.H., Bennett, G.N.& San, K.Y. (1994). Effect of modified glucose uptake using genetic engineering techniques on high-level recombinant protein production in *Escherichia coli* dense cultures. *Biotechnol Bioeng*, Vol. 44, pp. 952-960

Chou, C.P. (2007). Engineering cell physiology to enhance recombinant protein production in *Escherichia coli. Appl Microbiol Biotechnol*, Vol. 76, pp. 521-532

Cohen, S.N., Chang, A.C., Boyer, H.W.& Helling, R.B. (1973). Construction of biologically functional bacterial plasmids *in vitro*. *Proc Natl Acad Sci U S A*, Vol. 70, pp. 3240-3244

Contiero, J., Beatty, C., Kumari, S., DeSanti, C., Strohl, W.& Wolfe, A. (2000). Effects of mutations in acetate metabolism on high-cell-density growth of *Escherichia coli. J Ind Microbiol Biotechnol*, Vol. 24, pp. 421-430

Cortay, J.C., Nègre, D., Galinier, A., Duclos, B., Perrière, G.& Cozzone, A.J. (1991). Regulation of the acetate operon in *Escherichia coli:* purification and functional characterization of the IclR repressor. *EMBO J*, Vol. 10, pp. 675-679

Cozzone, A.J. (1998). Regulation of acetate metabolism by protein phosphorylation in enteric bacteria. *Annu Rev Microbiol*, Vol. 52, pp. 127-164

De Anda, R., Lara, A.R., Hernández, V., Hernández-Montalvo, V., Gosset, G., Bolívar, F.& Ramírez, O.T. (2006). Replacement of the glucose phosphotransferase transport system by galactose permease reduces acetate accumulation and improves process performance of *Escherichia coli* for recombinant protein production without impairment of growth rate. *Metab Eng*, Vol. 8, pp. 281-290

de Marco, A. (2009). Strategies for successful recombinant expression of disulfide bond-dependent proteins in *Escherichia coli. Microb Cell Fact*, Vol. 8, pp. 26

De Mey, M., Lequeux, G.J., Beauprez, J.J., Maertens, J., Van Horen, E., Soetaert, W.K., Vanrolleghem, P.A.& Vandamme, E.J. (2007a). Comparison of different strategies to reduce acetate formation in *Escherichia coli. Biotechnol Prog*, Vol. 23, pp. 1053-1063

De Mey, M., Lequeux, G.J., Beauprez, J.J., Maertens, J., Waegeman, H.J., Van Bogaert, I.N., Foulquie-Moreno, M.R., Charlier, D., Soetaert, W.K., Vanrolleghem, P.A.& Vandamme, E.J. (2010). Transient metabolic modeling of Escherichia coli MG1655 and MG1655 DeltaackA-pta, DeltapoxB Deltapppc ppc-p37 for recombinant beta-galactosidase production. *J Ind Microbiol Biotechnol*, Vol. 37, pp. 793-803

De Mey, M., Maeseneire, S.D., Soetaert, W.& Vandamme, E. (2007b). Minimizing acetate formation in E. *coli* fermentations. *J Ind Microbiol Biotechnol*, Vol. 34, pp. 689-700

Demain, A.L.& Vaishnav, P. (2009). Production of recombinant proteins by microbes and higher organisms. *Biotechnol Adv*, Vol. 27, pp. 297-306

Diaz-Ricci, J.C., Regan, L.& Bailey, J.E. (1991). Effect of alteration of the acetic acid synthesis pathway on the fermentation pattern of *Escherichia coli. Biotechnol Bioeng*, Vol. 38, pp. 1318-1324

Dittrich, C.R., Vadali, R.V., Bennett, G.N.& San, K.-Y. (2005). Redistribution of metabolic fluxes in the central aerobic metabolic pathway of E. *coli* mutant strains with deletion of the ackA-pta and poxB pathways for the synthesis of isoamyl acetate. *Biotechnol Prog*, Vol. 21, pp. 627-631

Dong, H., Nilsson, L.& Kurland, C.G. (1995). Gratuitous overexpression of genes in *Escherichia coli* leads to growth inhibition and ribosome destruction. *J Bacteriol*, Vol. 177, pp. 1497-1504

Durocher, Y.& Butler, M. (2009). Expression systems for therapeutic glycoprotein production. *Curr Opin Biotechnol*, Vol. 20, pp. 700-707

Eiteman, M.A.& Altman, E. (2006). Overcoming acetate in *Escherichia coli* recombinant protein fermentations. *Trends in Biotechnology*, Vol. 24, pp. 530-533

El-Mansi, E.M.& Holms, W.H. (1989). Control of carbon flux to acetate excretion during growth of *Escherichia coli* in batch and continuous cultures. *J Gen Microbiol*, Vol. 135, pp. 2875-2883

El-Mansi, M., Cozzone, A.J., Shiloach, J.& Eikmanns, B.J. (2006). Control of carbon flux through enzymes of central and intermediary metabolism during growth of *Escherichia coli* on acetate. *Curr Opin Microbiol*, Vol. 9, pp. 173-179

Farmer, W.R.& Liao, J.C. (1997). Reduction of aerobic acetate production by *Escherichia coli*. *Appl Environ Microbiol*, Vol. 63, pp. 3205-3210

Ferrer-Miralles, N., Domingo-Espín, J., Corchero, J.L., Vázquez, E.& Villaverde, A. (2009). Microbial factories for recombinant pharmaceuticals. *Microb Cell Fact*, Vol. 8, pp. 17

Fischer, E.& Sauer, U. (2003). A novel metabolic cycle catalyzes glucose oxidation and anaplerosis in hungry *Escherichia coli*. *J Biol Chem*, Vol. 278, pp. 46446-46451

Fisher, A.C., Haitjema, C.H., Guarino, C., Çelik, E., Endicott, C.E., Reading, C.A., Merritt, J.H., Ptak, A.C., Zhang, S.& DeLisa, M.P. (2011). Production of secretory and extracellular N-linked glycoproteins in *Escherichia coli*. *Appl Environ Microbiol*, Vol. 77, pp. 871-881

Fuchs, C., Koster, D., Wiebusch, S., Mahr, K., Eisbrenner, G.& Markl, H. (2002). Scale-up of dialysis fermentation for high cell density cultivation of *Escherichia coli*. *Journal of Biotechnology*, Vol. 93, pp. 243-251

Gill, R.T., Valdes, J.J.& Bentley, W.E. (2000). A comparative study of global stress gene regulation in response to overexpression of recombinant proteins in *Escherichia coli*. *Metab Eng*, Vol. 2, pp. 178-189

Global Industry Analysts, I. (February 09, 2011). Global Industrial Enzymes, In: *PRWeb*, July 20, 2011, Available from:
http://www.prweb.com/releases/industrial_enzymes/proteases_carbohydrases/prweb8121185.htm.

Gottesman, S. (1984). Bacterial regulation: global regulatory networks. *Annu Rev Genet*, Vol. 18, pp. 415-441

Gottesman, S. (1989). Genetics of proteolysis in *Escherichia coli*. *Annu Rev Genet*, Vol. 23, pp. 163-198

Gottesman, S. (1996). Proteases and their targets in *Escherichia coli*. *Annu Rev Genet*, Vol. 30, pp. 465-506

Gottesman, S.& Maurizi, M.R. (1992). Regulation by proteolysis: energy-dependent proteases and their targets. *Microbiol Rev*, Vol. 56, pp. 592-621

Hodgson, J. (1994). The changing bulk biocatalyst market. *Nat Biotechnol*, Vol. 12, pp. 789-790

Hoffmann, F.& Rinas, U. (2004). Stress induced by recombinant protein production in *Escherichia coli*. *Adv Biochem Eng Biotechnol*, Vol. 89, pp. 73-92

Holms, W.H. (1986). The central metabolic pathways of *Escherichia coli*: relationship between flux and control at a branch point, efficiency of conversion to biomass, and excretion of acetate. *Curr Top Cell Regul*, Vol. 28, pp. 69-105

Iuchi, S.& Lin, E.C. (1988). arcA (dye), a global regulatory gene in Escherichia coli mediating repression of enzymes in aerobic pathways. Proc Natl Acad Sci U S A, Vol. 85, pp. 1888-1892

Iuchi, S., Matsuda, Z., Fujiwara, T.& Lin, E.C. (1990). The arcB gene of Escherichia coli encodes a sensor-regulator protein for anaerobic repression of the arc modulon. Mol Microbiol, Vol. 4, pp. 715-727

Jensen, E.B.& Carlsen, S. (1990). Production of recombinant human growth hormone in Escherichia coli: expression of different precursors and physiological effects of glucose, acetate, and salts. Biotechnol Bioeng, Vol. 36, pp. 1-11

Jong, W.S.P., Saurí, A.& Luirink, J. (2010). Extracellular production of recombinant proteins using bacterial autotransporters. Curr Opin Biotechnol, Vol. 21, pp. 646-652

Jürgen, B., Breitenstein, A., Urlacher, V., Büttner, K., Lin, H., Hecker, M., Schweder, T.& Neubauer, P. (2010). Quality control of inclusion bodies in Escherichia coli. Microb Cell Fact, Vol. 9, pp. 41

Keseler, I.M., Collado-Vides, J., Santos-Zavaleta, A., Peralta-Gil, M., Gama-Castro, S., Muniz-Rascado, L., Bonavides-Martinez, C., Paley, S., Krummenacker, M., Altman, T., Kaipa, P., Spaulding, A., Pacheco, J., Latendresse, M., Fulcher, C., Sarker, M., Shearer, A.G., Mackie, A., Paulsen, I., Gunsalus, R.P.& Karp, P.D. (2011). EcoCyc: a comprehensive database of Escherichia coli biology. Nucleic Acids Research, Vol. 39, pp. D583-D590

Ko, C.-H., Tsai, C.-H., Lin, P.-H., Chang, K.-C., Tu, J., Wang, Y.-N.& Yang, C.-Y. (2010). Characterization and pulp refining activity of a Paenibacillus campinasensis cellulase expressed in Escherichia coli. Bioresour Technol, Vol.,

Kurland, C.G.& Dong, H. (1996). Bacterial growth inhibition by overproduction of protein. Mol Microbiol, Vol. 21, pp. 1-4

Lotti, M., Porro, D.& Srienc, F. (2004). Recombinant proteins and host cell physiology. J Biotechnol, Vol. 109, pp. 1-2

Maharjan, R.P., Yu, P.-L., Seeto, S.& Ferenci, T. (2005). The role of isocitrate lyase and the glyoxylate cycle in Escherichia coli growing under glucose limitation. Res Microbiol, Vol. 156, pp. 178-183

Makrides, S.C. (1996). Strategies for achieving high-level expression of genes in Escherichia coli. Microbiol Rev, Vol. 60, pp. 512-538

March, J.C., Eiteman, M.A.& Altman, E. (2002). Expression of an anaplerotic enzyme, pyruvate carboxylase, improves recombinant protein production in Escherichia coli. Appl Environ Microbiol, Vol. 68, pp. 5620-5624

Martinez-Antonio, A.& Collado-Vides, J. (2003). Identifying global regulators in transcriptional regulatory networks in bacteria. Curr Opin Microbiol, Vol. 6, pp. 482-489

Nakano, K., Rischke, M., Sato, S.& Märkl, H. (1997). Influence of acetic acid on the growth of Escherichia coli K12 during high-cell-density cultivation in a dialysis reactor. Appl Microbiol Biotechnol, Vol. 48, pp. 597-601

Noack, D., Roth, M., Geuther, R., Muller, G., Undisz, K., Hoffmeier, C.& Gaspar, S. (1981). Maintenance and genetic stability of vector plasmids pBR322 and pBR325 in Escherichia coli K12 strains grown in a chemostat. Mol Gen Genet, Vol. 184, pp. 121-124

Noronha, S.B., Yeh, H.J., Spande, T.F.& Shiloach, J. (2000). Investigation of the TCA cycle and the glyoxylate shunt in *Escherichia coli* BL21 and JM109 using (13)C-NMR/MS. *Biotechnol Bioeng*, Vol. 68, pp. 316-327

Parsell, D.A.& Sauer, R.T. (1989). Induction of a heat shock-like response by unfolded protein in *Escherichia coli:* dependence on protein level not protein degradation. *Genes Dev*, Vol. 3, pp. 1226-1232

Perrenoud, A.& Sauer, U. (2005). Impact of global transcriptional regulation by ArcA, ArcB, Cra, Crp, Cya, Fnr, and Mlc on glucose catabolism in *Escherichia coli*. *J Bacteriol*, Vol. 187, pp. 3171-3179

Phue, J.-N., Noronha, S.B., Ritabrata, Wolfe, A.J.& Shiloach, J. (2005). Glucose metabolism at high density growth of E. *coli* B and E. *coli* K: differences in metabolic pathways are responsible for efficient glucose utilization in E. *coli* B as determined by microarrays and Northern blot analyses. *Biotechnol Bioeng*, Vol. 90, pp. 805-820

Rittinger, K., Negre, D., Divita, G., Scarabel, M., Bonod-Bidaud, C., Goody, R.S., Cozzone, A.J.& Cortay, J.C. (1996). *Escherichia coli* isocitrate dehydrogenase kinase/phosphatase. *Eur J Biochem*, Vol. 237, pp. 247-254

Ryu, K., Kim, K.-H., Yoo, S.-Y., Lee, E.-Y., Lim, K.-H., Min, M.-K., Kim, H., Choi, S.I.& Seong, B.L. (2010). Production and characterization of active hepatitis C virus RNA-dependent RNA polymerase. *Protein Expr Purif*, Vol. 71, pp. 147-152

Sahdev, S., Khattar, S.K.& Saini, K.S. (2008). Production of active eukaryotic proteins through bacterial expression systems: a review of the existing biotechnology strategies. *Mol Cell Biochem*, Vol. 307, pp. 249-264

Salmon, K.A., Hung, S.-p., Steffen, N.R., Krupp, R., Baldi, P., Hatfield, G.W.& Gunsalus, R.P. (2005). Global gene expression profiling in *Escherichia coli* K12: effects of oxygen availability and ArcA. *J Biol Chem*, Vol. 280, pp. 15084-15096

Shiloach, J., Kaufman, J., Guillard, A.S.& Fass, R. (1996). Effect of glucose supply strategy on acetate accumulation, growth, and recombinant protein production by *Escherichia coli* BL21 (λDE3) and *Escherichia coli* JM109. *Biotechnol Bioeng*, Vol. 49, pp. 421-428

Siguenza, R., Flores, N., Hernandez, G., Mart\'inez, A., Bolivar, F.& Valle, F. (1999). Kinetic characterization in batch and continuous culture of *Escherichia coli* mutants affected in phosphoenolpyruvate metabolism: differences in acetic acid production. *World J Microbiol Biotechnol*, Vol. 15, pp. 587-592

Striedner, G., Pfaffenzeller, I., Markus, L., Nemecek, S., Grabherr, R.& Bayer, K. (2010). Plasmid-free T7-based *Escherichia coli* expression systems. *Biotechnol Bioeng*, Vol. 105, pp. 786-794

Szymanski, C.M., Yao, R., Ewing, C.P., Trust, T.J.& Guerry, P. (1999). Evidence for a system of general protein glycosylation in *Campylobacter jejuni*. *Mol Microbiol*, Vol. 32, pp. 1022-1030

Terpe, K. (2006). Overview of bacterial expression systems for heterologous protein production: from molecular and biochemical fundamentals to commercial systems. *Appl Microbiol Biotechnol*, Vol. 72, pp. 211-222

Tseng, T.-T., Tyler, B.M.& Setubal, J.C. (2009). Protein secretion systems in bacterial-host associations, and their description in the Gene Ontology. *BMC Microbiol*, Vol. 9, pp. S2

Varma, A., Boesch, B.W.& Palsson, B.O. (1993a). Biochemical production capabilities of *Escherichia coli*. *Biotechnol Bioeng*, Vol. 42, pp. 59-73

Varma, A., Boesch, B.W.& Palsson, B.O. (1993b). Stoichiometric interpretation of *Escherichia coli* glucose catabolism under various oxygenation rates. *Appl Environ Microbiol*, Vol. 59, pp. 2465-2473

Ventura, S.& Villaverde, A. (2006). Protein quality in bacterial inclusion bodies. *Trends Biotechnol*, Vol. 24, pp. 179-185

Wacker, M., Linton, D., Hitchen, P.G., Nita-Lazar, M., Haslam, S.M., North, S.J., Panico, M., Morris, H.R., Dell, A., Wren, B.W.& Aebi, M. (2002). N-linked glycosylation in *Campylobacter jejuni* and its functional transfer into E. *coli*. *Science*, Vol. 298, pp. 1790-1793

Waegeman, H., Beauprez, J., Moens, H., Maertens, J., De Mey, M., Foulquie-Moreno, M.R., Heijnen, J.J., Charlier, D.& Soetaert, W. (2011a). Effect of iclR and arcA knockouts on biomass formation and metabolic fluxes in Escherichia coli K12 and its implications on understanding the metabolism of Escherichia coli BL21 (DE3). *BMC Microbiol*, Vol. 11, pp. 70

Waegeman, H., De Lausnay, S., Beauprez, J., Maertens, J., De Mey, M.& Soetaert, W. (2011b). Increasing recombinant protein production in *Escherichia coli* K12 through metabolic engineering. *New Biotechnology*, submitted

Waegeman, H., Maertens, J., Beauprez, J., De Mey, M.& Soetaert, W. (2011c). Effect of *iclR* and *arcA* deletion on physiology and metabolic fluxes in *Escherichia coli* BL21(DE3). *Biotechnology Progress*, in press

Waegeman, H.& Soetaert, W. (2011). Increasing recombinant protein production in *Escherichia coli* through metabolic and genetic engineering. *Journal of Industrial Microbiology and Biotechnology*, accepted

Walsh, G. (2010). Post-translational modifications of protein biopharmaceuticals. *Drug Discov Today*, Vol. 15, pp. 773-780

Wong, M.S., Wu, S., Causey, T.B., Bennett, G.N.& San, K.-Y. (2008). Reduction of acetate accumulation in *Escherichia coli* cultures for increased recombinant protein production. *Metab Eng*, Vol. 10, pp. 97-108

Yamamoto, K.& Ishihama, A. (2003). Two different modes of transcription repression of the *Escherichia coli* acetate operon by IclR. *Mol Microbiol*, Vol. 47, pp. 183-194

Yang, Y.T., Aristidou, A.A., San, K.Y.& Bennett, G.N. (1999). Metabolic flux analysis of *Escherichia coli* deficient in the acetate production pathway and expressing the *Bacillus subtilis* acetolactate synthase. *Metab Eng*, Vol. 1, pp. 26-34

8

Approaches for Improving Protein Production in Multiple Protease-Deficient *Bacillus subtilis* Host Strains

Takeko Kodama[1*], Kenji Manabe[1], Yasushi Kageyama[1], Shenghao Liu[1],
Katsutoshi Ara[1], Katsuya Ozaki[1] and Junichi Sekiguchi[2]
[1]Biological Science Laboratories, Kao Corporation,
[2]Department of Bioscience and Textile Technology,
Interdisciplinary Graduate School of Science and Technology, Shinshu University,
Japan

1. Introduction

Bacillus subtilis is a Gram-positive, nonpathogenic organism which is widely used as a host for enzyme production, due to its ability to secrete large amounts of proteins into the growth medium (Simonen et al., 1993; Westers et al., 2004). The secretion of a target protein leads to the natural separation of the product from cell components, which simplifies downstream processing of the protein. Accordingly, there has been a great deal of research performed regarding protein production in *B. subtilis* (Simonen et al., 1993; Westers et al., 2004). Nevertheless, the yields of heterologous protein obtained from this organism are often insufficient (Harwood, 1992). Several bottlenecks in the *B. subtilis* secretion pathway have been reported, including poor targeting to the translocase, degradation of the secretory protein, and incorrect folding (Westers et al., 2004). One of the major bottlenecks involves the degradation of the produced protein by extracellular proteases; therefore, inactivation of extracellular proteases is essential for improvement of protein production with *B. subtilis* as the host.

2. Inhibition of proteolysis of heterologous and nature proteins after the translocation process by inactivation of multiple proteases

Eight extracellular proteases have been identified in *B. subtilis* to date, which are encoded by the following genes: *aprE* (Stahl et al., 1984; Wong et al., 1984), *bpr* (Sloma et al., 1990b; Wu et al., 1990), *epr* (Bruckner et al., 1990; Sloma et al., 1988), *mpr* (Rufo et al., 1990; Sloma et al., 1990a), *nprB* (Tran et al., 1991), *nprE* (Yang et al., 1984), *vpr* (Sloma et al., 1991), and *wprA* (Margot et al., 1996). Deletions in the *aprE* (encoding subtilisin, alkaline protease) and *nprE* (encoding neutral protease) genes were the first such mutations, whose mutants show lower activities of extracellular proteases (Sloma et al., 1991). In addition, a deletion mutation in the *epr* gene resulted in low protease activity in the culture supernatant. *wprA* encodes a 96-kDa protein that is processed to the CWBP23 propeptide and CWBP52 mature protease, forming a complex associated with the cell wall (Margot et al., 1996). This complex was also

A

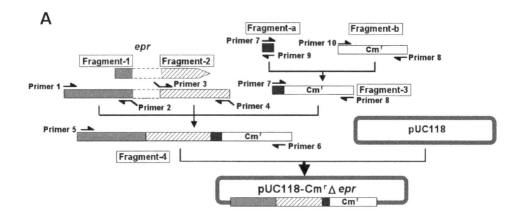

B

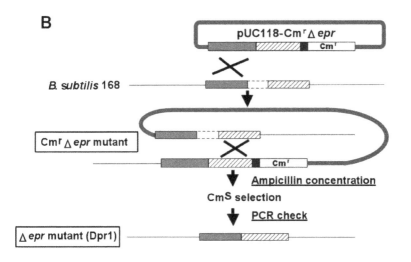

Fig. 1. Strategy for construction of a Δepr mutant. (A) Construction of a chloramphenicol-resistance (Cmʳ) plasmid, pUC118-CmʳΔepr. Fragment 3 was amplified with fragment a (containing the repU promoter of pUB110) and fragment b (containing the chloramphenicol resistance gene of pC194), and primers 7 and 8. Fragment 4 was amplified with fragments 1, 2, 3, and primers 5 and 6. The amplified fragment 4 was prepared by blunting and kination, and then cloned into the *Sma*I site of pUC118 to generate pUC118-CmʳΔepr. (B) Construction of the Δepr mutant. *B. subtilis* 168 cells were transformed with pUC118-CmʳΔepr, followed by selection for chloramphenicol resistance, obtaining CmʳΔepr. To obtain a Δepr mutant (deleted chloramphenicol-resistance cassette), the ampicillin concentration method was used (Kodama et al., 2007a). The chloramphenicol-sensitive (Cmˢ) Δepr mutant was confirmed by PCR using primers 5 and 4.

found in the culture supernatant of *B. subtilis* WB600 (Babe et al., 1998; Wu et al., 1991). Whether it is present in the cell wall or in the culture medium is therefore a critical factor in the degradation of heterologous proteins (Lee et al., 2000). Strains with deletion mutations in multiple extracellular proteases have since been constructed with extracellular protease activities of less than 0.5%, compared to the parental strain (Wu et al., 1991). It was recently reported that an eight protease-deficient strain, WB800, was a useful host for the production of various heterologous proteins (Murashima et al., 2002; L. Westers et al., 2006). However, the use of *B. subtilis* as a host has remained limited to bulk industrial enzyme production. Further optimization is necessary to develop production systems for heterologous proteins. This chapter focuses on the inhibition of proteolysis of secreted proteins after the translocation process by inactivation of multiple proteases which are extracellular (AprE, Bpr, Epr, Mpr, NprB, NprE, Vpr, and WprA), leaked outside from intracellular (AprX), and membrane-bound (HtrA and HtrB).

2.1 The intracellular protease, AprX is involved in degradation of a heterologous protein

In *B. subtilis*, extracellular protease-deficient mutants have been used in attempts to increase the productivity of heterologous proteins. We detected the protease activity of AprX using protease zymography in the culture medium at the late stationary growth phase. Construction of multiple-protease-deficient mutant without antibiotic-resistance markers and the effect of AprX on the heterologous protein production are descrived in detail in the following sections.

2.1.1 Construction of an eight-extracellular-protease-deficient mutant by marker-free deletion in *B. subtilis*

Antibiotic-resistance marker genes were used to create new bacterial strains. However, the number of markers available for use in *B. subtilis* and other bacteria is limited. We used the "ampicillin concentration" method for the creation of eight-extracellular-protease-deficient mutant with marker-free deletion (Fig. 1, Kodama et al., 2007a). Recently, several useful methods were developed to produce unmarked mutations in *B. subtilis* (Liu et al., 2008; Morimoto et al., 2008; Morimoto et al., 2009). These systems are more convenient for the introduction of multiple mutations.

2.1.2 Detection of AprX activity in the culture supernatant with protease zymography

Zymography has been used to detect proteolytic enzymes after electrophoretic separation in gels. Recently, the activities of some proteases, including Vpr have been detected by fibrin zymography of the extracellular proteins of *B. subtilis* (Murashima et al., 2002; L. Westers et al., 2006). The supernatant proteins of *B. subtilis* culture in modified 2xL broth (Kodama et al., 2007b) at 8 h (exponential growth phase), 25 h (early stationary phase), 50 h (mid-stationary phase), and 75 h (late stationary phase) (Fig. 2A) were analyzed by gelatin zymography (Fig. 2B). The resulting zymogram shows the protease activities as clear bands (or zones). In the exponential growth phase (8 h), no protease activity was detected (Fig. 2B, lane 1). However, protease activity increased during the stationary phase, and was highest at 75 h of the late stationary phase (Fig. 2B, lane 4). We examined the zymogram profile of the supernatant from the eight-extracellular-protease-deficient mutant (Dpr8) at 75 h, and found one clear band in the zymogram (Fig. 2C, lane 2). Protease activity disappeared in the aprX mutant at 75 h (Fig. 2C, lane 3). All of the protease activities completely disappeared in the KA8AX strain, in which nine genes (eight extracellular protease genes and *aprX*) were disrupted (Fig. 2C, lane 4). These results support the idea that the protease is serine protease AprX. To determine the serine and metal protease activities of this protease by zymography, PMSF and EDTA (2 mM each) were added to the supernatant of the 75 h culture of Dpr8 (Fig. 2D). EDTA decreased the activity of the protease slightly, whereas 2 mM PMSF completely inhibited the protease activity (Fig. 2D). These results suggest that the gelatin-degrading protease from the supernatant of Dpr8 culture is AprX. To determine whether AprX is the gelatin-degrading protease in the supernatant of the Dpr8 culture at 75 h, the AprX-FLAG fusion protein was constructed. The fusion gene was expressed with the original promoter and ribosomal binding site. On a zymographic gel, the activity bands corresponding to AprX-FLAG from both 168/AprX-FLAG and Dpr8/AprX-FLAG strains were located at slightly larger positions in size than those of AprX (Fig. 3). The size of the band corresponded to the size of the FLAG peptide. These results indicate that the activity of AprX is detectable as a single band by gelatin zymography of the supernatant of a 75 h culture of *B. subtilis* strains.

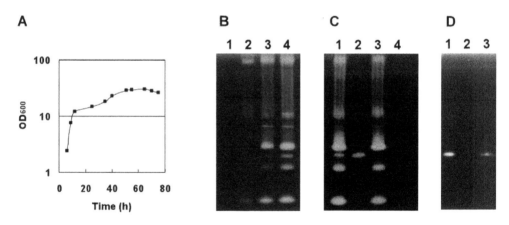

Fig. 2. (A) Cells from the wild-type were cultured in modified 2xL broth at 30°C. (B) Protein samples were prepared from the supernatants of the wild-type, cultured for various incubation times. Lane 1, 8 h culture; lane 2, 25 h; lane 3, 50 h; lane 4, 75 h. (C). Protein samples were prepared from the supernatants of the protease-deficient mutants after a 75 h culture. Lane 1, 168; lane 2, Dpr8; lane 3, AprXdd; lane 4, KA8AX. (D) PMSF or EDTA (2 mM each) was added to the supernatants of Dpr8 after a 75 h culture. Lane 1, Control (no addition); lane 2, addition of 2 mM PMSF; lane 3, addition of 2 mM EDTA. The samples were analyzed on SDS-12% polyacrylamide gels with 0.1% (w/v) gelatin. Proteins from the culture supernatants (equivalent to 0.3 μl) were applied to each lane for panels B, C, and D.

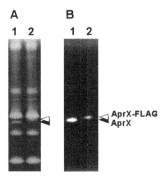

Fig. 3. Zymography of AprX-FLAG proteases. (A) Lane 1, 168; lane 2, 168/AprX-FLAG. (B) Lane 1, Dpr8; lane 2, Dpr8/AprX-FLAG. Proteins from the supernatants of the 75-h cultures (equivalent to 0.3 μl) were applied to the lanes for panels A and B. Arrowheads indicate the positions of AprX (closed symbol) and AprX-FLAG (open symbol).

2.1.3 Intracellular AprX leaked to the culture medium during the late stationary phase

It has been supposed that AprX is a serine protease belonging to the subtilase superfamily, and that it is an intracellular protease, because a canonical signal sequence for secretion has not been found in this protease (Valbuzzi et al.; 1999). However, AprX was detected in the culture medium by gelatin zymography (Fig. 3). *aprX* is transcribed during the stationary phase, and the regulator of SinR exerts negative effect on its transcription directly or indirectly (Valbuzzi et al.; 1999). However, *aprX* is not essential for either growth or sporulation (Valbuzzi et al.; 1999). As a result, the function of AprX has remained poorly understood. The Western blotting of AprX-FLAG from the intracellular fraction showed that the expression of AprX-FLAG began at 25 h, and that the expression level markedly increased after 50 h (Fig. 4). Our results agreed with a previous report that *aprX* is transcribed during the stationary phase. In contrast, a weak AprX-FLAG expression was detected in the supernatant only in the late stationary phase at 75 h (Fig. 4). This result agreed with the zymogram pattern of wild-type AprX (Fig. 2B). A slight decrease in cell density was observed after 50 h for the wild-type (Fig. 2A) and 168/AprX-FLAG strains (data not shown). The bands corresponding to AprX-FLAG from both the intra- and extracellular fractions were located at the same position. This result suggests that there is insufficient secretion of AprX, due to the absence of an obvious signal sequence. These observations also suggest that AprX is localized intracellularly by nature, and is leaked to the culture medium during the late stationary phase due to cell lysis.

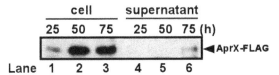

Fig. 4. Western blot analysis of AprX-FLAG. Western blot analysis was carried out to detect AprX-FLAG in the 168/AprX-FLAG strain with the anti-FLAG antibody. Proteins of cells (lanes 1-3) and supernatants (lanes 4-6) from 168/AprX-FLAG (0.02 OD600 units) were prepared as described in Materials and Methods. The arrowhead indicates the position of AprX-FLAG. The times of harvest of cells and supernatants are shown at the top.

2.1.4 AprX involved in degradation of the α-amylase-A522-PreS2 hybrid protein

AprX in the supernatant was able to degrade gelatin. Therefore, we considered that AprX may affect the production of secreted proteins. pTUBE522-preS2 has already been developed for the extracellular production of small peptides of the human hepatitis B virus preS2 antigen (42 amino acids) fused with *B. subtilis* α-amylase (deleting the C-terminal region to construct a peptide carrier) (Honda et al., 1993). To confirm the effect of AprX on the degradation of heterologous proteins, we examined the production of α-amylase-A522-PreS2 as a model of heterologous proteins, in multiple-protease-deficient *B. subtilis* strains. Cells carrying pTUBE522-preS2 were cultured in modified 2xL broth for 25, 50, and 75 h. α-amylase-A522-PreS2 in the supernatants from the cultures of Dpr7, Dpr8, and KA8AX strains was analyzed by Western blotting with the anti-PreS2 antibody (Fig. 5A). The Dpr7 strain lacked seven extracellular proteases (AprE deficiency excluded). No positive band corresponding to α-amylase-A522-PreS2 was detected at any phase for Dpr7 (pTUBE522-preS2) (Fig. 5A, lanes 1-3). Dpr8 (pTUBE522-preS2) produced α-amylase-A522-PreS2 at detectable levels, and production of the hybrid protein attained high levels after 50 h (Fig. 5A, lane 5). However, when AprX was produced in the supernatant of Dpr8 (pTUBE522-preS2) at 75 h, the amount of α-amylase-A522-PreS2 decreased markedly (Fig. 6A, lane 6). As expected, the degradation of α-amylase-A522-PreS2 was markedly inhibited in KA8AX (pTUBE522-preS2) at 75 h, with the relative amount of the hybrid protein produced by this strain being 1.8-times higher than that of Dpr8 at 50 h (Fig. 5A, lanes 5 and 9; Fig. 5B). KA8AX produced α-amylase-A522-PreS2 up to 80 mg/L, which is at least eightfold higher than the amount produced by the improved strain in a previous study (Lee et al., 2000; Fig. 5, lane 9). We also examined the degradation of α-amylase-A522-PreS2 by AprX protease. First, we prepared AprX protease from KA8AX (pDG-AprX) that was grown in a medium containing 1 mM IPTG for 4 h. The overexpression of AprX was confirmed by gelatin zymography (Fig. 6A, lane 2). Afterwards, the α-amylase-A522-PreS2 protein prepared from the supernatant of KA8AX (pTUBE522-preS2) at 75 h was mixed with AprX protease, and the mixture was incubated at 37°C for 60 min. The degradation of α-amylase-A522-PreS2 was analyzed by Western blotting using the anti-PreS2 antibody. H₂O and intracellular proteins extracted from KA8AX (pDG-AprX) cells cultured without IPTG did not decrease the amount of the α-amylase-A522-PreS2 protein (Figs. 6B, C, lanes 2 and 4), but intracellular proteins extracted from KA8AX (pDG-AprX) cells cultured with 1 mM IPTG decreased the amount to 70% (Figs. 6B, C, lanes 5 and 6).

These results indicate that the AprX protease directly degraded the α-amylase-A522-PreS2 protein. One bottleneck of the production of α-amylase-A522-PreS2 was partially solved by the disruption of eight extracellular proteases and AprX, as shown in this chapter. However, the supernatant from the KA8AX culture at 75 h contained not only a small amount of α-amylase-A522-PreS2, but also a large amount of α-amylase protein (determined by Western blotting with the anti-α-amylase antibody; data not shown). On other hand, no PreS2 peptide was detected by Western blotting with the anti-PreS2 antibody (Fig. 5A). These results indicate that the degradation of α-amylase-A522-PreS2 was not inhibited completely in the KA8AX strain, and that there were as yet unidentified protease(s) involved in the proteolysis of the PreS2 region. Therefore, there is still room for improving the inhibition of hybrid protein degradation. It has been reported that IspA (Isp) was identified as a major intracellular serine protease (Koide et al., 1986). We evaluated the inhibition of the degradation of α-amylase-A522-PreS2 by the inactivation of IspA in the KA8AX strain. However, the productivity of α-amylase-A522-PreS2 in the ten-protease deficient mutant was almost same as that in the KA8AX strain. A search in the GenoList database for *B. subtilis* 168 genome (http://genodb.pasteur.fr/cgi-bin/WebObjects/GenoList) of proteases and peptidases revealed the presence of 31 known and 11 putative proteases, and 38 known and 12 putative peptidases, respectively. Section 2.2 describes the investigation of membrane-bound proteases involved in protein degradation.

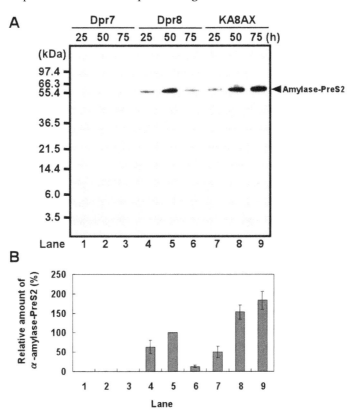

Fig. 5. Western blot analysis of the α-amylase-A522-PreS2 hybrid protein in the extracellular fractions of Dpr7, Dpr8, and KA8AX. (A) Western blot analysis was carried out to detect α-amylase-A522-PreS2 with the anti-PreS2 antibody. Culture supernatants from Dpr7 (lanes 1-3), Dpr8 (lanes 4-6), and KA8AX (lanes 7-9) were collected after 25, 50 h, and 75 h of cultivation, and subjected to Tricine-SDS-PAGE and Western blotting, as described in the Materials and Methods. Proteins from the culture supernatants (equivalent to 1 μl) were applied to each lane. The arrowhead indicates the position of α-amylase-A522-PreS2. The times of harvest of supernatants are shown at the top. (B) The relative α-amylase-A522-PreS2 protein amounts were compared on the basis of band intensities on Western blots (the amount of α-amylase-A522-PreS2 at 50 h in the Dpr8 strain was set to 100%). The presented results are the average of three individual experiments. Error bars correspond to the standard errors of the means (SEM). Lane numbers in panel A correspond to those in panel B.

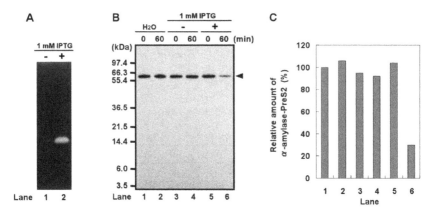

Fig. 6. The α-amylase-A522-PreS2 hybrid protein was degraded by AprX. (A) Zymography of supernatants from the KA8AX(pDG-AprX) strain. Lane 1, without IPTG; lane 2, with IPTG. (B) Western blot analysis of degradation of α-amylase-A522-PreS2 by AprX. AprX from KA8AX (pDG-AprX) mutant cells, cultured for 4 h with or without 1 mM IPTG was prepared as described in the Material and Methods. α-Amylase-A522-PreS2 from 10 μl supernatants of the KA8AX (pTUBE522-preS2) mutant (at 75 h cultivation) was mixed with 10 μl of AprX solution. After incubation at 37°C for 60 min, PMSF (final concentration, 10 mM) was added to the samples to stop the reaction. Western blot analysis was carried out to detect α-amylase-A522-PreS2 with the anti-PreS2 antibody; +, addition of 1 mM IPTG (AprX); -, no addition. The reaction mixture (equivalent to 1 μl) was applied to each lane. The arrowhead indicates the position of α-amylase-A522-PreS2. (C) The relative amounts of α-amylase-A522-PreS2 were obtained by comparing the band intensities on Western blots (the α-amylase-A522-PreS2 amount in lane 1 was set as 100%). Lanes 1 to 6 in panel C correspond to lanes 1 to 6 in panel B.

2.2 The effect of HtrA and HtrB on the degradation of secreted proteins

In this section we describe the effects of membrane-bound proteases and a two-component system on degradation of secreted proteins, and transcriptional regulation of the membrane-bound protease genes.

2.2.1 Cell envelope-associated quality control proteases

In *B. subtilis*, the accumulation of misfolded proteins at the membrane-cell wall interface is sensed by the CssR–CssS two-component system, which consists of the membrane-embedded sensor kinase, CssS and the response regulator, CssR (Hyyryläinen et al., 2001). This system responds to general protein secretion stresses, which can be triggered by either homologous (e.g., overproduction of LipA) or heterologous (e.g., overproduction of AmyQ and hIL-3) proteins, and consequently activates the transcription of the monocistronic *htrA* and *htrB* genes (Darmon et al., 2002; H. Westers et al., 2006; Hyyryläinen et al., 2007). HtrA and HtrB are membrane-bound serine proteases that are responsible for the degradation of misfolded proteins, and can thereby rescue the cell from a lethal accumulation of misfolded proteins in the cell envelope. In addition, HtrA has a dual localization, because it can be detected in the membrane-associated cellular fraction as well as the growth medium. Therefore, HtrA has a chaperone-like activity that might assist misfolded proteins in recovering their conformation, while also targeting unsuccessful protein for degradation (Antelmann et al., 2003). Induction of *htrA* and *htrB* expressions is responsive to secretion stress in a manner dependent on the CssRS two-component system. In addition, *htrA* and *htrB* expressions are negatively autoregulated and reciprocally cross-regulated (Noone et al., 2000, Noone et al., 2001). Therefore, the absence of HtrA leads to the increased synthesis of HtrB, and vice versa (Noone et al., 2001).

2.2.2 High-level lipase A (LipA) production in eleven proteases mutant

We examined the production of lipase A (LipA) of *B. subtilis* (van Pouderoyen et al., 2001), as a valuable model for industrial enzyme production, in a nine-protease-deficient *B. subtilis* strain. Therefore, we constructed the pHLApm plasmid, in which LipA with the promoter and ribosomal binding site of an alkaline cellulase gene, *egl-237* (Hakamada et al., 2000) was cloned into pHY300PLK (Takara, Japan). LipA was overproduced in *B. subtilis*. Cells carrying pHLApm were cultured in modified 2xL broth for 12, 24, 36, 48, 60, and 75 h. The productivity of LipA in the supernatants from cultures of the 168 and Dpr9 (in which nine genes encoding eight extracellular proteases and AprX were precisely and completely deleted from the chromosome) strains was calculated based on the activity of LipA (Fig. 7). In 24 h cultivation, the production level of the LipA in 168 and Dpr9 could be obtained at 860 mg/L, an excellent yield which is 1.4-times higher than that of previously reported (Lesuisse et al., 1993). After 24 h, the amount of LipA markedly decreased in the 168 strain (Fig. 7). In contrast, degradation of LipA in the Dpr9 was effectively inhibited, compared with the 168 strain. However, after 36 h, the production of LipA in Dpr9 was reduced by approximately 10% (Fig. 7). These results showed that LipA was also degraded in the Dpr9 strain. Overproduction of both homologous (LipA) and heterologous (AmyQ and hIL-3) proteins induces the expression of *htrA* and *htrB* by the CssRS system (Darmon et al., 2002; H. Westers et al., 2006). From the currently available data, it seems most likely that limitation of both proteases of HtrA and HtrB improved the yield of heterologous proteins (Vitikainen, M., H. L. et al., 2005). To confirm the effect of HtrA and HtrB on the degradation of secreted proteins, we examined the production of LipA of *B. subtilis* in the *htrA* and/or *htrB* deficient *B. subtilis* strains. We constructed Dpr9ΔhtrA,

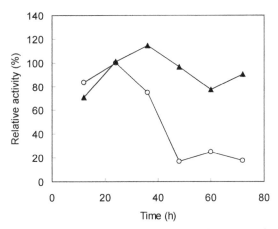

Fig. 7. Time course of LipA activity in the Dpr9 mutant. Cells were cultured in modified 2xL broth at 30°C. The accumulation of LipA in the culture medium was measured at various incubation times. Open circles, wild type strain; closed triangles, Dpr9 mutant.

Dpr9ΔhtrB, and Dpr9ΔhtrA/B (with eleven inactivated proteases), and evaluated each strain for the production of LipA. No effect on LipA production was observed in Dpr9ΔhtrA and Dpr9ΔhtrB. However, the production of LipA by the Dpr9ΔhtrA/B strain was at 1100 mg/L, which is 1.2-times higher than that of the Dpr9 strain (Fig. 8). These results suggest that inactivation of both *htrA* and *htrB*, as well as the nine proteases, has improved the productivity of *B. subtilis* for the production of LipA.

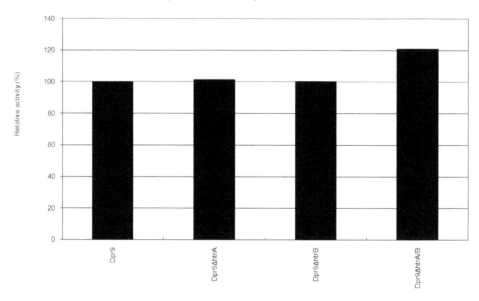

Fig. 8. Enhanced productivity of LipA in the absence of both *htrA* and *htrB*. Cells were cultured in modified 2xL broth at 30°C. The accumulation of LipA in the culture medium was measured at 48 h. The relative activities of LipA are shown (the amount of Dpr9 was set to 100%).

2.2.3 Transcriptional regulation of *htrB* and *htrA* by reciprocal cross regulation

We predicted that there was no difference between the productivities of LipA in the Dpr9ΔhtrA and Dpr9ΔhtrB strains, because the inactivation of either *htrA* or *htrB* results in a compensating overexpression of the other gene (Noone et al., 2001). To confirm that the overexpressions of *htrA* and *htrB* are caused by the inactivation of the other gene, we examined the level of expression of the *htrB-lacZ* fusion for the Dpr9ΔhtrA mutant, as well as the similar expression of the *htrA-lacZ* fusion for the Dpr9ΔhtrB mutant. Cells carrying pHY300PLK (control) and pHLApm (LipA overexpression) were cultured in modified 2xL broth for 48 h. As shown in Table 1, Dpr9ΔhtrA cells harboring pHLApm transcribed *htrB-lacZ* at a 4-fold increased level, compared with Dpr9 harbouring pHLApm (from 0.51 to 2.30 U). Similarly, a 10-fold increase in the *htrA-lacZ* expression level was observed in the Dpr9ΔhtrB mutant (from 0.41 to 4.26 U). The expressions of *htrB-lacZ* and *htrA-lacZ* also demonstrated reciprocal cross regulation in cells carrying pHY300PLK. These observations suggest that the overexpression of *htrB* in Dpr9ΔhtrA and of *htrA* in Dpr9ΔhtrB might affect LipA production. The expression level of *htrB-lacZ* in LipA-producing Dpr9 was 2.4-times higher than that of non-LipA-producing Dpr9 (Table 1). There was almost no change in the expression level of *htrA-lacZ*, between Dpr9 cells harboring pHLApm and Dpr9 cells harbouring pHY300PLK. The expression of the *htrB-lacZ* reporter gene fusion has previously been shown to be more sensitive to secretion stress than the *htrA-lacZ* reporter gene fusion (Hyyryläinen et al., 2001). These results suggest that Dpr9 produced LipA in weak response to secretion stress.

Expressed gene	Strain	Plasmid	Expression [a]
htrB-lacZ	Dpr9	pHY300PLK	0.22±0.01
	Dpr9ΔhtrA	pHY300PLK	1.46± 0.11
	Dpr9	pHYLApm	0.51± 0.11
	Dpr9ΔhtrA	pHYLApm	2.30± 0.02
htrA-lacZ	Dpr9	pHY300PLK	0.38± 0.03
	Dpr9ΔhtrB	pHY300PLK	1.29± 0.01
	Dpr9	pHYLApm	0.41± 0.03
	Dpr9ΔhtrB	pHYLApm	4.26± 0.02

[a] One activity unit is defined as 1 nmol of O-nitrophenyl-ß-D-galactopyranoside hydrolysed per min per µg of OD_{600}. The results presented are the average of three individual experiments. Plus/minus values represent standard deviations.

Table 1. Expression of transcriptional fusions between the *htrA* and *htrB* promoters and *lacZ* reporter gene in various genetic backgrounds.

3. Conclusion

This chapter focused on biotechnological approaches to optimization of heterologous protein and enzyme production by multiple protease-deficient mutations in *B. subtilis*. Section 2.2 described the identification of AprX protease using gelatin zymography and the effects of AprX on heterologous protein production. The nine-protease-deficient KA8AX strain (lacking nine genes encoding eight extracellular proteases and AprX) effectively prevented proteolysis of α-amylase-A522-PreS2 [PreS2 antigen of human hepatitis B virus (HBV) fused with the N-terminal 522 amino acids of *B. subtilis* α-amylase] in the late stationary growth phase and improved the yield of the fusion protein. In addition, AprX was detected in the culture medium due to leakage on cell lysis during the late stationary growth phase. Section 2.3 described that the inactivation of nine-proteases and both *htrA* and *htrB* (resulting the Dpr9ΔhtrA/B mutant) improved the productivity of LipA in *B. subtilis*. In particular, the productivity of the LipA in the Dpr9ΔhtrA/B strain was 1100 mg/L, an optimal yield which is 1.8-times higher than that of previously reported. There was no difference in the productivities of LipA in the Dpr9ΔhtrA and Dpr9ΔhtrB strains, compared with that of Dpr9. Because the transcriptions of *htrA* and *htrB* are controlled by reciprocal cross regulation, overexpression of *htrB* in the Dpr9ΔhtrA strain and of *htrA* in the Dpr9ΔhtrB strain might affect LipA production. The previous approach for effective protein production was to generate a strain which has the inactivation of eight extracellular proteases in *B. subtilis* as the host. We reported that AprX leaked outside of cells, and HtrA/HtrB membrane-bond proteases of *B. subtilis* were also key proteases involved in the degradation of natural and heterologous proteins. In addition, nine- or eleven-protease-deficient strains of *B. subtilis* were helpful in improving protein productivity. Our findings, described in this chapter should contribute to the generation of hosts to be further optimized for protein production.

4. Acknowledgment

We would like to thank Mr. Keiji Endo, Mr. Kazuhisa Sawada, Dr. Koji Nakamura, Dr. Yasutaro Fujita, Dr. Fujio Kawamura, and Dr. Naotake Ogasawara for useful advice and discussions, and Dr. Hiroshi Kakeshita and Dr. Kunio Yamane for their generous gift of plasmid of pTUBE522-PreS2, and for useful advice and discussions. This work was supported by the New Energy and Industrial Technology Development Organization (NEDO).

5. References

Antelmann, H.; Darmon, E.; Noone, D.; Veening, J.W.; Westers, H.; Bron, S.; Kuipers, O.P.; Devine, K.M.; Hecker, M. & van Dijl, J.M. (2003). The extracellular proteome of *Bacillus subtilis* under secretion stress conditions. *Molecular Microbiology*. Vol. 49, No. 1, pp. 143–156, ISSN 0950-382X

Babe, L. M. & B. Schmidt. (1998). Purification and biochemical analysis of WprA, a 52-kDa serine protease secreted by *B. subtilis* as an active complex with its 23-kDa propeptide. *Biochimica et Biophysica Acta*, Vol. 1386, No. 1, pp. 211–219, ISSN 0304-4165

Bruckner, R.; Shoseyov, O. & Doi, R. H. (1990). Multiple active forms of a novel serine protease from *Bacillus subtilis*. *Molecular and General Genetics MGG*, Vol. 221, No. 3, pp. 486-490, ISSN 0026-8925

Darmon, E.; Noone, D.; Masson, A.; Bron S.; Kuipers, O.P.; Devine, K.M. & van Dijl, J.M. (2002). A novel class of heat and secretion stressresponsive genes is controlled by the autoregulated CssRS twocomponent system of *Bacillus subtilis*. *Journal of Bacteriology*, Vol. 184, No. 20, pp. 5661–5671, ISSN 0021-9193

Hakamada, Y.; Hatada, Y.; Koike, K.; Yoshimatsu, T.; Kawai, S.; Kobayashi, T. & Ito, S. (2000). Deduced amino acid sequence and possible catalytic residues of a thermostable, alkaline cellulase from an alkaliphilic *Bacillus* strain. *Bioscience Biotechnology and Biochemistry*, Vol. 64, No. 11, pp. 2281-2289, ISSN 0916-8451

Harwood, C.R. (1992). *Bacillus subtilis* and its relatives: molecular biological and industrial workhorses. *Trends in Biotechnology*, Vol. 10, No. 7, pp. 247-256, ISSN 0167-7799

Honda, K., Fujieda, H., Ogawa, K., Imai, M., Yamamoto, H., Ikeda, T. & Yamane, K. (1993). Extracellular production of human hepatitis B virus preS2 antigen as hybrid proteins with *Bacillus subtilis* α-amylases in high-salt-concentration media. *Applied Microbiology and Biotechnology*, Vol. 40, No. 2-3, pp. 341-347, ISSN 0175-7598

Hyyryläinen, H.L.; Bolhuis, A.; Darmon, E.; Muukkonen, L.; Koski, P.; Vitikainen, M.; Sarvas, M.; Prágai, Z.; Bron, S.; van Dijl, J.M. & Kontinen V.P. (2001). A novel two-component regulatory system in *Bacillus subtilis* for the survival of severe secretion stress. *Molecular Microbiology*, Vol. 41, No. 5, pp. 1159–1172, ISSN 0950-382X

Hyyryläinen, H.L.; Pietiäinen, M.; Lundén, T.; Ekman, A.; Gardemeister, M.; Murtomäki-Repo, S.; Antelmann, H.; Hecker, M.; Valmu, L.; Sarvas, M. & Kontinen, V.P. (2007). The density of negative charge in the cell wall influences two-component signal transduction in *Bacillus subtilis*. *Microbiology*. Vol. 153, No. 7, pp. 2126-2136, ISSN 1350-0872

Kodama, T.; Endo, K.; Sawada, K.; Ara, K.; Ozaki, K.; Kakeshita, H.; Yamane, K. & Sekiguchi, J. (2007)a. *Bacillus subtilis* AprX Involved in Degradation of a Heterologous Protein During the Late Stationary Growth Phase. *Journal of Bioscience and Bioengineering*, Vol. 104, No. 2, pp. 135-143, ISSN 1389- 1723

Kodama, T.; Endo, K.; Ara, K.; Ozaki, K.; Kakeshita, H.; Yamane, K. & Sekiguchi, J. (2007)b. Effect of the *Bacillus subtilis spo0A* mutation on cell wall lytic enzymes and extracellular proteases, and prevention of cell lysis. *Journal of Bioscience and Bioengineering*, Vol. 103, No. 1, pp. 13-21, ISSN 1389- 1723

Koide, Y.; Nakamura, A.; Uozumi, T & Beppu, T. (1986). Cloning and sequencing of the major intracellular serine protease gene of *Bacillus subtilis. Journal of Bacteriology*, Vol. 167, No. 1, pp110-116, ISSN 0021-9193

Lee, S.J.; Kim, D.M.; Bae, K.H.; Byun, S.M. & Chung, J.H. (2000). Enhancement of secretion and extracellular stability of staphylokinase in *Bacillus subtilis* by *wprA* gene disruption. *Applied and Environmental Microbiology*, Vol. 66, No. 2, pp. 476-480, ISSN 0099-2240

Lesuisse, E.; Schanck, K. & Colson, C. (1993). Purification and preliminary characterization of the extracellular lipase of *Bacillus subtilis* 168, an extremely basic pH-tolerant enzyme. *European Journal of Biochemistry*, Vol. 216, No. 1, pp.155-160, ISSN 0014-2956

Liu, S.; Endo, K.; Ara, K.; Ozaki, K. & Ogasawara, N. (2008). Introduction of marker-free deletions in *Bacillus subtilis* using the AraR repressor and the ara promoter. *Microbiology*, Vol. 154, No. 9, 2562-2570, ISSN 1350-0872

Margot, P. & Karamata, D. (1996). The *wprA* gene of *Bacillus subtilis* 168, expressed during exponential growth, encodes a cell-wall-associated protease. *Microbiology*, Vol. 142, No. 12, pp. 3437-3444, ISSN 1350-0872

Morimoto, T.; Kadoya, R.; Endo, K.; Tohata, M.; Sawada, K.; Liu S.; Ozawa, T.; Kodama, T.; Kakeshita, H.; Kageyama, Y.; Manabe, K.; Kanaya, S.; Ara, K.; Ozaki, K. & Ogasawara, N. (2008). Enhanced recombinant protein productivity by genome reduction in *Bacillus subtilis. DNA Research*, Vol. 15, No. 2, pp. 73-81, ISSN 1340-2838

Morimoto, T.; Ara, K.; Ozaki, K & Ogasawara, N. (2009). A new simple method to introduce marker-free deletions in the *Bacillus subtilis* genome. *Genes & Genetic Systems*, Vol. 84, No. 4, pp. 315-318, ISSN 1341-7568

Murashima, K.; Chen, C.L.; Kosugi, A.; Tamaru, Y.; Doi, R.H. & Wong, S.L. (2002). Heterologous production of *Clostridium cellulovorans engB*, using protease-deficient *Bacillus subtilis*, and preparation of active recombinant cellulosomes. *Journal of Bacteriology*, Vol. 184, No.1, 76-81, ISSN 0021-9193

Noone, D.; Howell, A. & Devine, K.M. (2000). Expression of *ykdA*, encoding a *Bacillus subtilis* homologue of HtrA, is heat shock inducible and negatively autoregulated. *Journal of Bacteriology*, Vol. 182, No. 6, pp. 1592-1599, ISSN 0021-9193

Noone, D.; Howell, A.; Collery, R. & Devine, K.M. (2001). YkdA and YvtA, HtrA-Like serine proteases in *Bacillus subtilis*, engage in negative autoregulation and reciprocal cross-regulation of *ykdA* and *yvtA* gene expression. *Journal of Bacteriology*, Vol. 183, No. 2, pp. 654–663, ISSN 0021-9193

Rufo, G. A.; Jr., Sullivan, B. J.; Sloma, A. & Pero, J. (1990). Isolation and characterization of a novel extracellular metalloprotease from *Bacillus subtilis. Journal of Bacteriology*, Vol. 172, No. 2, pp. 1019-1023, ISSN 0021-9193

Simonen, M. & Palva, I. (1993). Protein secretion in *Bacillus* species, *Microbiological Reviews*. Vol.57, No.1, pp. 109-137, ISSN 0146-0749

Sloma, A.; Ally, A., Ally, D. & Pero, J. (1988). Gene encoding a minor extracellular protease in *Bacillus subtilis. Journal of Bacteriology*, Vol. 170, No. 12, pp. 5557-5563, ISSN 0021-9193

Sloma, A.; Rudolph, C. F.; Rufo, G. A., Jr.; Sullivan, B. J.; Theriault, K. A.; Ally, D. & Pero, J. (1990). Gene encoding a novel extracellular metalloprotease in *Bacillus subtilis. Journal of Bacteriology*, Vol. 172, No. 2, pp. 1024-1029, ISSN 0021-9193

Sloma, A.; Rufo, G. A.; Jr., Rudolph, C. F.; Sullivan, B. J.; Theriault, K. A. & Pero, J. (1990). Bacillopeptidase F of *Bacillus subtilis*: purification of the protein and cloning of the gene. *Journal of Bacteriology*, Vol. 172, No. 3, pp. 1470-1477, ISSN 0021-9193, Erratum in: *Journal of Bacteriology*, Vol. 172, No. 9, pp. 5520-5521, ISSN 0021-9193

Sloma, A., Rufo, G. A., Jr., Theriault, K. A., Dwyer, M., Wilson, S. W., & Pero, J. (1991). Cloning and characterization of the gene for an additional extracellular serine protease of *Bacillus subtilis*. *Journal of Bacteriology*, Vol. 173, No. 21, pp. 6889-6895, ISSN 0021-9193

Stahl, M. L. & Ferrari, E (1984). Replacement of the *Bacillus subtilis* subtilisin structural gene with an in vitro-derived deletion mutation. *Journal of Bacteriology*, Vol. 158, No. 2, pp. 411-418, ISSN 0021-9193

Tran, L.; Wu, X. C. & Wong, S. L. (1991). Cloning and expression of a novel protease gene encoding an extracellular neutral protease from *Bacillus subtilis*. *Journal of Bacteriology*, Vol. 173, No. 20, pp. 6364-6372, ISSN 0021-9193

Valbuzzi, A.; Ferrari, E.; & Albertini, A.M. (1999). A novel member of the subtilisin-like protease family from *Bacillus subtilis*. *Microbiology*, Vol. 145, No. 11, pp. 3121-3127, ISSN 1350-0872

van Pouderoyen, G.; Eggert, T.; Jaeger, K.E. & Dijkstra, B.W. (2001). The crystal structure of *Bacillus subtilis* lipase: a minimal alpha/beta hydrolase fold enzyme. *Journal of Molecular Biology*, Vol. 309, No. 1, pp. 215-226, ISSN 0022-2836

Vitikainen, M.; Hyyrylainen H. L.; Kivimaki A.; Kontinen V. P. & Sarvas M. (2005). Secretion of heterologous proteins in *Bacillus subtilis* can be improved by engineering cell components affecting post-translocational protein folding and degradation. *Journal of Applied Microbiology*, Vol. 99, No.2, pp. 363-375, ISSN 1364-5072

Westers, L.; Westers, H.; & Quax, W.J. (2004). *Bacillus subtilis* as cell factory for pharmaceutical proteins: a biotechnological approach to optimize the host organism. *Biochimica et Biophysica Acta*, Vol. 1694, Issues 1-3, pp. 299-310, ISSN 0304-4165

Westers, H.; Westers, L.; Darmon, E.; van Dijl, J.M.; Quax, W.J. & Zanen, G. (2006). The CssRS two-component regulatory system controls a general secretion stress response in *Bacillus subtilis*. *FEBS Journal*, Vol. 273, No. 16, pp. 3816–3827, ISSN 1742-464X

Westers, L.; Dijkstra, D.S.; Westers, H.; van Dijl, J.M. & Quax, W.J. (2006). Secretion of functional human interleukin-3 from *Bacillus subtilis*. *Journal of Biotechnology*, Vol. 123, No. 2, pp. 211-24, ISSN 0168-1656

Wong, S.L.; Price, C. W.; Goldfarb, D. S. & Doi, R. H. (1984). The subtilisin E gene of *Bacillus subtilis* is transcribed from a sigma 37 promoter in vivo. *Proceedings of the National Academy of Sciences of the United States of America*, Vol. 81, No. 4, pp. 1184-1188, ISSN 0027-8424

Wu, X. C.; Nathoo, S.; Pang, A. S.; Carne, T. & Wong, S. L. (1990). Cloning, genetic organization, and characterization of a structural gene encoding bacillopeptidase F from *Bacillus subtilis*. *The Journal of Biological Chemistry*, Vol. 265, No. 12, pp. 6845-6850, ISSN 0021-9258

Wu, X. C.; W. Lee, L. Tran, & S. L. Wong. (1991). Engineering a *Bacillus subtilis* expression-secretion system with a strain deficient in six extracellular proteases. *Journal of Bacteriology*, Vol. 173, No. 16, pp. 4952–4958, ISSN 0021-9193

Yang, M.Y.; Ferrari, E. & Henner, D. J. (1984). Cloning of the neutral protease gene of *Bacillus subtilis* and the use of the cloned gene to create an in vitro-derived deletion mutation. *Journal of Bacteriology*, Vol. 160, No. 1, pp. 15-21, ISSN 0021-9193

9

The Development of Cell-Free Protein Expression Systems and Their Application in the Research on Antibiotics Targeting Ribosome

Witold Szaflarski, Michał Nowicki and Maciej Zabel
Department of Histology and Embryology, Poznań University of Medical Sciences,
Poland

1. Introduction

There is a little doubt that increasing developments of protein synthesis are in high demand. Not only proteins are participants in all biochemical processes of the living cell, continually accelerating advances in proteomics, (i.e. the science of proteins and their reciprocal interactions in the cell) are increasingly underscoring the need to perfect techniques that facilitate the production of specified proteins at an industrial scale that meets the necessary standards of purification (Kim and Kim 2009). Investigations that have built the foundation for such protein production have largely originated from discoveries in the middle of the last century. Such advances firstly elucidated new cellular environments of protein production. Subsequent developments focused on the specificity of protein synthesis and the general efficiency of production has been developed largely by genomic analysis and genetic recombination.

Several *in vitro* systems of protein synthesis are commercially available worldwide. Many of these methods are categorized according to the derivation of their extracts, from either prokaryotic cells such as *Escherichia coli* (*E. coli*) or, alternatively, eukaryotic cells such as wheat germ or rabbit reticulocytes. While such extracts can be enriched by cofactors that enhance the efficacy of protein biosynthesis, there are obvious limitations to such systems.

An important criterion involves also simplicity of the system and its potential application: (1) simple systems, such as synthesis of phenylalanine homopolymer (*poly(U)-dependent poly(Phe) expression*) are generally applied in studies that analyze protein biosynthesis itself and on factors which block the process, i.e. antibiotics. This is in contrast to (2) the complex systems that are able to link transcription and translation into a single system.

The most advanced cell-free system based on the application of semi-permeable membrane allowing the concentration of reaction compartment during the work with ribosomes. Such membrane separates the feeding compartment where energy-rich molecules are deposed and can be moved to the reaction compartment with a simple diffusion. Moreover, such a feeding compartment is a suitable space where by-products potentially interfering with the biosynthesis can be deposed.

Recently, many of different cell-free based systems are available and the customer can select the most suitable for the specific application. Here, we described the most popular systems and we demonstrated how these systems can be utilized to study interactions between antibiotics and the ribosome.

1.1 The beginning of cell-free protein synthesis

In 1950s, several research teams independently demonstrated that protein biosynthesis can take place even after disintegration of the cell membrane (Siekievitz and Zamecnik 1951; Borsook et al. 1950; Winnick 1950; Gale and Folkes 1954). Thus, the isolated cytoplasm has been found to comprise the entire set of components necessary to conduct protein biosynthesis. As first, Zamecnik prepared fully active cell-free system based on mitochondrium-isolated ribosomes from an animal (Littlefield et al. 1955; Keller and Littlefield 1957). The team further demonstrated that the reactions were dependent on the supply of high energy molecules, such as ATP and GTP. The first *in vitro* systems of protein synthesis based on isolated bacterial ribosomes were designed independently by two teams, German (Schachtschabel and Zillig 1959) and American (Lamborg and Zamecnik 1960). However, both of them were only capable of translating endogenous mRNAs, what was their main limitation. Nevertheless, this discovery provided a proof that extracellular biosynthesis was possible at all and consequently it provided a new approach to synthesize proteins and to study molecular mechanisms of protein biosynthesis. An open nature of the *in vitro* systems was very attractive especially to the latter approach.

The discovery of protein expression systems on the template of exogenous mRNA molecules significantly extended applications of extracellular protein biosynthesis. The achievement took place in 1961 in the laboratory of Nirenberg and Matthaei (Nirenberg and Matthaei 1961). A short incubation at the physiological temperature of around 37°C proved sufficient to remove endogenous mRNA molecules from ribosomes. Free ribosomes obtained from the procedure were subsequently used for protein synthesis on the template of exogenous mRNA molecules. Of great importance, the ribosomes could be "programmed" by synthetic mRNA molecules. The technique of Nirenberg became the classical system of extracellular protein synthesis and, taking advantage of it, its originator deciphered the genetic code, for which he received the Nobel prize in 1968. In the subsequent systems, additional procedures of purifying ribosomes from endogenous mRNA molecules were applied to DEAE cellulose, permitting the separation ribosomes from free nucleic acids *via* chromatography.

Incubation of ribosomes, preceding the proper protein biosynthesis and conducted in the same manner as in the technique of Nirenberg, was later successfully applied in eukaryotic *in vitro* systems. Extracts of animal cells enriched with purified ribosomes conducted efficient protein biosynthesis. The technique was again successful using the template of exogenous mRNA molecules (Schreier and Staehelin 1973). During approximately the same timeframe, investigators applied this capacity to extracts of wheat germs and, of great interest, found that the endogenous as opposed to exogenous expression of mRNA molecules manifested naturally low levels of protein (Marcus, Efron, and Weeks 1974; Roberts and Paterson 1973; Anderson, Straus, and Dudock 1983). Other techniques of eliminating endogenous mRNA were based on application of calcium ion-dependent bacterial RNAse, used to augment the efficiency of protein expression system in lysates of erythrocytes (Jackson and Hunt 1983; Merrick 1983; Pelham and Jackson 1976) as well as in other lysates originating from animal cells (Henshaw and Panniers 1983). In the 1980s, it was subsequently found that bypassing the expression of endogenous mRNA molecules significantly improved the efficiency of extracellular protein expression systems.

1.2 Simple systems based on synthesis of protein homopolymers

In such systems, the principal homopolymeric system involves synthesis of polyalanine on the template of poly(U) chain (*poly(U)-dependent poly(Phe) synthesis*). A buffered medium containing free ribosomes and the remaining components necessary for a translation reaction with the template of poly(U), the polyuridine homopolymer is added. The poly(U) template has the capacity to bind a ribosome without involvement of the Shine-Dalgarno sequence (sequence on mRNA which binds to the region of 16S rRNA) also has the ability to "program" the ribosome for synthesis of poly(Phe). Efficiency of the optimised systems of polyphenyloalanine synthesis may reach 300 amino acid incorporations per ribosome, which represents a significant achievement allowing for a unitemporal and complete analysis of all protein biosynthesis components (Szaflarski et al. 2008). In contrast, the first attempts of the type, performed in 1950s and 1960s resulted in merely 2-5 amino acid incorporations per ribosome.

The homopolymeric system was prepared by isolation of two cellular fractions, which were subsequently enriched in high-energy molecules, free amino acides and poly(U)-mRNA, providing the template. The fractions were obtained from bacterial extracts, which were fractionated by centrifugation (for a detailed description of ribosome isolation see (Blaha et al. 2000)). The so-called fraction S30 (obtained by centrifugation at approximately 30,000 rpm for 24 h) was rich in ribosomes and was used to purify free ribosome subunits in a sucrose gradient (centrifugation at around 45,000-60,000 rpm for 15 h). Ribosomes prepared in this manner were incubated at the temperature of 37°C in a buffer containing, for instance, Mg^{2+} ions at the concentration of 4.5 mM in order to obtain complete correct 70S ribosome structure capable of performing protein synthesis.

S100 fraction was obtained from supernatant of the S30 fraction and it provides the source of protein factors indispensable to conduct translation (i.a., initiation factors: IF1, IF2, IF3, specific aminoacyl-tRNA synthetases, elongation factors: EF-Tu, EF-G, EF-Tu).

The reaction of polyphenylanaline synthesis represents a simple and widely used technique in several varieties. Instead of a poly(U) template, a poly(A) template can be used, enabling the synthesis of polylysine. Unfortunately, however, the polymer was poorly soluble in water; this property markedly restricts applicability of the system at a broader scale. Nevertheless, application of certain detergents permits its application in studies as seen in previous experiments conducted with the functional analysis of two antibiotics (pactamycin and edein), representing inhibitors of protein synthesis (Dinos et al. 2004). Here, the incorporation of near-cognate lysine instead of phenylalanine on the template of poly(U) can be precisely measured using double radioisotope labeling. If any antibiotic impacts on the translation accuracy (for example aminoglycoside paromomycin) it can be confirmed by detection of higher incorporation of lysine. Followed that technique edein was found to be an error-prone antibiotic in contrast to pactamycin which did not induce any miscoding (Dinos et al. 2004).

1.3 Biosynthesis of protein in a couple transcription-translation system

Nirenberg and Matthaei (Matthaei and Nirenberg 1961) again were the first to describe the DNA dependence of the bacterial extracellular synthesis of protein. The dependence was corroborated by synthesis of a protein on the template of endogenous DNA molecules. Another group of investigators extended synthesis of endogenous proteins by application of exogenous DNA, which originated from a bacteriophage (Byrne et al. 1964; Wood and Berg 1962). Unfortunately, the systems manifested a relatively poor efficiency using either endogenous or viral DNA. Moreover, they were accompanied by a non-specific expression of cellular and bacteriophage proteins. However, the continuing improvements of the joint transcription and translation system resulted in its dissemination; with it ultimately becoming a significant laboratory tool (Lederman and Zubay 1967; DeVries and Zubay 1967).

In the improved system, suggested by Zubay, a preliminary bacterial extract was subjected to incubation in order to degrade mRNA and DNA molecules by cellular nucleases (Zubay 1973). The system gained popularity due to the ease of its preparation, stability of components and a relatively high efficiency. In the system designed by Gold and Schweiger ribosomes were isolated from cellular extracts to their homogenous form and so prepared ribosomes were supplemented with a cytoplasmic fraction, cleared of nucleic acids by ion-exchange chromatography (Schweiger and Gold 1969, 1969, 1970). Such a preparation of components for extracellular protein synthesis produced a remarkable reduction in the non-specific expression of protein. The troublesome procedure, however, remained the disadvantage of the system.

2. Contemporary systems of the cell-free protein expression

The 1980s and 1990s witnessed development of the *in vitro* systems in the form of optimization of cellular extracts, including the application of bacterial strains lacking the genes that code enzymes of endonuclease type (RNases) (Zaniewski, Petkaites, and Deutscher 1984). Such reaction mixes allowed researchers to keep bacteria in the reactive mix for a much longer period of time. Contemporary systems are characterized by a high mRNA level even after 24 h of the reaction (Iskakova et al. 2006). In 1990s the developing methods of genetic engineering and bioinformatics produced tremendous advances inside *in vitro* systems. One tremendous achievement was the ability to obtain data on structure of mRNA transcripts and on the effect of the structure on the efficiency of biosynthesis. The spatial structure of mRNA and primarily the sequence located at the 5' terminus, proved to be very important (Graentzdoerfer et al. 2002) due to its ability to fold secondary structures covering Shine-Dalgarno sequence.

The several years of studies on structure and function of individual elements of mRNA sequence resulted in a design of the optimum expression vector for the *in vitro* protein expression systems. This has been exemplified by pIVEX (*In Vitro EXpression*) plasmid (Betton 2003). This plasmid is characterized by its ability to form secondary structures at the level of mRNA, particularly within the Shine-Dalgarno system and AUG initiation codon.

Due to this, both fragments of mRNA sequence are exposed and they easily bind to a ribosome. An integral part of the vector also includes the sequence of the bacteriophage promoter, T7, which permits transcription of a given gene using T7 polymerase. Two types of vectors are commercially available, including: (1) those containing His-tag sequences and (2) Strep-tag, located on the amine or carboxylic terminus of the protein, which allows for an easy and rapid purification of the biosynthesis products.

2.1 *In vitro* systems based on application of a semipermeable membrane

At present, the *in vitro* expression systems enjoy wide application due to the ease of performing the reaction without the need to apply sophisticated equipment. However, they exhibit low efficiency, not exceeding few tens of product nanograms in 50 μl reaction. The solution which markedly increased protein expression level involved application of a semipermeable membrane, used for the first time by Spirin (Baranov and Spirin 1993). This produced a significant increase volume of the so-called feeding mix, containing free amino acids, ribonucleotides and high energy molecules (mainly ATP and GTP), securing in parallel high concentration of ribosomes with the yield produced in the reaction compartment (Fig. 1, A and B). Application of a semipermeable membrane further enabled a significant extension of the biosynthetic reaction since it could be continuously supplied by inflow of indispensable reactants from the feeding mix (Fig. 1B). The maximum duration of conducting the reaction averaged at approximately 30-50 h, with *plateau* of reaction product being reached following around 30 h at 30°C. The system produced a remarkable yield. For the first time milligram quantities of protein per 1 ml of the reaction were obtained. The example involved expression of GFP (*Green Fluorescence Protein*), the synthesis of which reached the level of around 5 mg in the course of a single 24 h (Fig. 2 A and B) reaction using the RTS (*Rapid Translation System*) (Iskakova et al. 2006). RTS is manufactured by Roche company and it has been designed in the basis of A. Spirin's patent (*U.S. Pat. No. 5,478,730*).

The RTS is based not only on the ingenious application of a semipermeable membrane but also coupling the transcription and translation reactions, used also in the earlier designed systems (Fig. 1A). Such an approach markedly abbreviated duration of the process and reduced formation of nonspecific products since only the gene present in the expression vector was undergoing transcription and, then, translation.

2.2 Advantages and drawbacks of RTS

The RTS system manifests several advantages. Due to release of ribosomes from the cell and provision of appropriate conditions for the translation reaction, toxic proteins can be produced. If using *in vivo* conditions, such toxicities factors would surely block living processes in the cell. Enrichment of free amino acids with their radioactively labeled substitutes permitted the effective labeling of nascent polypeptides.

The open nature of RTS systems and other *in vitro* techniques provided a handy tool for studies on protein biosynthesis itself and on molecules such as antibiotics, which would otherwise, of course block protein production. An interesting approach within RTS involves screening analysis of known or potential antibiotics. Such an approach allows for a very rapid determination of the inhibitory concentration of a given antibiotic (determination of IC_{50}) or preliminary analysis of the mechanism of antibiotic action.

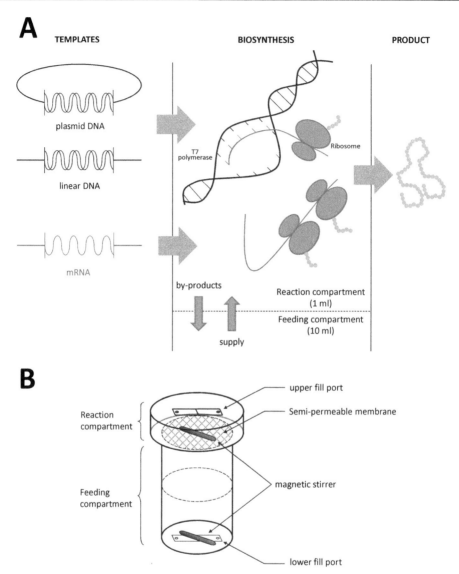

Fig. 1. Production of proteins in RTS system. (A) Principles of the process; coupled transcription-translation reaction runs on one of three templates in the reaction compartment supplied by energy rich-components and amino acids from the feeding compartment. (B) RTS reaction chamber with the semipermeable membrane separating reaction and feeding compartments.

A typical marker protein used in studies on *in vitro* systems is GFP protein, due to the ease of estimating the total product quantity and the fraction of active molecules (the phenomenon of fluorescence). The GFP protein is almost ideal for this purpose since it manifests a characteristic structure (Fig. 2A). The structure involves a barrel formed by 11 β structures and an active centre, the chromophore, located inside the molecule. The β structures serve to protect the chromophore from the access of water molecules that could otherwise block the fluorescence of the reaction (Yang, Moss, and Phillips 1996). The appearance of inappropriate amino acids in GFP disturbs its structure and allows for penetration of its interior by water molecules.

Using this system, it was possible to demonstrate that the *in vitro* system fully confirmed the role of aminoglycosides as inducers of translation errors (Szaflarski et al. 2008). In the presence of one of the aminoglycosides (streptomycin) expression of total GFP and active fraction (i.e. GFP molecules inducing fluorescence) were followed by SDS polyacrylamide gel electrophoresis (SDS PAGE) and native PAGE, respectively. The relations between results read by these two techniques determined whether antibiotic (e.g. streptomycin) impacted on the fidelity of translation reaction. If the active fraction of GFP decreases faster than the total expression it means that full-size polypeptide chains are produced on the ribosome however their protein folding failed due to increasing number of wrong amino acids in the GFP and this protein was inactive (Fig. 2C). In view of the literature on aminoglycoside character, this provided evidence for introduction of erroneous amino acids to GFP molecule. This technique was demonstrated to be suitable to discriminate opposite effects of edeine and pactamycin acting on the ribosome (Dinos et al. 2004).

Aminoglycosides were also tested in another *in vitro* synthesis system: the poly(U)-dependent translation of polyphenylalanine (Szaflarski et al. 2008). In contrast to results obtained in the RTS system, they failed to block the ribosome. However, following addition of an additional leucine, it was found to be introduced to the polyphenylalanine chain already at low concentration of streptomycin (around 1 µM). This reflected the similarity of leucine codon and phenylalanine codon (UUC *vs.* UUU) and, thereafter, in situations inducing elevated probability of translation errors, leucine was introduced instead of phenylalanine. Thus, the observation did not allow a direct comparison between the two *in vitro* translation systems. The RTS system resembles more closely the natural conditions and, therefore, is more sensitive to action of antibiotics due to higher number of potential targets.

Nevertheless, extracellular protein biosynthesis is linked to disadvantages which for several years have been successively eliminated. In the course of studies the systems such as RTS was found to support expression of relatively high amounts of protein but around half of the proteins were found to be biologically inactive (Iskakova et al. 2006). However, application of specific translation factors as EF-4 allowed reaching 100% efficiency of RTS system (Qin et al. 2006).

The causes of lowered activity of proteins inside *in vitro* systems may be multiple but the most probable one involves application of the bacteriophage polymerase T7, which is exceedingly rapid. The transcription and translation are strictly interrelated during *in vivo* conditions with elimination of the free space on mRNA between polymerase and the ribosome. This prevents against development of spatial structures in mRNA molecule and it does not allow for a precocious termination of translation. Application of T7 polymerase disturbs the natural interrelationship between the polymerase and the ribosome, which may lead to errors at the level of translation or to incorporation of inappropriate amino acids to the growing polypeptide chain. The solution worked out by involved genetic recombination of the polymerase T7 of such a type that the enzyme contained two point mutations that decreased the rate of activity (He et al. 1997). The results proved that despite the lowered efficiency of the total biosynthesis the content of active proteins increased to almost 100%. A similar effect was obtained decreasing temperature of the reaction. Most probably this reflected the fact that, as compared to ribosomes, polymerase is much more temperature sensitive, producing a slower rate of transcription and a similar pace of translation (Lewicki et al. 1993).

3. Energy consumption and its regeneration inside in vitro systems

Protein biosynthesis represents a process of particular energetic requirements. In biological systems the energy is obtained from hydrolysis of high energy bonds. For introduction of a single amino acid to the growing polypeptide chain, the cell sacrifices as many as 10 high energy bonds which is equivalent to hydrolysis of 10 molecules of ATP or GTP, each characterized by bonding energy of ΔG^0 = -6 kcal/mol. The extreme energetic requirement of a cell supporting protein biosynthesis explains development of sophisticated systems which control energy loss. However, the systems retain the Achilles heel of contemporary systems of protein biosynthesis *in vitro* in which their output remains seriously restricted by excessive uncontrolled leaks of energy: usually not more than 5% of energy is expended to support current protein biosynthesis while the remaining energy is wasted in uncontrolled biochemical reactions. It should be borne in mind that injuring the cell we introduce an extreme chaos to its metabolism. In contrast to *in vitro* conditions, in the *in vivo* conditions in a bacterial cell as much as 70% of energy can be directed to support protein biosynthesis (Szaflarski and Nierhaus 2007).

Therefore, in the techniques of protein biosynthesis *in vitro* a continuous replenishment of high energy compounds (i.a. ATP and GTP), necessary for efficacious transcription and translation reactions, continues to pose an enormous challenge. The earliest to be designed system of replenishing the high energy compounds involved enrichment of the cellular extract with millimolar concentrations of phosphoenolpyruvate (PEP), a derivative of pyruvic acid, and with pyruvate kinase, which catalyzes transfer of a phosphate group from phosphoenolpyruvate to AMP and ADP, yielding ADP and ATP, respectively. Also GMP and GDP represent substrates for the kinase. PEP is distinguished among all biologically active compounds by its content of the energetically most valuable bond: the phosphoester bond contained in the compound carries the energy of ΔG^0=-12 kcal/mol. Nevertheless, the PEP-based system carries also an extreme disadvantage: the by-product formed during regeneration of the high energy molecules involves orthophosphoric acid. The acid lowers pH of the reaction and, which is even more important, it binds magnesium ions (Mg^{2+}), markedly reducing their level in the reaction. Mg^{2+} stabilizes ribosome structure by its interaction with rRNA, therefore their reduced level results in a disturbed ribosome structure and a reduced efficiency of protein biosynthesis.

One of the ways in which the lowered concentration of Mg^{2+} ions can be avoided involves transformation of orthophosphoric to acetylphosphate using pyruvate oxidase and a defined prosthetic group (TPP or FAD). The reaction requires an access of molecular oxygen, the availability of which is restricted inside *in vitro* systems, particularly when the reaction is conducted in a few milliliter volumes. Nevertheless, the application of pyruvate oxidase has significantly extend the duration of the effective protein biosynthesis reaction which has markedly increased efficiency (Kim and Swartz 2000).

However, the above approach still hardly can be considered ideal. First of all, the restricted access of molecular oxygen markedly reduces the potential for utilization of the system on a larger scale. At present, the solution widely applied involves application of a combined system, based on the traditional PEP/pyruvate kinase approach with acetyl phosphate synthesis by acetyl-CoA, which allows for an effective elimination of free phosphoric acid during synthesis of acetylphosphate from acetyl-CoA (Jewett and Swartz 2004). Application of the system provided a breakthrough and permitted milligram quantities of the produced protein in a volume of just one milliliter (Iskakova et al. 2006).

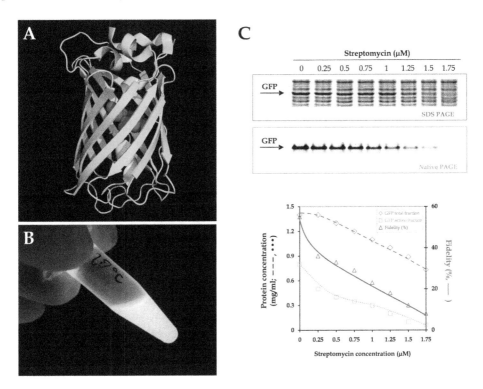

Fig. 2. Expression of GFP as a reporter protein in the presence of antibiotic streptomycin. (A) The molecular structure of GFP with the internal chromophore as red (coordinates based on PDB acc. no. 2B3Q). (B) The luminescence of GFP seen under UV lamp. (C) Parallel analysis of GFP total expression (SDS PAGE) and its activity measured as its luminescence (native PAGE). The graph demonstrates the fidelity (red line) of translation as the ratio between total expression of GFP (blue dashed line) and the active fraction of GFP (green dotted line). Increasing concentrations of streptomycin caused dramatic decrease in the ratio of the active GFP to the total protein. Techniqual and experimental details see in Dinos et al. 2004; Szaflarski et al. 2008; Qin et al. 2006.

4. Biotechnological application of extracellular protein synthesis systems

4.1 Efficient expression of several proteins and their screening analysis in parallel

The extracellular protein expression provided the base for many automated techniques of high-throughput expression and screening of several proteins in parallel (Angenendt et al. 2004; Spirin 2004). In such a system, using a single 96- or 384-well plate, multiple genes can be copied, transcribed and translated in parallel, providing substrates for subsequent high-throughput analysis of protein functions. The extensive scale of expression as well as the rate of analysis warrant that the technique deserves to be considered in proteomics and protein engineering based on screening analysis of gene and protein libraries as well as of entire genomes and proteomes (He and Taussig 2007, 2008). The *in vitro* systems of protein expression permit synthesis of protein population in a single reaction, which represents an ideal and economic solution in complete screening analysis of proteins within a given gene library (Chandra and Srivastava).

Such solutions can be proved by the technology known as *in vitro* expression cloning (IVEC) (King et al. 1997). In the technique a large genomic library is preliminarily subcloned to groups each containing 50-100 plasmids placed in the standard 96-well plate. They provide a template for expression in an *in vitro* system. The plasmid-containing genes which yield protein products are subsequently transferred by cloning to expression vectors, which allow for synthesis of milligram quantities of proteins in RTS type *in vitro* systems.

The IVEC technique can further be improved by combining it with gene cloning and amplification using the PCR reaction, thus eliminating the time-consuming cloning of the genes to plasmids (Gocke and Yu 2009). Preparation of the appropriate primers containing promoter sequences and Shine-Dalgarno sequences has facilitated protein synthesis directly from products of the PCR reaction (Rungpragayphan, Nakano, and Yamane 2003). The example includes an application of extracellular protein expression system for screening analysis of the entire *Arabidopsis thaliana* genome in order to identify new genes and products of their expression (Sawasaki et al. 2002).

Systems of *in vitro* translation have found application also in medical studies. The protein truncated test (PTT) has been worked out in order to identify open reading frames (ORF) (Roest et al. 1993). Detection of precocious translation termination within an ORF may reflect mutation at the level of DNA, which provides grounds for distinguishing another genetically-conditioned disease. Other applications are linked to production and analysis inside *in vitro* systems of potential vaccines which, even if obtained on bacterial or animal ribosomes, manifest the same immunological variables as those obtained in cultures of human cells (Kanter et al. 2007). At present, new proteins representing potential anti-malaria vaccines have been fully worked out in the systems of extracellular protein expression (Tsuboi et al. 2008). Also the expression of virus-like particles (VLP) has been characterized inside *in vitro* systems The examples include the phage protein, MS2 and C-terminal fragment of the protein core in the hepatitis B virus (HBV), the biological activity of which has been identical to the forms obtained *in vivo* (Bundy, Franciszkowicz, and Swartz 2008).

4.2 Production of proteins "resistant" to expression

Considering the open character of *in vitro* systems due to the absence of biological membranes, the systems could have been applied with excellent results for production of toxic or membraneous proteins, which manifest poor expression in *in vivo* systems (Jackson et al. 2004). At present the *in vitro* systems allow for expression at a large scale of membraneous proteins in modified extracts of *E. coli* (Klammt et al. 2006). Due to application in the *in vitro* systems of techniques allowing for formation of disulphide bridges, production of antibodies also has become possible. The example involves immunoglobulin G (IgG), which when obtained in the system has proven to be fully biologically active, manifesting affinity to antigen and stability identical to the natural antibody (Frey et al. 2008).

4.3 Analysis of molecular interactions

Extracellular protein expression systems markedly facilitate the molecular analysis of interactions between protein and X substance, where X may involve another protein, DNA, RNA or a ligand (Jackson et al. 2004). In order to identify the interaction, one of the reactants must be labeled (a protein, nucleotide or ligand) and placed in a system in which protein, the other reactant is synthesized. Then, the arising complex is isolated from the reaction mix using immunoprecipitation (Derbigny et al. 2000) or it may be directly analyzed in agarose or polyacrylamide gels (Lee and Chang 1995).

4.4 Protein display technologies

The essence of protein display technologies involves establishing a link between genetic information (genotype) and function of an unknown protein (phenotype) in the protein library. The principal technique involves a ribosome display (He and Taussig 1997; Hanes and Pluckthun 1997). Elimination of the STOP codon in mRNA permitted to obtain stable complexes of mRNA-ribosome-protein. Thus, a kind of a frozen structure was obtained, from the threshold of genetic world and proteomics. Subsequently, binding of the protein formed on the ribosome to a defined ligand (which may involve also DNA or RNA) resulted in development of an informational link between a given ligand and the sequence of protein mRNA. Then, a given mRNA-ribosome-protein-ligand complex can be isolated by affinity chromatography from the medium containing also other ligands while mRNA sequences are identified by reverse transcription and DNA sequencing. In order to amplify efficacy of the system, the process is conducted in a cyclic manner, i.e., the isolated mRNAs are independently amplified and added again to the mixture of ribosomes and ligands, enabling a more effective selection of an individual specific ligand. In combination with methods of genetic engineering, including mutagenesis, the protein display technologies can be applied not only in proteomics but also in molecular evolution studies. Now, the processes of interactions between DNA, RNA and protein, which took millions of years of evolution may be analyzed in the laboratory and their rate may be multiplied by selective amplification of DNA.

5. Conclusion

Cell-free systems will be optimized and improved according to their expression yield, protein specificity ("difficult proteins") and protein folding. They will be more broadly applied in protein microarrays technology where can be utilized for the analysis of protein-protein interaction. Furthermore, protein technologies based on cell-free biosynthesis will be applied for protein engineering in order to synthesis specific antibodies or enzymes, as well as for production of proteins for crystallisation.

The "post-genome" research requires comprehensive tool which will allow determination of structure, function and specific location of the proteins in the network of proteomes. It has to be performed effectively, quickly and on the multiple platform where large number of proteins can be analyzed in the same time. Based on cell-free systems such analysis is possible especially in the comparison to traditional cell-based systems where their miniaturization is rather impossible.

6. Acknowledgements

This study was supported by the Polish Ministry of Science and Higher Education (grants no. 0172/B/P01/2009/36). Authors thank to Dr. Jan Jaroszewski for his help in lingual edition of the manuscript. Witold Szaflarski thanks to Prof. Knud H. Nierhaus for his long-time powerful mentoring.

7. References

Anderson, C. W.; Straus, J. W., & Dudock, B. S. (1983). Preparation of a cell-free protein-synthesizing system from wheat germ. *Methods Enzymol*, Vol. 101, pp. 635-44.

Angenendt, P.; Nyarsik, L.; Szaflarski, W.; Glokler, J.; Nierhaus, K. H.; Lehrach, H.; Cahill, D. J., & Lueking, A. (2004). Cell-free protein expression and functional assay in nanowell chip format. *Anal Chem*, Vol. 76, No. 7, pp. 1844-1849.

Baranov, V. I., & Spirin, A. S. (1993). Gene expression in cell-free system on preparative scale. *Methods Enzymol*, Vol. 217, pp. 123-42.

Betton, J. M. (2003). Rapid translation system (RTS): a promising alternative for recombinant protein production. *Curr Protein Pept Sci*, Vol. 4, No. 1, pp. 73-80.

Blaha, G.; Stelzl, U.; Spahn, C.M.T.; Agrawal, Rajendra K; Frank, Joachim, & Nierhaus, Knud H. (2000). Preparation of functional ribosomal complexes and the effect of buffer conditions on tRNA positions observed by cryoelectron microscopy. *Methods Enzymol*, Vol. 317, pp. 292-309.

Borsook, H.; Deasy, C. L.; Haagensmit, A. J.; Keighley, G., & Lowy, P. H. (1950). Incorporation *in vitro* of labeled amino acids into rat diaphragm proteins. *J Biol Chem*, Vol. 186, No. 1, pp. 309-315.

Bundy, B. C.; Franciszkowicz, M. J., & Swartz, J. R. (2008). Escherichia coli-based cell-free synthesis of virus-like particles. *Biotechnol Bioeng*, Vol. 100, No. 1, pp. 28-37.

Byrne, R.; Levin, J. G.; Bladen, H. A., & Nirenberg, M. W. (1964). The in Vitro Formation of a DNA-Ribosome Complex. *Proc Natl Acad Sci U S A*, Vol. 52, pp. 140-8.

Chandra, H., & Srivastava, S. Cell-free synthesis-based protein microarrays and their applications. *Proteomics*, Vol. 10, No. 4, pp. 717-30.

Derbigny, W. A.; Kim, S. K.; Caughman, G. B., & O'Callaghan, D. J. (2000). The EICP22 protein of equine herpesvirus 1 physically interacts with the immediate-early protein and with itself to form dimers and higher-order complexes. *J Virol*, Vol. 74, No. 3, pp. 1425-35.

DeVries, J. K., & Zubay, G. (1967). DNA-directed peptide synthesis. II. The synthesis of the alpha-fragment of the enzyme beta-galactosidase. *Proc Natl Acad Sci U S A*, Vol. 57, No. 4, pp. 1010-2.

Dinos, G.; Wilson, D. N.; Teraoka, Y.; Szaflarski, W.; Fucini, P.; Kalpaxis, D., & Nierhaus, K. H. (2004). Dissecting the ribosomal inhibition mechanisms of edeine and pactamycin: the universally conserved residues G693 and C795 regulate P-site RNA binding. *Mol Cell*, Vol. 13, No. 1, pp. 113-24.

Frey, S.; Haslbeck, M.; Hainzl, O., & Buchner, J. (2008). Synthesis and characterization of a functional intact IgG in a prokaryotic cell-free expression system. *Biol Chem*, Vol. 389, No. 1, pp. 37-45.

Gale, E. F., & Folkes, J. P. (1954). Effect of nucleic acids on protein synthesis and amino-acid incorporation in disrupted staphylococcal cells. *Nature*, Vol. 173, No. 4417, pp. 1223-1227.

Gocke, C. B., & Yu, H. (2009). Identification of SUMO targets through in vitro expression cloning. *Methods Mol Biol*, Vol. 497, No., pp. 51-61.

Graentzdoerfer, A; Watzele, M; Buchberger, B; Wizemann, S; Metzler, T; Mutter, W, & Nemetz, C. 2002. Optimization of the Translation Initiation Region of Prokaryotic Expression Vectors: High Yield In Vitro Protein Expression and mRNA folding. In *Cell-Free Translation Systems*, edited by A. S. Spirin: Springer.

Hanes, J., & Pluckthun, A. (1997). In vitro selection and evolution of functional proteins by using ribosome display. *Proc Natl Acad Sci USA*, Vol. 94, No. 10, pp. 4937-4942.

He, B; Rong, M; Lyakhov, D; Gartenstein, H; Diaz, G; Castagna, R; McAllister, W T, & Durbin, R K. (1997). Rapid mutagenesis and purification of phage RNA polymerases. *Protein Express Purif*, Vol. 9, No., pp. 142-151.

He, M., & Taussig, M. J. (1997). Antibody-ribosome-mRNA (ARM) complexes as efficient selection particles for in vitro display and evolution of antibody combining sites. *Nucleic Acids Res*, Vol. 25, No. 24, pp. 5132-4.

He, M., & Taussig, M. J. (2007). Rapid discovery of protein interactions by cell-free protein technologies. *Biochem Soc Trans*, Vol. 35, No. Pt 5, pp. 962-5.

He, M., & Taussig, M. J. (2008). Production of protein arrays by cell-free systems. *Methods Mol Biol*, Vol. 484, pp. 207-15.

Henshaw, E. C., & Panniers, R. (1983). Translational systems prepared from the Ehrlich ascites tumor cell. *Methods Enzymol*, Vol. 101, pp. 616-29.

Iskakova, Madina B.; Szaflarski, Witold; Dreyfus, Marc; Remme, Jaanus, & Nierhaus, Knud H. (2006). Troubleshooting coupled *in vitro* transcription–translation system derived from *Escherichia coli* cells: synthesis of high-yield fully active proteins. *Nucl. Acids Res.*, Vol., pp. e135.

Jackson, A. M.; Boutell, J.; Cooley, N., & He, M. (2004). Cell-free protein synthesis for proteomics. *Brief Funct Genomic Proteomic*, Vol. 2, No. 4, pp. 308-19.

Jackson, R. J., & Hunt, T. (1983). Preparation and use of nuclease-treated rabbit reticulocyte lysates for the translation of eukaryotic messenger RNA. *Methods Enzymol*, Vol. 96, pp. 50-74.

Jewett, M. C., & Swartz, J. R. (2004). Mimicking the *Escherichia coli* cytoplasmic environment activates long-lived and efficient cell-free protein synthesis. *Biotechnol Bioeng*, Vol. 86, No. 1, pp. 19-26.

Kanter, G.; Yang, J.; Voloshin, A.; Levy, S.; Swartz, J. R., & Levy, R. (2007). Cell-free production of scFv fusion proteins: an efficient approach for personalized lymphoma vaccines. *Blood*, Vol. 109, No. 8, pp. 3393-9.

Keller, E. B., & Littlefield, J. W. (1957). Incorporation of ^{14}C-amino acids into ribonucleoprotein particles from the Ehrlich mouse ascites tumor. *J. Biol. Chem.*, Vol. 224, No. 1, pp. 13-30.

Kim, D. M., & Swartz, J. R. (2000). Prolonging cell-free protein synthesis by selective reagent additions. *Biotechnol Prog*, Vol. 16, No. 3, pp. 385-390.

Kim, H. C., & Kim, D. M. (2009). Methods for energizing cell-free protein synthesis. *J Biosci Bioeng*, Vol. 108, No. 1, pp. 1-4.

King, R. W.; Lustig, K. D.; Stukenberg, P. T.; McGarry, T. J., & Kirschner, M. W. (1997). Expression cloning in the test tube. *Science*, Vol. 277, No. 5328, pp. 973-4.

Klammt, C.; Schwarz, D.; Lohr, F.; Schneider, B.; Dotsch, V., & Bernhard, F. (2006). Cell-free expression as an emerging technique for the large scale production of integral membrane protein. *Febs J*, Vol. 273, No. 18, pp. 4141-53.

Lamborg, M. R., & Zamecnik, P. C. (1960). Amino acid incorporation into protein by extracts of E. coli. *Biochim Biophys Acta*, Vol. 42, No., pp. 206-11.

Lederman, M., & Zubay, G. (1967). DNA-directed peptide synthesis. 1. A comparison of T2 and *Escherichi coli* DNA-directed peptide synthesis in two cell-free systems. *Biochim Biophys Acta*, Vol. 149, No. 1, pp. 253-258.

Lee, H. J., & Chang, C. (1995). Identification of human TR2 orphan receptor response element in the transcriptional initiation site of the simian virus 40 major late promoter. *J Biol Chem*, Vol. 270, No. 10, pp. 5434-40.

Lewicki, B. T. U.; Margus, T.; Remme, J., & Nierhaus, K. H. (1993). Coupling of rRNA transcription and ribosomal assembly *in vivo* - formation of active ribosomal subunits in *Escherichia coli* requires transcription of rRNA genes by host RNA polymerase which cannot be replaced by bacteriophage-T7 RNA polymerase. *J Mol Biol*, Vol. 231, pp. 581-593.

Littlefield, J. W.; Keller, E. B.; Gross, J., & Zamecnik, P. C. (1955). Studies on cytoplasmic ribonucleoprotein particles from the liver of the rat. *J. Biol. Chem.*, Vol. 217, No. 1, pp. 111-123.

Marcus, A.; Efron, D., & Weeks, D. P. (1974). The wheat embryo cell-free system. *Methods Enzymol*, Vol. 30, No. 0, pp. 749-54.

Matthaei, J. H., & Nirenberg, M. W. (1961). Characteristics and stabilization of DNAase-sensitive protein synthesis in E. coli extracts. *Proc Natl Acad Sci U S A*, Vol. 47, No., pp. 1580-1588.

Merrick, W. C. (1983). Translation of exogenous mRNAs in reticulocyte lysates. *Methods Enzymol*, Vol. 101, pp. 606-15.

Nirenberg, M.W., & Matthaei, J.M. (1961). The dependance of cell-free protein synthesis in E. coli upon naturally occuring or synthetic polyribonucleotides. *Proc Natl Acad Sci U S A*, Vol. 47, pp. 1588-1602.

Pelham, H. R. B., & Jackson, R. J. . (1976). An efficient mRNA-dependent translation system from reticulocyte lysates. *Eur J Biochem*, Vol. 67, pp. 247-256.

Qin, Y.; Polacek, N.; Vesper, O.; Staub, E.; Einfeldt, E.; Wilson, D. N., & Nierhaus, K. H. (2006). The highly conserved LepA is a ribosomal elongation factor that back-translocates the ribosome. *Cell*, Vol. 127, No. 4, pp. 721-733.

Roberts, B E, & Paterson, B M (1973). Efficient translation of TMV RNA and rabbit globin 9S RNA in a cell-free system from commercial wheatgerm. *Proc Natl Acad Sci U S A*, Vol. 70, No. 8, pp. 2330-2334.

Roest, P. A.; Roberts, R. G.; Sugino, S.; van Ommen, G. J., & den Dunnen, J. T. (1993). Protein truncation test (PTT) for rapid detection of translation-terminating mutations. *Hum Mol Genet*, Vol. 2, No. 10, pp. 1719-21.

Rungpragayphan, S.; Nakano, H., & Yamane, T. (2003). PCR-linked in vitro expression: a novel system for high-throughput construction and screening of protein libraries. *FEBS Lett*, Vol. 540, No. 1-3, pp. 147-50.

Sawasaki, T.; Ogasawara, T.; Morishita, R., & Endo, Y. (2002). A cell-free protein synthesis system for high-throughput proteomics. *Proc Natl Acad Sci U S A*, Vol. 99, No. 23, pp. 14652-7.

Schachtschabel, D., & Zillig, W. (1959). [Investigations on the biosynthesis of proteins. I. Synthesis of radiocarbon labeled amino acids in proteins of cell free nucleoprotein-enzyme-system of Escherichia coli.]. *Hoppe Seylers Z Physiol Chem*, Vol. 314, No. 4-6, pp. 262-75.

Schreier, MH, & Staehelin, T. (1973). Translation of rabbit hemoglobin meessenger RNA in vitro with purified and partially purified components from brain or liver of different species. *Proc Natl Acad Sci U S A*, Vol. 70, No., pp. 462-465.

Schweiger, M., & Gold, L. M. (1969). Bacteriophage T4 DNA-dependent in vitro synthesis of lysozyme. *Proc Natl Acad Sci U S A*, Vol. 63, No. 4, pp. 1351-8.

Schweiger, M., & Gold, L. M. (1969). DNA-dependent in vitro synthesis of bacteriophage enzymes. *Cold Spring Harb Symp Quant Biol*, Vol. 34, No., pp. 763-6.

Schweiger, M., & Gold, L. M. (1970). Escherichia coli and Bacillus subtilis phage deoxyribonucleic acid-directed deoxycytidylate deaminase synthesis in Escherichia coli extracts. *J Biol Chem*, Vol. 245, No. 19, pp. 5022-5.

Siekievitz, P, & Zamecnik, P C. (1951). In vitro incorporation of l-C14-DL-alanine into proteins of rat-liver granular fractions. *Fed Proc*, Vol. 10, No., pp. 246-247.

Spirin, A. S. (2004). High-throughput cell-free systems for synthesis of functionally active proteins. *Trends Biotechnol*, Vol. 22, No. 10, pp. 538-45.

Szaflarski, W., & Nierhaus, K. H. (2007). Question 7: optimized energy consumption for protein synthesis. *Orig Life Evol Biosph*, Vol. 37, No. 4-5, pp. 423-8.

Szaflarski, W.; Vesper, O.; Teraoka, Y.; Plitta, B.; Wilson, D. N., & Nierhaus, K. H. (2008). New features of the ribosome and ribosomal inhibitors: non-enzymatic recycling, misreading and back-translocation. *J Mol Biol*, Vol. 380, No. 1, pp. 193-205.

Tsuboi, T.; Takeo, S.; Iriko, H.; Jin, L.; Tsuchimochi, M.; Matsuda, S.; Han, E. T.; Otsuki, H.; Kaneko, O.; Sattabongkot, J.; Udomsangpetch, R.; Sawasaki, T.; Torii, M., & Endo, Y. (2008). Wheat germ cell-free system-based production of malaria proteins for discovery of novel vaccine candidates. *Infect Immun*, Vol. 76, No. 4, pp. 1702-8.

Winnick, T. (1950). Studies on the mechanism of protein synthesis in embryonic and tumor tissues. II. Inactivation of fetal rat liver homogenates by dialysis, and reactivation by the adenylic acid system. *Arch Biochem*, Vol. 28, No. 3, pp. 338-447.

Wood, W. B., & Berg, P. (1962). The effect of enzymatically synthesized ribonucleic acid on amino acid incorporation by a soluble protein-ribosome system from Escherichia coli. *Proc Natl Acad Sci U S A*, Vol. 48, No., pp. 94-104.

Yang, F.; Moss, L. G., & Phillips, G. N., Jr. (1996). The molecular structure of green fluorescent protein. *Nat Biotechnol*, Vol. 14, No. 10, pp. 1246-51.

Zaniewski, R; Petkaites, E, & Deutscher, M P. (1984). A multiple mutant of *Escherichia coli* lacking the exoribonucleases RNase II, RNase D, and RNase BN. *J Biol Chem*, Vol. 259, No. 19, pp. 11651-11653.

Zubay, G. (1973). *In vitro* synthesis of protein in microbial systems. *Annu Rev Genet*, Vol. 7, No., pp. 267-287.

Part 3

Molecular Biotechnology and Genetic Engineering

Effect of Environmental Stresses on S-Layer Production in *Lactobacillus acidophilus* ATCC 4356

Moj Khaleghi[1] and Rouha Kasra Kermanshahi[2]
[1]Department of Biology, Faculty of Sciences, Shahid Bahonar University, Kerman,
[2]Department of Biology, Faculty of Sciences, Alzahra University, Tehran,
Iran

1. Introduction

The gastrointestinal tract (GTI) is the organ with the largest surface area in the human body, having in an adult between 150 and 200 m^2 (Holzapfel et al., 1998). Interesting in this context is the fact that a huge number of microorganisms live and interact with the host in the stomach and gut. The GIT of an adult human is estimated to harbour about 10^{13}–10^{14} viable bacteria, i.e. 10 times the total number of eukaryotic cells in all tissues of man's body (Holzapfel et al., 1998; Velez et al., 2007). In the gastrointestinal tract, the bacteria are affected both by the physiological conditions (such as low pH, bile salt and enzymes) and by other microorganisms which exist in the GIT. Because of the presence of enzymes, salts and acids in the gastric juice, the environmental conditions in the stomach are destructive to a number of microorganisms (Holzapfel et al., 1998). The microbial community in the gastrointestinal tract is complex and consists of several hundred species, of which lactic acid bacteria constitute a minor proportion. Lactic acid bacteria (LAB) are Gram-positive bacteria which excrete lactic acid as a main fermentation product into the medium. This biochemical definition associates lactic acid bacteria of different phylogenetic branches of bacterial evolution: the "low GC" taxa, e.g. *Enterococcus*, *Lactobacillus*, *Lactococcus*, *Leuconostoc*, *Pediococcus* and *Streptococcus*, and the "high GC" genus *Bifidobacterium*. Species of these genera can be found in gastrointestinal tract of man and animal (Klaenhammer et al., 2005; Klein et al., 1998; Mathur & Singh, 2005). They have also been involved since time immemorial in food processing and food preservation, and are applied in particular for the manufacturing of dairy products, fermented meat, vegetables, bread and ensilage (Pouwels et al., 1998). Several lactic acid bacteria have the potential to promote the health of the host or to prevent and treat diseases. Such bacteria are referred to as probiotic, meaning 'for life' (Latin *pro*=for and *biotic*=life) (Herich & Levkut, 2002; Marteau & Rambaud, 2002; Mercenier et al., 2002; Ouwehand et al., 2003). The strains of LAB used as probiotics usually belong to species of the genera *Lactobacillus*, *Enterococcus*, and *Bifidobacterium* (Klein et al., 1998). The effects of probiotic microorganisms on the host have been discussed extensively in the literature. It has been proposed that probiotics possess several advantageous properties, such as antagonistic actions, production of antimicrobial substances, modulation of immune responses, and impact on the metabolic activities of the gut (Marteau & Rambaud, 2002;

O'Toole & Cooney, 2008; Sanders & Klaenhammer, 2001). *In vitro* and animal studies have further shown inhibitory effects of probiotic bacteria to be mediated by their interference with the adhesion of gastrointestinal pathogens or with toxins produced by the pathogenic microorganisms (O'Toole & Cooney, 2008; Sullivan & Nord, 2002). However, beneficial influence of probiotics has been demonstrated in clinical studies (De Roos & Katan, 2000; Kos et al., 2008; Mishra et al., 2008; O'Toole & Cooney, 2008; Park et al., 2008; Resta-Lenert & Barrett, 2003; Sanders & Klaenhammer, 2001). Some possible health effects include immune system stimulation, cholesterol lowering, prevention and treatment of diarrhea, prevention of cancer recurrence, improvement in lactose intolerance, and reduction of allergy (De Roos & Katan, 2000; Mercenier et al., 2002; Reid, 1999; Sanders & Klaenhammer, 2001). The most common probiotic strains belong to two genera, *Lactobacillus* and *Bifidobacterium* (Sanders & Klaenhammer, 2001; Sullivan & Nord, 2002). Some strains of *Lactobacillus* are found as natural commensals of the GIT, the oral cavity and the female uro-genital tract of animals and humans (Pouwels et al., 1998). Also, Lactobacilli are of considerable technological and commercial importance because of their role in the manufacturing and preservation of many fermented food products (Schar-Zammaretti et al., 2005). *Lactobacillus acidophilus* is one of the major species of the genus *Lactobacillus* found in human and animal intestines (Frece et al., 2005). Nowadays, *Lactobacillus acidophilus* strains are commonly used as probiotic (Sanders & Klaenhammer, 2001), and marketed as capsules, powders, enriched yogurts, yogurt-like products, and milk (De Roos & Katan, 2000). Adhesion to intestinal epithelial cells is an important prerequisite for colonization of probiotic strains in the gastrointestinal tract (Altermann et al., 2004; Kos et al., 2003). This is mediated either non-specifically by physico-chemical factors (such as hydrophobicity) or, specifically, by adhesive bacterial surface molecules and epithelial receptor molecules (such as S-layer, fibronectin and mucin-binding proteins) (Holzapfel et al., 1998).

2. Probiotics and stress

In the intestinal tracts of mammals and avians, species of the genera *Lactobacillus*, *Enterococcus*, *Streptococcus* and *Bifidobacterium* are the dominant indigenous lactic micro-biota. Commonly recovered *Lactobacillus* isolates from the human gastrointestinal tract include *L. acidophilus*, *L. salivarius*, *L. casei*, *L. plantarum*, *L. fermentum*, *L. brevis* and *L. reuteri*. The findings that colonization by lactobacilli and other lactic acid bacteria improves infection resistance of the host have led to the production and consumption of probiotics (Kosin & Rakshit, 2006). The probiotics must resist multiple stresses including the GIT conditions and food processing. An important attribute for certain probiotic bacteria functions are survival and growth in the intestinal tract (Sanders & Klaenhammer, 2001). The gastric juice contains hydrochloric acid, which induces an extremely low pH. The fasting pH in the stomach is approximately 1.5, while the pH increases to between 3.0 and 5.0 when food is eaten (Cotter & Hill, 2003). In the intestine, the conditions are less extreme because of in the intestine pH is higher than the stomach, but the bacteria still have to endure bile and pancreatic juices (intestinal fluids). Other factors affecting microbial life in this environment are immunoglobulins, defensins, a continuously regenerating epithelium, peristaltic movement of intestinal content, and a viscous mucus layer (Dunne et al., 2001; Tannock et al., 1999). Moreover, probiotics used in food technology are exposed to various adverse conditions during processing, such as temperature changes, acidity, osmotic and oxidative stress (Kosin & Rakshit, 2006). Such stresses may reduce the physiological activity of the cells and readily kill the cells. Once the cells have survived the stresses, they can colonize and grow to adequate numbers to provide the beneficial effect to the host. Survival mechanisms exhibited by bacteria when confronted with stress are generally referred to as the stress response (De Angelis & Gobbetti, 2004; Jan et al., 2001; Kim et al., 2001). However, the ability to survive passage through the intestinal tract and potentially establish residence there is considered as an important feature. The degree of retention is likely dependent on the ability of the bacteria to interact with eukaryotic cell surfaces or with the mucosal layer surrounding these cells (Altermann et al., 2004). Several factors contribute to the interaction of Lactobacilli with the host tissues, such as cell surface hydrophobicity (Vadillo-Rodriguez et al., 2005; Van der Mei et al., 2003), autoaggregation (Kos et al., 2003), lipoteichoic acids (Granato et al., 1999) and external surface proteins (such as S-layer, fibronectin and mucin-binding proteins) (Altermann et al., 2004; Avall-Jaaskelainen & Palva, 2005; Frece et al., 2005; Kos et al., 2003; Kosin & Rakshit, 2006; Pouwels et al., 1998; Velez et al., 2007; Ventura et al., 2002). Surface layer (S-layer) has been identified as the outermost structure of cell envelope in numerous organisms from the domains Bacteria (in both Gram-positive and Gram-negative Eubacteria) and Archaea (Debabov, 2004; Sara & Sleytr, 2000). S-layer proteins are non-covalently bound to the cell wall and assemble into surface layers with high degrees of positional order often completely covering the cell wall, and can be disintegrated into monomers by denaturing agents such as urea or guanidine hydrochloride (Avall-Jaaskelainen & Palva, 2005; Engelhardt & Peters, 1998; Lortal et al., 1992; Sleytr et al., 2001). The S-layer has been detected in a few species of the genus *Lactobacillus* (Avall-Jaaskelainen & Palva, 2005). Lactobacilli surface layer proteins (S-layers) are generally monomolecular crystalline arrays exhibiting a morphologically similar, oblique lattice structure and representing 10-15% of the total protein content of the bacterial cell wall (Avall-Jaaskelainen & Palva, 2005; Jakava-Viljanen et al., 2002). Several reports have appeared in which functions of S-layer are described or assumed (Boot et al., 1993; Jakava-Viljanen et al., 2002; Sara & Sleytr, 2000). However, no general function has been identified for S-layer proteins, but several lactobacillar S-layers have been identified as putative adhesions with affinity for intestinal epithelial cells, extracellular matrices and/ or to lipoteichoic acid (LTA) of other bacterial species (Avall-Jaaskelainen & Palva, 2005; Buck et al., 2005; Frece et al., 2005; Garrote et al., 2004; Hynonen et al., 2002; Kos et al., 2003; Velez et al., 2007; Vidgren et al., 1992). Of the various roles proposed for the bacterial S-layer, it is a protective sheath against hostile environment (Avall-Jaaskelainen & Palva, 2005; Frece et al., 2005; Khaleghi et al., 2010, 2011; Kos et al., 2003; Schar-Zammaretti et al., 2005). Adhesive S-layers have a role in inhibition of adhesiveness of pathogenic bacteria and thus can contribute to probiotic effects of lactobacilli. To date, several lactobacilli S-layer protein-encoding genes have been cloned, sequenced and deposited in GenBank (Avall-Jaaskelainen & Palva, 2005; Velez et al., 2007). The presence of multiple S-layer protein genes seems to be quite common for bacteria (Avall-Jaaskelainen & Palva, 2005; Ben-Jacob et al., 2000; Boot & Pouwels, 1996; Jakava-Viljanen et al., 2002). Multiple S-layer genes have been identified in the genomes of *L. acidophilus, L. amylovorus, L. gallinarum, L. crispatus, L. brevis, L. gasseri and L. johnsonii* (Avall-Jaaskelainen & Palva, 2005; Boot & Pouwels, 1996; Jakava-Viljanen et al., 2002; Ventura et al., 2002). There is also increasing evidence that S-layer-carrying bacteria may use S-layer variation, by expressing alternative S-layer protein genes, for adaptation to different stress factors such as the immune response of the host for pathogens and drastic changes in the environmental conditions for nonpathogens (Boot & Pouwels, 1996; Frece et al., 2005; Jakava-Viljanen et al., 2002; Pouwels et al., 1998; Sara & Sleytr, 2000). According to Pouwels' study (1998), phase variation or antigenic variation, as a result of inversion of the *slp* segment, might enable *Lactobacillus acidophilus* bacteria to better adhere to specific regions of the mucosa. Variation in S-layer gene expression as a response to environmental changes has also been described in *Geobacillus stearothermophilus, Bacillus anthracis,* and *Campylobacter fetus* (Boot & Pouwels, 1996; Mignot et al., 2002).

There is some evidence that the surface properties of microorganisms are dependent on the growth conditions and the composition of the fermentation medium (Schar-Zammaretti et al., 2005; Waar et al., 2002; Dufrene & Rouxhet, 1996; Millsap et al., 1997). Schar-Zammaretti (2005) suggested that S-layer protein is preferentially expressed under different fermentation media. Furthermore, it has been shown that the S-layer production is changed with the change in medium (such as bile salt, penicillin G) (Khaleghi et al., 2010, 2011).

The aim of this study was to gain more knowledge about S-layer production and *slpA* gene expression in different growth conditions (pH and temperature) in *Lactobacillus acidophilus* ATCC 4356. Moreover, the reassembly of S-layer subunits was studied under these stresses.

2.1 S-layer protein, *slp*A expression and stresses

The S-layer proteins of Lactobacilli are relatively small, 25 kDa to 71 kDa in size (Avall-Jaaskelainen & Palva, 2005), whereas the molecular masses of S-layers in other bacterial species range up to 200 kDa (Sara & Sleytr, 2000). The Lactobacillar S-layers are highly basic proteins with calculated isoelectric point values ranging from 9.35 to 10.4. Yet, all the other S-layer proteins characterized are weakly acidic (Avall-Jaaskelainen & Palva, 2005). Evidence shows that the S-layer protein is important for *Lactobacillus acidophilus* (Frece et al., 2005; Khaleghi et al., 2010, 2011; Kim et al., 2001; Toba et al., 1995).

Lactobacillus acidophilus strains isolated from humans and animals, which belong to DNA homology groups A, are reported to possess a *slpA* gene, while the strains which belong to the DNA homology groups B appear not to have an *slpA* gene (Boot et al., 1993). According to Boots' study (1995, 1996, 1996c), there are two S-layer protein encoding genes, *slpA* and *slpB*, in *Lactobacillus acidophilus* ATCC 4356; of the two, *slpA* is active and *slpB* is silent in normal growth conditions. The two S-protein genes are located 6-kb apart on the chromosome, in a reverse orientation relative to each other. The *slpA* gene is interchanged with the *slpB* gene through inversion of a chromosomal fragment in a fraction of a culture (0.3% of the cell growth under laboratory conditions). Thus, it seems that S-layer variation of non-pathogenic lactobacilli has the same function as S-layer variation for pathogenic organisms such as *Campylobacter fetus*, namely to circumvent an immune response of the infected host (Boot & Pouwels, 1996; Boot et al., 1995, 1996b, 1996c).

Therefore, the present study investigated the effects of some stresses on the S-layer production, reassembly of S-layer subunits, and *slpA* gene expression.

2.1.1 Materials and methods

2.1.1.1 Stress conditions

To study the effect of heat and pH stresses on S-layer production and *slpA* gene expression, the *Lactobacillus acidophilus* ATCC 4356 was cultivated in MRS broth (Merck) for heat stress (30, 45, 50 and 55 ºC) (Kim et al., 2001); for pH stress, MRS broth was adjusted to pH 3-7 (adjusted with HCl and NaOH) (Jan et al., 2001; Lorca et al., 1998). The pH and temperature of the control culture (MRS broth) were 6.5 and 37 °C, respectively, according to manufacturer's recommendation and as described previously (Boot et al., 1993; Silva et al., 2005; Smit et al., 2001).

2.1.1.2 Isolation of S-layer

Lactobacillus acidophilus ATCC 4356 was obtained from the Germany Type Culture Collection and was cultivated anaerobically (in jar with Anaerocult A-strip, Merck) in MRS broth (Merck) at 37 °C (Boot et al., 1993; Smit et al., 2001). *Lactobacillus casei* ATCC 393 was used as negative control for isolation of S-layer.

For isolation of S-layer and total RNA, the recommended optical density is 0.7 at 695 nm (the end of log phase) (Boot et al., 1993; Smit et al., 2001) and 0.2-0.4 at 600 nm (mid-log phase), respectively (Boot et al., 1995). But in this study, we compared *slp*A gene expression and S-layer production at the same time. In addition, the S-layer production was compared in OD_{600}= 0.4 and OD_{600}= 0.7. Therefore, S-layer protein and total RNA were isolated in exponentially growing cells ($OD_{600} \approx 0.4$). Also, S-layer protein was extracted at OD_{600}= 0.7. For extraction of S-layer, *Lactobacillus acidophilus* ATCC 4356 was cultivated anaerobically in MRS broth at 37 °C. In general, 100 ml of pre-warmed MRS medium (under stress conditions and control) was inoculated 1: 100 (v/v) with an overnight culture and cultivated until the optical density at 600 nm reached 0.4 and 0.7. Cells were harvested by centrifugation at 15000 ×g for 15 min at 4 °C. The cells were washed twice with 100 ml of ice-cold water. The cell pellet was extracted with 0.1 volume of 4 M guanidine hydrochloride (pH 7) for one hour at 37 °C and centrifuged at 18000 ×g for 15 min. The supernatant, containing S-layer protein monomers, was dialyzed against water at 4 °C for 16-24 h (Boot et al., 1993). The dialyzed extracts were analyzed by SDS-PAGE (Smit et al., 2001). SDS-PAGE of protein samples was carried out using Precision Plus Protein Standard [low molecular weight marker (10-250 kDa) - Biorad]. The samples were run on 12% polyacrylamide gel at 100 V. Protein bands were visualized by Coomassie blue staining. Protein concentration was determined according to Bradford's method (Bradford, 1976). For normalization of the measured absorption values, BSA (Merck) was used.

2.1.1.3 Reassembly of S-layer monomers and transmission electron microscopy (TEM)

S-layer self-assembly subunits were studied by the negative staining technique. To prepare the TEM samples, several droplets of the dialyzed protein were pipetted onto the carbon-coated grids and left for 1-16 h to immobilize the proteinaceous structures. Samples were washed once with distilled water and then stained with 2% uranyl acetate for two minutes (Avall-Jaaskelainen et al., 2002; Smit et al., 2001). The grids were dried by nitrogen flow and studied by Zeiss/CEM 902 a transmission electron microscopy (TEM) at 60 kV.

2.1.1.4 Isolation of total RNA

For isolation of total RNA, *L. acidophilus* ATCC 4356 cells were grown in MRS broth (under stress conditions and control) until they reached an optical density of approximately OD_{600}=0.4. The cells were subsequently harvested by centrifugation (5000 ×g for 10 min at 4°C) and washed with an ice-cold TE buffer (Boot et al., 1995). The total RNA was isolated using a protective RNeasy Minikit (Qiagen) according to the manufacturer's recommendations, and then treated with deoxyribonuclease I (DNase I, RNase-free; Fermentas) at 37°C for 30 min according to manufacturer's recommendations.

2.1.1.5 RT-PCR

The reverse transcription (RT) of the RNA samples was performed with 150 ng of total RNA and 0.5 µg of Oligo dT primer using a First Strand cDNA Synthesis kit (Fermentas) at 42°C for 60 min, as recommended by the manufacturer. Forward and reverse primers were designed for the *slpA* gene of *Lactobacillus acidophilus* ATCC 4356 as follows: *slpA* forward (5'-TGG CCG TTC TTG AAT GTG TA-3') and *slpA* reverse (5'-ACA TCA ACG CTG CAA ACA TC-3'). These primers generated a 154 bp PCR product in the PCR reaction.

16S rRNA was used as the internal control gene based on previously reported primers (Trotha et al., 2001) that generate a 370 bp PCR product.

The final volume of the PCR reaction was 25 µl with the following components: 1 µl cDNA (≈ 7.5 ng), 1 µl (100 pmol/µl) from each primer, 0.5 µl dNTPs mix, 0.5 µl $MgCl_2$, and 0.25 µl (5 U/µl) *Taq* DNA polymerase (Fermentas). The Mastercycler (Eppendorf) was programmed as follows: initial denaturation for 5 min at 94°C; 30 cycles at 94°C for 45 sec, 54°C for 30 sec, 72°C for 30 sec, and a final extension at 72°C for 8 min. The PCR products (and 50 bp DNA ladder, Fermentas) were separated on a 1% agarose gel and visualized by ethidium bromide staining.

2.1.1.6 Statistical assessment

All the experiments and measurements were repeated at least three times. All the statistical analyses were performed using SPSS and Excel 2003 software. All the experimental results were analyzed using mean descriptive statistics, the correlation coefficient, and a single-factorial analysis of variance. A value of $P<0.05$ was regarded as statistically significant.

2.1.2 Results

The growth curve of *Lactobacillus acidophilus* ATCC 4356 showed that it took approximately 8 h to reach $OD_{600}= 0.4$ and approximately 14 h to reach $OD_{600}= 0.7$ (Data not shown). It is important to know that *Lactobacillus acidophilus* ATCC 4356 was live and grew after 14 h under stress conditions. However, the results indicated that *Lactobacillus acidophilus* was not live in 50 °C, 55 °C, pH 3 and 4 after 14 h (Table 1). In this study, pH 5, 6, and 7, as well as temperatures 30 °C and 45 °C were chosen as the stress conditions.

Culture condition	Cell count (CFU/ml)
Control*	9.6×10^9
pH 3	No growth
pH 4	No growth
pH 5	8.38×10^6
pH 6	7.24×10^8
pH 7	9.03×10^9
30 °C	8.98×10^8
45 °C	3.63×10^7
50 °C	No growth
55 °C	No growth
*pH 6.5 & 37 °C.	

Table 1. Viable cell counts of *Lactobacillus acidophilus* ATCC 4356 under stress conditions and the control after 14 h of inoculation.

The surface proteins of *Lactobacills acidophilus* ATCC 4356 were extracted by treatment of whole cells with 4M guanidine hydrochloride, and analyzed by SDS-PAGE. One dominant band of 43-46 kDa, which is known as the S-protein (Boot et al., 1993; Smit et al., 2001) and a few faint bands were visible on gel (Fig. 1b). In the mid-log-phase (OD_{600}=0.4), S-protein production was low in control group, so the 43-46 kDa band was not seen on gel clearly (Fig. 1a, c; lane 1). However, in the control group with OD_{600}=0.7, a 43-46 kDa band was visible on SDS-PAGE gel (Fig. 1b, d; lane 1).

Under stress conditions (OD_{600}=0.4 & 0.7), S-protein band was visible and the band became sharper in pH 5 and 45 °C (Fig. 1). It seemed that S-protein bands were not different in pH 6, 7 and the control (Fig. 1b).

No protein bands (43-46 kDa) were visible on SDS-PAGE gel from isolated protein of *Lactobacillus casei* ATCC 393.

To determine the total proteins, the Bradford method was used. The total proteins were compared between the control and the group under stress conditions. In the case of pH 5 and 45 °C, total protein content was higher than others (Fig. 2). Moreover, the total protein production level was lowest at 30 °C ($p < 0.001$). In pH 6, 7 and control, the protein content was almost similar. After comparing the results of total protein analysis under stress and control conditions, the range of difference in protein content was similar in OD_{600} = 0.4 and 0.7.

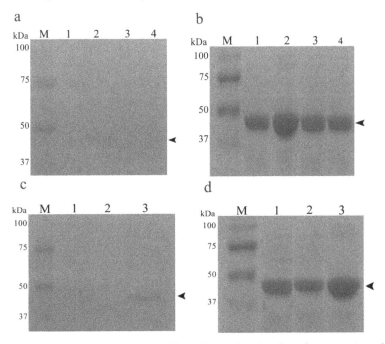

Fig. 1. SDS-PAGE gel (12% polyacrylamide) analysis of isolated surface proteins of *Lactobacillus acidophilus* ATCC 4356 at: **a)** mid-log phase (OD_{600}= 0.4) and **b)** exponential growth phase (OD_{600}= 0.7) under pH stress (lane 1, control; lane 2, pH 5; lane 3, pH 6; lane 4, pH 7). **c)** mid-log phase (OD_{600}= 0.4) and **d)** exponential growth phase (OD_{600}= 0.7) under heat stress (lane 1, control; lane 2, 30 °c; lane 3, 45 °c). M, Protein marker.

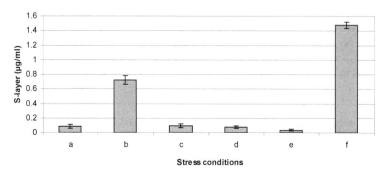

Fig. 2. Extracted surface proteins (extracted in mid-log phase) were compared in control and the group under stress conditions by Bradford method. (a) Control (pH 6.5 & 37 °C); (b) pH 5; (c) pH 6; (d) pH 7; (e) 30 °C; (f) 45 °C. Error bars represent standard deviations of the mean values of results from three independent experiments.

To assess the change in S-layer protein content of the cell wall under stress conditions by transmission electron microscopy (TEM), we chose 45 °C in which S-layer production was highest (Fig. 2). In the electron microscopy study, the presence of the S-layer on the outer surface of *Lactobacillus acidophilus* ATCC 4356 was clearly demonstrated (Fig. 3). In particular, the bacterial surface was completely covered with an S-layer in the control (Fig. 3a, b), but an excess of S-layer protein was found at the both ends of the bacterial cell under stress condition (Fig. 3c, d).

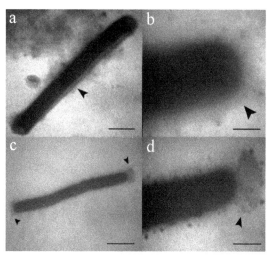

Fig. 3. Electron microscopic images of *Lactobacillus acidophilus* ATCC 4356. **a, b:** *Lactobacillus acidophilus* ATCC 4356 was completely covered by the S-layer in the control. **c, d:** an excess of S-layer was found at the both ends of the bacterial cell at 45°c. S-layer protein was indicated by black arrow (scale bar (a, c) = 2.5 µm and scale bar (b, d) = 0.6 µm).

The crystallization of S-layer was investigated by TEM. S-protein, which was isolated by guanidine hydrochloride, aggregated readily upon removal of the salt by dialysis, and formed a white precipitate. Analysis of these precipitates by TEM showed that they were composed exclusively of crystalline lattice (Fig. 4). It seems that the reassembly of S-layer subunits was similar in the group under stress condition (45°c) and control. The results indicated that S-layer has two-fold (p2) symmetry with a periodicity of 11.3 and 5.5 nm in the control. After comparing the lattice parameters, we found that they were similar (under stress condition and control).

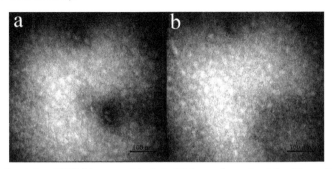

Fig. 4. Negatively stained TEM image of isolated S-layer from *Lactobacillus acidophilus* ATCC 4356. **a)** The control (37 °c); **b)** 45 °c (scale bar = 100nm).

The results indicated that the stress influenced *slp*A gene expression. Interestingly, the *slp*A gene expression increased in pH 5 and 45 °C. Under pH stress, comparison of the *slp*A gene expression showed that in the pH 6 and 7, the *slp*A gene expression was lower than that in the control (Fig. 5a) ($p < 0.001$). In addition, the *slp*A gene expression decreased at 30 °C, and the *slp*A gene expression was highest at 45 °C (Fig. 5b) ($p < 0.001$). However, major differences in the *slp*A gene expression were observed between the control and the group under stress conditions (Fig. 5).

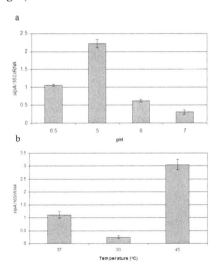

Fig. 5. Comparison of *slp*A gene expression under control and stress conditions (a: pH stress; b: heat stress) as compared to expression of housekeeping gene (*16S RNA*) in the same reaction to normalize the data. Error bars represent standard deviations of the mean values of results from three independent experiments.

3. Discussion

To investigate the effects of pH and heat stresses on S-layer production and *slpA* gene expression (OD_{600} = 0.4 and 0.7) in *Lactobacillus acidophilus* ATCC 4356, we studied the survival of bacteria under stress conditions for 14 h. It was found that in pH 3, 4 and the temperatures 50-55 °C, *Lactobacillus acidophilus* ATCC 4356 could not survive. Then, pH 5, 6, and 7, as well as temperatures 30 °C and 45 °C were chosen as the stress conditions. SDS-PAGE gel (12% polyacrylamide) of S-layer protein showed that a single dominant band (43-46 kDa) was visible (Fig. 1). According to the previous studies, *Lactobacillus acidophilus* ATCC 4356 has S-layer protein with 43 kDa molecular weight (Boot et al., 1993; Smit et al., 2001). Also, *Lactobacillus acidophilus* ATCC 4356 has been used as positive control for S-layer protein in other studies (Fitzsimons et al., 2003; Frece et al., 2005). In our study, S-layer extraction was carried out according to Boot's method (Boot et al., 1993) in *Lactobacillus acidophilus* ATCC 4356 and *Lactobacillus casei* ATCC 393 (as negative control), and the results were compared together. As S-layer is the outermost structure of cell envelope in *Lactobacillus acidophilus* ATCC 4356, the extracted protein was only from cell wall without lysis of bacterial cells. In SDS-PAGE gel, there was one dominant band of 43-46 kDa and a few faint bands which were not 43 kDa. But there was no protein band in negative control. Therefore, we confirmed that these proteins are S-layer proteins. The S-layer protein was isolated with 4 M guanidine hydrochloride. During the mid-log phase (OD_{600} = 0.4), the S-layer protein production was low in the control group, but it was clearly visible in OD_{600} = 0.7 (Fig. 1, lane 1). Under stress conditions, the production of S-layer protein increased at pH 5 and 45°C. However, the increase in S-protein production was found in OD_{600} = 0.4 and 0.7 (Fig. 1, 2). Evidence showed that S-layer production was increased under stress conditions (Khaleghi et al., 2010, 2011) and medium components (Schar-Zammaretti et al., 2005). It was found that S-layer proteins were present during all growth phases of *Lactobacillus acidophilus* M92 under heat stress (Frece et al., 2005). This suggested that the S-layer protein is preferentially expressed under conditions which are not optimal for bacterial growth. It has been proposed that S-layer plays a role as a protective sheath in *Lactobacillus acidophilus* ATCC 4356. In addition, some studies identified that the S-layer proteins of Lactobacilli were important for hydrophobicity, autoaggregation and adherence of this bacteria to different host surfaces (Frece et al., 2005; Greene & Klaenhammer, 1994; Hynonen et al., 2002; Kos et al., 2003; Pelletier et al., 1997; Sillanpaa et al., 2000; Toba et al., 1995; Vadillo-Rodriguez et al., 2004, 2005; Van der Mei et al., 2003).

Transmission electron microscopic analysis showed that *Lactobacillus acidophilus* ATCC 4356 was completely covered by S-layer at 37 °C. Also, S-layer covered the bacterial cells at 45 °C, but an excess of S-protein was found at the both ends of the bacterial cell (Fig. 3). It seems that the S-layer protein has a protective role for *Lactobacillus acidophilus* ATCC 4356. Previously, the excess of S-layer was found at the site of separation of the two daughter cells in *Clostridium thermosaccharolyticum*, which prevented the exposure of the newly synthesized parts of the cell wall to the environment. It has been proposed that several bacteria produce an excess of S-protein to ensure complete coverage of the cell wall during cell division, and either store excess S-layer protein in the peptidoglycan layer or secrete it into the environment (Boot & Pouwels, 1996).

Because of the adhesion role of S-layer to the epithelial cells in Lactobacilli, it is important to investigate of the self-assembly ability of S-protein monomers under stress conditions. One of these stresses is heat stress that *Lactobacillus* is encountered it during food processing. According to the investigation, the TEM images showed that dialyzed S-protein monomers were able to recrystalized at 45 °C as same as 37 °C. The S-layer structure was an oblique lattice with p2 symmetry, and its parameters were respectively 11.3 and 5.5 nm. This finding was corresponded to the lattice parameters in *Lactobacillus acidophilus* ATCC 4356 (Smit et al., 2001), *Lactobacillus helveticus* ATCC 12046 (Lortal et al., 1992) and *Lactobacillus brevis* ATCC 14869 (Jakava-Viljanen et al., 2002).

According to the investigation of *slp*A gene expression, it was found that the *slp*A gene expression increased at 45 °C and pH 5. It proves that S-layer is very important for *Lactobacillus acidophilus* ATCC 4356, and is a protective sheath for the bacteria. In the pH 6, 7 and 37 °C, the *slp*A gene expression was lower than that in the control. It is not known why the *slp*A gene expression was different from the S-protein production in pH 6 and 7. It has been suggested that, either a little inversion has happened on *slp*A and *slp*B in unfavorable growth conditions or the presence of HCl and NaOH, used for adjusting of pH, can influence or block *slp*A gene expression. Another explanation could be that the S-layer mRNA has a relatively long half-life of 15 min and it can be repeatedly translated. As S-layer proteins represent 10-15% of the total amount of proteins in *Lactobacillus* cells, their transcription and secretion mechanisms must be efficient and tightly regulated. Multiple promoters precede several S-layer genes (Boot & Pouwels, 1996) including S-layer genes of *Lactobacillus acidophilus* (Boot et al., 1996a) and *Lactobacillus brevis* (Vidgren et al., 1992) and are likely to ensure efficient transcription of these genes. Also, the half-lives of mRNA-encoding lactobacillar S-layer proteins are relatively high, approximately 15 min, which enables efficient protein translation (Boot et al., 1996a). In addition to the actively transcribed S-layer protein gene (*slp*A), *Lactobacillus acidophilus* has also the silent *slp*B gene. The inversion of the *slp* segment causes an interchange of the active and the silent S-layer genes, which resembles a mechanism of phase variation in expression of bacterial surface antigen (Avall-Jaaskelainen & Palva, 2005; Boot & Pouwels, 1996; Boot et al., 1996b; Pouwels et al., 1998). It was found that the frequency of inversion was high (1/300), yet all attempts to demonstrate expression of the *slp*B gene have so far been unsuccessful (Boot et al., 1996b; Pouwels et al., 1998). Phase variation or antigenic variation, as a result of inversion of the *slp* segment, might enable *Lactobacillus acidophilus* bacteria to better adhere to specific regions of mucosa (Pouwels et al., 1998).

4. Conclusion

In conclusion, we found that environmental conditions influenced the S-layer protein and *slp*A gene expression. Nevertheless, it seems that high temperature (45 °C) did not influence the self-assembly of S-layer monomers.

For future investigations, the *slp*B gene expression and adhesion of *Lactobacillus acidophilus* to the epithelial cells should be studied under stress and control conditions.

5. Acknowledgments

This work was supported by the Graduate Studies Office and Research Office of the University of Isfahan and International Center for Science, High Technology and Environmental Sciences.

6. References

Altermann, E., Buck, B. L., Cano, R. & Klaenhammer, T. R. (2004). Identification and phenotypic characterization of the cell-division protein CdpA. *Gene*,Vol. 342, No. 1, (September 2004), pp. 189-197, ISSN 0378-1119

Avall-Jaaskelainen, S., Kyla-Nikkila, K., Kahala, M., Miikkulainen-Lahti, T. & Palva, A. (2002). Surface display of foreign epitoes on the *Lactobacillus brevis* S-layer. *Applied and Environmental Microbiology*, Vol. 68, No. 12, (December 2002), pp. 5943-5951, ISSN 0099-2240

Avall-Jaaskelainen, S. & Palva, A. (2005). *Lactobacillus* surface layers and their applications. *FEMS Microbiology Review*, Vol. 29, No. 3, (August 2005), pp. 511-529, ISSN 0168-6445

Ben-Jacob, E., Cohen, I., Golding, I., Gutnick, D. L., Tcherpakov, M., Helbing, D. & Ron, I. G. (2000). Bacterial cooperative organization under antibiotic stress. *Physica A*, Vol. 282, No. 1-2, (February 2000), pp. 247-282, ISSN 0378-4371

Boot, H. J., Kolen, C. P. A. M., van Noort, J. M. & Pouwels, P. H. (1993). S-Layer Protein of *Lactobacillus acidophilus* ATCC 4356: purification, expression in *Escherichia coli*, and nucleotide sequence of the corresponding genet. *The Journal of Bacteriology*, Vol. 175, No. 19, (October 1993), pp. 6089-6096, ISSN 0021-9193

Boot, H. J., Kolen, C. P. A. M. & Pouwels, P. H. (1995). Identification, cloning, and nucleotide sequence of a silent S-layer protein gene of *Lactobacillus acidophilus* ATCC 4356 which as extensive similarity with the S-layer protein gene of this species. *The Journal of Bacteriology*, Vol. 177, No. 24, (December 1995), pp. 7222-7230, ISSN 0021-9193

Boot, H. J. & Pouwels, P. H. (1996). Expression, secretion and antigenic variation of bacterial S-layer proteins. *Molecular Micribiology*, Vol. 21, No. 6, (September 1996), pp. 1117-1123, ISSN 0950-382X

Boot, H. J., Kolen, C. P. A. M., Andreadaki, F. J., Leer, R. J. & Pouwels, P. H. (1996a). The *Lactobacillus acidophilus* S-Layer protein gene expression site comprises two consensus promoter sequences, one of which directs transcription of stable mRNA. *The Journal of Bacteriology*, Vol. 178, No. 18, (September 1996), pp. 5388-5394, ISSN 0021-9193

Boot, H. J., Kolen, C. P. A. M & Pouwels, P. H. (1996b). Interchange of the active and silent S-layer protein genes of *Lactobacillus acidophilus* by inversion of the chromosomal *slp* segment. *Molecular Micribiology*, Vol. 21, No. 4, (August 1996), pp. 799-809, ISSN 0950-382X

Boot, H. J. Kolen, C. P. A. M., Pot, B., Kersters, K. & Pouwels, P. H. (1996c). The presence of two S-layer protein-encoding genes is conserved among species related to *Lactobacillus acidophilus*. *Microbiology*, Vol. 142, No. 9, (September 1996), pp. 2375-2384, ISSN 0302-8933

Bradford, M. M. (1976). A rapid & sensitive method for the quantification of microgram quantities of protein utilizing the principle of dye-binding. *Analytical Biochemistry*, Vol. 72, No.1-2, (January 1976), pp. 248-254, ISSN 0003-2697

Buck, B. L., Altermann, E., Svingerud, T. & Klaenhammer, T. (2005). Functional analysis of putative adhesion factors in *Lactobacillus acidophilus* NCFM. *Applied and Environmental Microbiology*, Vol. 71, No. 12, (December 2005), pp. 8344-8351, ISSN 0099-2240

Cotter, P. D. & Hill, C. (2003). Surviving the Acid Test: Responses of Gram- Positive Bacteria to Low pH. *Microbiology and Molecular Biology Reviews*, Vol. 67, No. 3, (September 2003), pp. 429-453, ISSN 1092-2172

De Angelis, M. &. Gobbetti, M. (2004). Environmental stress responses in *Lactobacillus*: A review. *Proteomics*, Vol. 4, No. 1, (January 2004), pp.106-122, ISSN 1615-9853

De Roos, N. M. & Katan, M. B. (2000). Effects of probiotic bacteria on diarrhea, lipid metabolism, and carcinogenesis: a review of papers published between 1988 and

1998. *The American Journal of Clinical Nutrition,* Vol. 71, No. 2, (February 2000), pp. 405-411, ISSN 0002-9165

Debabov, V. G. (2004). Bacterial and archaeal S-layers as a subject of nanobiotechnology. *Journal of Molecular Biology,* Vol. 38, No. 4, (January 2004), pp. 482-493, ISSN 0026-8933

Dufrene, Y. F. & Rouxhet, P. G. (1996). Surface composition, surface properties, and adhesiveness of *Azospirillum brasilense* — variation during growth. *Canadian Journal of Microbiology,* Vol. 42, No. 6, (June 1996), pp. 548-556, ISSN 0008-4166

Dunne, C., O'Mahony, L., Murphy, L., Thornton, G. & et al. (2001). In vitro selection criteria for probiotic bacteria of human origin: correlation with in vivo findings 1-4. *The American Journal of Clinical Nutrition,* Vol. 73, No. 2, (February 2001), pp. 386s-392s, ISSN 0002-9165

Engelhardt, H. &. Peters, J. (1998). Structural research on surface layers: a focus on stability, surface layer homology domains, and surface layer-cell wall interactions. *Journal of Structural Biology,* Vol. 124, No. 2-3, (December 1998), pp. 276-302, ISSN 1047-8477

Fitzsimons, N. A., Akermans, A. D. L., de Vos, W. M. & Vaughan, E. E. (2003). Bacterial gene expression detected in human faeces by reverse transcription-PCR. *Journal of Microbiological Methods,* Vol. 55, No. 1, (April 2003), pp. 133-140, ISSN 0167-7012

Frece, J., Kos, B., Svetec, I. K., Zgaga, Z., Mrsa, V. & Suskovic, J. (2005). Importance of S-layer proteins in probiotc activity of *Lactobacillus acidophilus* M92. *Journal of Applied Microbiology,* Vol. 98, No.2, (February 2005), pp. 285-292, ISSN 1364-5072

Garrote, G. L., Delfederico, L., Bibiloni, R., Abraham, A. G. & et al. (2004). Lactobacilli isolated from kefir grains: evidence of the presence of S-layer proteins. *Journal of Dairy Research,* Vol. 71, No. 2, (March 2004), pp. 222-230, ISSN 0022-0299

Granato, D., Perotti, F., Masserey, I., Rouvet, M., Golliard, M. & et al. (1999). Cell surface-associated lipoteichoic acid acts as an adhesion factor for attachment of *Lactobacillus johnsonii* La1 to human enterocyte-like Caco-2 cells. *Applied and Environmental Microbiology,* Vol. 65, No. 3, (March 1999), pp. 1071-1077, ISSN 0099-2240

Greene, J. D. & Klaenhammer, T. R. (1994). Factors involved in adherence of lactobacilli to human Caco-2 cells. *Applied and Environmental Microbiology,* Vol. 60, No. 12, (December 1994), pp. 4487-4494, ISSN 0099-2240

Herich, R. &. Levkut, M. (2002). Lactic acid bacteria, probiotics and immune system. *Veterinarni Medmedicina-Czech,* Vol. 47, No. 6, (June 2002), pp. 169-180, ISSN 0375-8427

Holzapfel, W. H., Haberer, P., Snel, J., Schillinger, U. & Huis in't Veld, J. H. J. (1998). Overview of gut flora and probiotics. *International Journal of Food Microbiology,* Vol. 41, No. 2, (May 1998), pp. 85-101, ISSN 0168-1605

Hynonen, U., Weterlund-Wikstrom, B., Palva, A. & Korhonen, T. K. (2002). Identification by flagellum display of an epithelial cell- and fibronectin-binding function in the SlpA surface protein of *Lactobacillus brevis. The Journal of Bacteriology,* Vol. 184, No. 12, (June 2002), pp. 3360-3367, ISSN 0021-9193

Jakava-Viljanen, M., Avall-Jaaskelainen, S., Messner, P., Sleytr, U. B. & Palval, A. (2002). Isolation of three new surface layer protein genes (*slp*) from *Lactobacillus brevis* ATCC 14869 and characterization of the change in their expression under aerated and anaerobic conditions. *The Journal of Bacteriology,* Vo. 184, No. 24, (December 2002), pp. 6786-6795, ISSN 0021-9193

Jan, G., Leverrier, P., Pichereau, V. & Boyaval, P. (2001). Changes in protein synthesis and morphology during acid adaptation of *Propionibacterium freudenreichii. Applied and Environmental Microbiology,* Vol. 67, No. 5, (May 2001), pp. 2029-2036, ISSN 0099-2240

Khaleghi, M., Kasra Kermanshahi, R., Yaghoobi, M. M., Zarkesh-Esfahani, S. H. & Baghizadeh, A. (2010). Assessment of bile salt effects on S-Layer production, *slp* gene expression and some physicochemical properties of *Lactobacillus acidophilus* ATCC 4356. *Journal of Microbiology and Biotechnology*, Vol. 20, No. 4, (April 2010), pp. 749-756, ISSN 1017-7825

Khaleghi, M., Kasra Kermanshahi, R. & Zarkesh-Esfahani, S. H. (2011). Effects of penicillin G on morphology and some physiological parameters of *Lactobacillus acidophilus* ATCC 4356. *Journal of Microbiology and Biotechnology*, Vol 21, No. 8, (August 2011), pp. 822-829, ISSN 1017-7825

Kim, W. S., Perl, L., Park, J. H., Tandianus, J. E. & Dunn, N. W. (2001). Assessment of stress response of the probiotic *Lactobacillus acidophilus*. *Current Microbiology*, Vol. 43, No. 5, (April 2001), pp. 346-350, ISSN 0343-8651

Klaenhammer, T. R., Barrangou, R., Buck, B. L., Azcarate-Peril, M. A. & Altermann, E. (2005). Genomic features of lactic acid bacteria effecting bioprocessing and health. *FEMS Microbiology Review*, Vol. 29, No. 3, (August 2005), pp. 393-409, ISSN 0168-6445

Klein, G., Pack, A., Bonaparte, C. & Reuter, G. (1998). Taxonomy and physiology of probiotic lactic acid bacteria. *International Journal of Food Microbiology*, Vol. 41, No. 2, (March 1998), pp. 103-125, ISSN 0168-1605

Kos, B., Suskovic, J., Vukovic, S., Simpraga, M., Frece, J. & Matosic, S. (2003). Adhesion and aggregation ability of probitic strain *Lactobacillus acidophilus* M92. *Journal of Applied Microbiology*, Vol. 94, No. 6, (January 2003), pp. 981-987, ISSN 1364-5072

Kos, B., Suskovic, J., Beganovic, J., Gjuracic, K., Frece, J., Iannaccone, C. & Caganella, F. (2008). Characterization of the three selected probiotic strains for the application in food industry. *World Journal of Microbiology Biotechnology*, Vol. 24, No. 5, (May 2008), pp. 699-707, ISSN 1573-0972

Kosin, B. & Rakshit, S. K. (2006). Microbial and processing criteria for production of probiotics: A Review. *Food Technology and Biotechnology*, Vol. 44, No. 3, (March 2006), pp. 371-379, ISSN 1330-9862

Lorca, G. L., Raya, R. R., Taranto, M. P. & de Valdez, G. F. (1998). Adaptive acid tolerance response in *Lactobacillus acidophilus*. *Biotecholoyg Letters*, Vol. 20, No. 3, (March 1998), pp. 239-241, ISSN 0141-5492

Lortal, S., van Heijenoort, J., Gruber, K. & Sleytr, U. B. (1992). S-layer of *Lactobacillus helveticus* ATCC 12046: isolation, chemical, characterization and re-formation after extraction with lithium chloride. *Microbiology*, Vo. 138, No. 3, (March 1992), pp. 611-618, ISSN 0022-1287

Marteau, P. & Rambaud, J. C. (2002). Probiotics and health: new facts and ideas. *Current Opinion in Biotechnology*, Vol. 13, No. 5, (October 2002), pp. 486-489, ISSN 0958-1669

Mathur, S. & Singh., R. (2005). Antibiotic resistance in food lactic acid bacteria- a review. *International Journal of Food Microbiology*, Vol. 105, No. 3, (December 2005), pp. 281-295, ISSN 0168-1605

Mercenier, A., Pavan, S. & Pot, B. (2003). Probiotics as biotherapeutic agents: present knowledge and future prospects. *Current Pharmaceutical Design*, Vol. 8, No. 2, (February 2003), pp. 99-110, ISSN 1381-6128

Mignot, T., Mesnage, S., Couture-Tosi, E., Mock, M. & Fouet, A. (2002). Developmental switch of S-layer protein synthesis in *Bacillus anthracis*. *Molecular Microbiology*, Vol. 43, N0. 6 (December, 2001), pp. 1615–1627, ISSN 0950-382X

Millsap, K. W., Reid, G. H., Van der Mei, C. & Busscher, H. J. (1997). Cluster analysis of genotypically characterized *Lactobacillus* species based on physicochemical cell surface properties and their relationship with adhesion to hexadecane. *Canadian Journal of Microbiology*, Vol. 43, No. 3, (March 1997), pp. 284–291, ISSN 0008-4166

Mishra, V. K., Mohammad, G. & Jha, A. (2008). Immunomodulation and anticancer potentials of yogurt probiotic. *EXCLI Journal*, Vol. 7, No. 1, (October 2008), pp. 177-184, ISSN 1611-2156

O'Toole, P. W. & Cooney, J. C. (2008). Probiotic bacteria influence the composition and function of the intestinal microbiota. *Interdisciplinary Perspectives on Infectious Diseases*, Article ID. 175285, (September 2008), pp. 1-6, ISSN 1687-7098

Ouwehand, A. C., Batsman, A. & Salminen, S. (2003). Probiotics for the skin: a new area of potential application. *Letters in Applied Microbiology*, Vol. 36, No. 5, (May 2003), pp. 327-331, ISSN 0266-8254

Park, Y. H., Kim, J. G., Shin, Y. W., Kim, H. S. & et al. (2008). Effects of *Lactobacillus acidophilus* 43121 and a mixture of *Lactobacillus casei* and *Bifidobacterium longum* on the serum cholesterol level and fecal sterol excretion in hypercholesteroemia-induced pigs. *Bioscience, Biotechnology, and Biochemistry*, Vol. 72, No. 2, (February 2008), pp. 595-600, ISSN 0916-8451

Pelletier, C., Bouley, C., Cayuela, C., Bouttier, S., Bourlioux, P. & Bellon-Fontaine, M. N. (1997). Cell surface characteristics of *Lactobacillus casei* subsp. *casei, Lactobacillus paracasei* subsp. *paracasei*, and *Lactobacillus rhamnosus* srtains. *Applied and Environmental Microbiology*, Vol. 63, No. 5, (May 1997), pp. 1725-1731, ISSN 0099-2240

Pouwels, P. H., Leer, R. J., Shaw, M., den Bak-Glashouwer, M-J. H. & et al. (1998). Lacic acid bacteria as antigen delivery vehicles for oral immunization purposes. *International Journal of Food Microbiology*, Vol. 41, No. 2, (March 1998), pp. 155-167, ISSN 0168-1605

Reid, G. (1999). The scientific basis for probiotic strains of *Lactobacillus*. *Applied and Environmental Microbiology*, Vol. 65, No. 9, (September 1999), pp. 3763-3766, ISSN 0099-2240

Resta-Lenert, S. & Barrett., K. E. (2003). Live probiotics protect intestinal epithelial cells from the effects of infection with enteroinvasive *Escherichia coli* (EIEC). *Gut*, Vol. 52, No. 7, (July 2003), pp. 988-997, ISSN 0017-5749

Sanders, M. E. & Klaenhammer, T. R. (2001). Invited Review: The scientific basis of *Lactobacillus acidophilus* NCFM functionality as a probiotic. *Journal of Dairy Science*, Vol. 84, No. 2, (February 2001), pp. 319-331, ISSN 0022-0302

Sara, M. & Sleytr, U. B. (2000). S-layer proteins. *The Journal of Bacteriology*, Vol. 182, No. 4, (February 2000), pp. 859-868, ISSN 0021-9193

Schar-Zammaretti, P., Dillmann, M-L., D'Amico, N., Affolter, M. & Ubbink, J. (2005). Influence of fermentation medium composition on physicochemical surface properties of *Lactobacillus acidophilus*. *Applied and Environmental Microbiology*, Vol. 71, No. 12, (December 2005), pp. 8165-8173, ISSN 0099-2240

Sillanpaa, J., Martinez, B., Antikainen, J., Toba, T., Kalkkinen, N., Tankka, S. & et al. (2000). Characterization of the collagen-binding S-layer protein CbsA of *Lactobacillus crispatus*. *The Journal of Bacteriology*, Vol. 182, No. 22, (November 2000), pp. 6440-6450, ISSN 0021-9193

Silva, J., Carvalho, A. S., Ferreira, R., Vitorino, R., Amado, F., Domingues, P., Teixeira, P. & Gibbs,P. A. (2005). Effect of the pH of growth on the survival of *Lactobacillus delbrueckii* subsp. *bulgaricus* to stress conditions during spray-drying. *Journal of Applied Microbiology*, Vol. 98, No. 3, (March 2005), pp. 775-782, ISSN 1364-5072

Sleytr, U. B., Sara, M., Pum, D. & Schuster, B. (2001). Caracterization and use of crystalline bacterial cell surface layers. *Progress in Surface Science*, Vol. 68, No. 7-8, (October 2001), pp. 231-278, ISSN 0079-6816

Smit, E., Oling, F., Demel, R., Martiez, B. & Pouwels, P. H. (2001). The S-layer protein of
    Lactobacillus acidophilus ATCC 4356: Identification and characterization of domains
    responsible for S-protein assembly and cell wall binding. *Journal of Molecular
    Biology*, Vol. 305, No. 2, (January 2001), pp. 245-257, ISSN 0022-2836

Sullivan, A. & Nord, C. E. (2002). Probiotics in human infections. *Journal of Antimicrobial
    Chemotherapy*, Vol. 50, No. 5, (October 2002), pp. 625-627, ISSN 0305-7453

Tannock, G. W., Tilsala-Timisjarvi, A., Rodtong, S., Ng, J., Munro, K. & Alatossava, T. (1999).
    Identification of *Lactobacillus* isolates from the gastrointestinal tract, silage, and
    yoghort by 16S-23S rRNA gene intergenic spacer region sequence comparisons.
    Applied and Environmental Microbiology, Vol. 65, No. 9, (September 1999), pp. 4264-
    4267, ISSN 0099-2240

Toba, T., Virkola, R., Westerlund, B., Bjorkman, Y., Sillanpaa, J., Vartio, T., Kalkkinen, N. &
    Korhonen, T. K. (1995). A collagen-binding S-layer protein in *Lactobacillus crispatus*.
    Applied and Environmental Microbiology, Vol. 61, No. 7, (July 1995), pp. 2467-2471,
    ISSN 0099-2240

Trotha, R., Konig, T. H. W. & Konig, B. (2001). Rapid ribosequencing-an effective diagnostic
    tool for detecting microbial infection. *Infection*, Vol. 29, No. 1, (January-February
    2001), pp. 12-16, ISSN 0019-9567

Vadillo-Rodriguez, V., Busscher, H. J., Norde, W., De Vries, J. & van der Mei, H. C. (2004).
    Dynamic cell surface hydrophobicity of *Lactobacillus* strains with and without
    surface layer proteins. *The Journal of Bacteriology*, Vol. 186, No. 19, (July 2004), pp.
    6647-6650, ISSN 0021-9193

Vadillo-Rodriguez, V., Busscher, H. J., van der Mei, H. C., De Vries, J. & Norde, W. (2005).
    Role of *Lactobacillus* cell surface hydrophobicity as probed by AFM in adhesion to
    surfaces at low and high ionic strength. *Colloids and Surfaces B: Biointerfaces*, Vol. 41,
    No. 1, (March 2005), pp. 33-41, ISSN 0144-8617

Van der Mei, H. C., Van der Belt-Gritter, B., Pouwels, P. H., Martinez, B. & Busscher, H.
    J.(2003). Cell surface hydrophobicity is conveyed by S-layer proteins - a study in
    recombinant lactobacilli. *Colloids and Surfaces B: Biointerfaces*, Vol. 28, No. 2-3, (April
    2003), pp. 127-134, ISSN 0144-8617

Velez, M. P., De Keersmaecker, S. C. J. & Vanderleyden, J. (2007). Adherence factors of
    Lactobacillus in the human gastrointestinal tract. *FEMS. Microbiology Letters*, Vol.
    276, No. 2, (September 2007), pp. 140-148, ISSN 1574-6968

Ventura, M., Jankovic, I., Walker, D. C., Pridmore., R. D. & Zink, R. (2002). Identification and
    characterization of novel surface proteins in *Lactobacillus johnsonii* and *Lactobacillus
    gasseri*. *Applied and Environmental Microbiology*, Vol. 68, No.12, (December 2002), pp.
    6172-6181, ISSN 0099-2240

Vidgren, G., Palva, I., Pakkanen, R., Lounatmaa, K. & Palva, A. (1992). S-layer protein gene
    of *Lactobacillus brevis*: cloning by polymerase chain reaction and determination of
    the nucleotide sequence. *The Journal of Bacteriology*, Vol. 174, No. 22, (November
    1992), pp. 7419-7427, ISSN 0021-9193

Waar, K., van der Mei, H. C., Harmsen, H. J. M., Degener, J. E. & Busscher, H. J. (2002).
    Adhesion to bile drain materials and physicochemical surface propertise of
    Enterococcus faecalis strains growth in the presence of bile. *Applied and Environmental
    Microbiology*, Vol. 68, No. 8, (August 2002), pp. 3855-3858, ISSN 0099-2240

11

Built-In Synthetic Gene Circuits in *Escherichia coli* – Methodology and Applications

Bei-Wen Ying and Tetsuya Yomo
Osaka University,
Japan

1. Introduction

Synthetic approaches are widely employed in the emerging research field of systems and synthetic biology, to learn the living organisms in a physical and systematic manner, such as, cellular dynamics and network interactions. Synthetic gene circuits potentially offer the insights into nature's underlying design principles (Hasty *et al*, 2002), and genetic reconstructions will give better understanding of naturally occurring functions (Sprinzak and Elowitz, 2005). Technical improvements in synthetic biology will provide not only engineering novelty for applications in biotechnology (McDaniel and Weiss, 2005) but also the fundamental understanding of living systems.

It is well-known that a library of the parts comprised in the gene circuits, which can be found in MIT Parts Registry (http://parts.mit.edu/), provides a variety for genetic reconstruction. As well, a new born organization (http://biobricks.org/) provides a platform (BioBrick™ parts) for scientists and engineers to work together. Current pioneer studies provided the successful examples of synthetic circuits working in the living cells, such as, the mutual inhibitory circuits functionally constructed in bacterial cells (Gardner *et al*, 2000), and with newly introduced biological functions (Kashiwagi *et al*, 2006). However, the reported cases generally do not include the vast majority of many failures. After defining a conceptual design as specifying how individual components are connected to accomplish the desired function, the next step is constructing the well-designed foreign circuit in living cells. So far, the strategies for construction of synthetic gene circuits are more of an art form than a well-established engineering discipline, mostly, in a "Plug and Play" manner (Haseltine and Arnold, 2007).

Carriers (vector) used for genetic construction are commonly limited in the plasmid, due to the advantageous of its efficiency and easy manipulation. Successful constructions have been reported to mimic a toggle switch in bacterial cells (Gardner *et al*, 2000), to build a synthetic predator-prey ecosystem (Balagadde *et al*, 2008), to address the dynamical property of positive feedback system (Maeda and Sano, 2006), to study the behaviour of the synthetic circuit under complex conditions: unregulated, repressed, activated, simultaneously repressed and activated (Guido *et al*, 2006). However, noise due to the copy number variation in plasmids is inevitable. As know, copy number variation is an important and widespread component within and between cell populations. For example, CNV can cause statistically significant changes in concentrations of RNA associated with growth rate changes in bacteria (Klappenbach *et al*, 2000; Stevenson and Schmidt, 2004); as well as, small-scale copy number variation can cause a dramatic, nonlinear change in gene expression from the theoretical study on various genetic modules (network motifs) (Mileyko *et al*, 2008). Thus, low-copy plasmids are utilized for generation of cellular function in the studies of demonstrating that negative auto-regulation speeds the response times of transcription networks (Rosenfeld *et al*, 2002), identifying heuristic rules for programming gene expression with combinatorial promoters (Cox *et al*, 2007), studying the biological networks and produce diverse phenotypes (Guet *et al*, 2002), *etc*. As well, combination of low-copy plasmid and genome has been applied to analyze the multistablity in lactose operon in bacterial (Ozbudak *et al*, 2004), to evaluate the fluctuation in gene regulation at the single cell level (Rosenfeld *et al*, 2005), and to study noise propagation (Pedraza and van Oudenaarden, 2005), and so on. Nevertheless, neither controlling the copy number of plasmid in a living cell nor keeping a constant copy number of plasmid in a growing cell population is easy.

Difficulties in synthetic approaches of genetic constructions are faced, in particular, as the fact that a stable construction is essential for steady phenotypic quantification. Practical methodology is required for the stable maintenance of the synthetic gene circuits in growing cells. As the genome is the most stable genetic circuit in living cells, insertion synthetic circuit into the genome will promise a best solution. Short fragment genome recombination of a reporter gene is widely applied, particularly, such as, the accurate prediction of the behaviour of gene circuits from component properties (Rosenfeld *et al*, 2007), and the study on intrinsic and extrinsic noise in a single cell level (Elowitz *et al*, 2002). It is becoming aware of the importance of genome integration of the synthetic gene networks.

Though the single copy of genome is the best choice for carrying the synthetic circuit stable along with the cell division and propagation, building a complex synthetic circuit, commonly comprised of a few genetic parts, into genome is not an easy job due to the flowing reasons. Inducing these parts into the genome one by one is time consuming, and the frequently repeated genomic construction process can potentially result in unexpected mutagenesis or stress-induced genomic recombination. The modified method introduced here reduces the frequency of recombination, and provides a time-saving approach for efficient synthetic construction on the bacterial genome. The availability of long insertions allows the easy artificial reconstruction of complicated networks on the genome. The examples of synthetic circuits constructing in *Escherichia coli* cells using the refined methods are described in detail. An assortment of synthetic circuits integrated into the genome working as design principles are shown. The switch-like response of the synthetic circuit sensitive to nutritional conditions is specially presented. Constructing synthetic gene circuits integrated in bacterial genome is to form a stable built-in artificial structure, and provides a powerful tool for the studies not only on the field of synthetic and systems biology based on bacteria but also on the applications potential for genetic engineering to achieve metabolic reconstruction.

2. Methodology: Genome-integration of foreign DNA sequences

As the classic methods for genome recombination, a number of general allele replacement methods have been used to inactivate bacterial chromosomal genes (Dabert and Smith, 1997;

Kato *et al*, 1998; Link *et al*, 1997; Posfai *et al*, 1999). These methods all require creating the gene disruption on a suitable plasmid before recombining it onto the chromosome, leading to its complexity in the methodology. A relatively simple method was developed by Wanner's group, a simple and highly efficient method to disrupt chromosomal genes in *Escherichia coli* in which PCR primers provide the homology to the targeted genes (Datsenko and Wanner, 2000). The procedure is based on the Red system that promotes a greatly enhanced rate of recombination over that exhibited by *recBC*, *sbcB*, or *recD* mutants when using linear DNA.

Elegant applications of Wanner's method have been reported, such as, the construction of single-gene knock-out mutants (Baba *et al*, 2006), construction of targeted single copy of lac fusions (Ellermeier *et al*, 2002), produce insertion alleles for about 2,000 genes systematic mutagenesis of *Escherichia coli* genome (Kang *et al*, 2004). Because of the limitation on the insertion length, the optimization on transformation procedure was performed to produce recombinant prophages carrying antibiotic resistance genes (Serra-Moreno *et al*, 2006). Wanner's method is very efficient on deletion mutation, even for quite long genome segments, whereas, insertion is limited within 2-3 Kbs technically. The requirement on constructing complicated networks is facing to the technical problem on the length limitation.

The methodology of genetic construction was recently published as the research article on a new protocol for more efficient integration of larger genetic circuits into the *Escherichia coli* chromosome. Complex synthetic circuits are commonly comprised of a few genetic parts. Inducing these parts into the genome one by one is time consuming, and the frequently repeated genomic construction process can potentially result in unexpected mutagenesis or stress-induced genomic recombination. The refined procedure introduced here shows the availability of the efficient artificial reconstruction of complex networks on the *Escherichia coli* genome, and provides a powerful tool for complex studies and analysis in synthetic and systems biology. Comparison between the genome integrated and the plasmid incorporated genes, reduced cell-to-cell variation was clearly observed in genome format. The method demonstrated that the integrated circuits show more stable gene expression than those on plasmids and so we feel this technique is an essential one for microbiologists to use.

2.1 Refined method

The method has been modified including medium, temperature, transformation, and selection, as described elsewhere (Ying *et al*, 2010). The synthetic sequences need to be wholly constructed on a plasmid in advance. Following PCR amplification and purification of the linear target sequence, transformation (electroporation) for genome replacement is performed, to introduce it into competent cells. To distinguish genomic recombinants from the original plasmid carriers, the target synthetic sequence encodes a different antibiotic resistant gene from the original plasmid. False transformants (*i.e.*, transformed colony) carrying the plasmid grow on both antibiotic plates; genomic recombinants grow only on the plate carrying the antibiotic whose resistant gene is encoded in the circuit, but not the one encoded in the plasmid. Dual antibiotics selection for positive transformants reduces the labour and cost of large-scale screening, and uncovers a high ratio of positive candidates on the colony PCR check. The steps of the refined method are described as follows, along with the schematic illustration of the process (Figure 1).

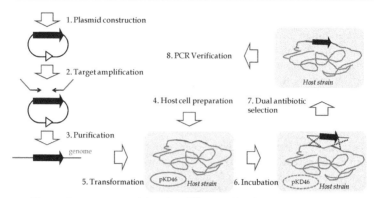

Fig. 1. Scheme of homologous recombination. The numbering steps are corresponding to the listed procedure of the refined method. Modified from the original paper (Ying *et al*, 2010).

- Construction of the synthetic sequence on a plasmid (often containing an Amp^R gene).
- PCR amplification of the target foreign DNA sequence, with the homogenous region corresponding to the recombination site.
- Clean-up (buffer exchange or gel extraction) using commercial kits.
- Digestion by the enzyme *DpnI* at 37°C for 2 h to remove the trace amount of the original plasmid.
- Clean-up and condensation of the target sequence. Any commercial kit is convenient.
- Transformation to the host strain containing the plasmid of pKD46, encoding the recombinase. Electroporation is crucial.
- Culturing in the rich medium (SOC) with 1 mM of arabinose, at 37°C for 2 h. Quiet incubation often increases the efficiency of transformation.
- Plating for antibiotic selection, incubation overnight at 37°C. Once using a slow growth strain, the additional incubation time is required.
- Strike the single colonies onto two plates, each with a different antibiotic, and incubate overnight at 37°C.
- Selection based on the difference of the clones between the two plates: positive candidates exhibited fast growth on the Gene^R (the antibiotics resistant gene different from Amp^R) plate, and slow or no growth on the Amp^R plate. This dual antibiotics screening on the plates promoted the final positive selection by colony PCR.
- Colony PCR for final confirmation. This step is essential to make sure that no unexpected recombination occurred in genome, particularly repeated homologous recombination have been performed.

2.2 High efficiency of recombination

Synthetic DNA sequences of various lengths (1 − 10 Kbs) have been inserted into the different sites on genome, such as, *intC*, *argG*, *glnA*, *leuB*, *ilvE*, *hisC* and *galK*. Comparatively short insertions result in accurate genome replacement. In contrast, longer insertions generally lead to fewer transformants and a worse outcome (*i.e.*, fewer positive colonies); nevertheless, usually there are still sufficient transformants for further selection (Table 1). Genome location (gene site) dependent efficiency of homologous recombination was noticed (unpublished data). The site of *galK* always gave the best score of successful recombination, regardless of the length of inserted sequences. The efficiency of successful recombination, based on the positive ratio of the colony PCR, depended on the insertion length. Previous studies provide myriad examples of short DNA fragment recombinations (Elowitz *et al*, 2002), but none of long ones. The ability of this methodology to successfully recombine a foreign DNA fragment of 9 to 10 Kbs indicates that it is possible to construct a complex synthetic gene circuit of a considerable length onto the bacterial genome.

Structure of Insertion	Length (Kbs)	[1]Transformants (per plate)	[2]Positive ratio (colony PCR)
Gene[R] (cat) for deletion	<1.5	hundreds	~100%
Gene A-Gene[R]	2-3	hundreds	70-100%
Gene A-Gene B-Gene[R]	3-4	20-100	70-100%
Gene A-Gene B-Gene C-Gene[R]	4-5	20-100	50-80%
Gene A-Gene B-...-Gene[R]	9-10	10-30	30-50%

Table 1. Recombination efficiency. Genes A, B, C, etc. represent an assortment of genes incorporated into the synthetic circuits. Gene[R] indicates the antibiotic resistance gene. [1]Number of clones grown on an agar plate plated with 300 µL of SOC transformation mixture. [2]Ratio of the number of positive clones to the total number of clones applied in the colony PCR test, for final confirmation. Modified from the original paper (Ying *et al*, 2010).

In addition, the homologous sequences of various lengths have been evaluated for the transformation efficiency at the identical genome location. In general, a 100-bp overlap, which can be easily generated by PCR amplification, could enable reasonably high genome recombination efficiency. Thought longer homologous sequences are supposed to give higher recombination efficiency, our test showed that length dependency was only present when the homologous sequence ranged from 50 to 150 bps, as an overlap of 300 to 500 bps decreased the number of transformants (Ying *et al*, 2010). We assume that the accessibility of the secondary structure, caused by the complex conformation of genome, for annealing possibly plays a role in successful recombination. According to the reported studies, genome integration has been generally carried out for 1 or 2 genes for each recombination step. If a relatively large fragment comprising 4 to 8 genes, four cycles of genome recombination procedure are needed. The reduced repetition of homologous recombination procedure of this refined protocol allows long DNA segments exchange at once. It greatly saves time and labour, and is supposed to contribute to the microbiological engineering.

2.3 Reduced cell-to-cell variation

Copy number variation in genetic materials can cause large variation within and between cell populations. As reported, it caused significant changes in RNA concentrations associated with growth rate changes (Klappenbach *et al*, 2000; Stevenson *et al*, 2004), as well as dramatic differences in gene expression compared to theory in various genetic modules (network motifs) (Mileyko *et al*, 2008). It is why combination of low-copy plasmids and genomes is currently applied in the studies on fluctuations in living cells (Ozbudak *et al*, 2004; Rosenfeld *et al*, 2005).

The relation between the copy number of the plasmid and the cell-to-cell variation in gene expression has been evaluated by an assortment of experiments using the flow cytometry, as previously reported (Ito *et al*, 2009). The results showed that the copy number of the plasmid was correlated to the protein abundance, which indicated that the variable in DNA copy number contributed to the cell-to-cell variation in gene expression. The copy number issue is important to be considered, when discussing the property of the target gene and its expression fluctuation. Genomic recombination seems to be the best choice for the studies on biological noise and phenotypic fluctuation, as the genome is a stable carrier of the constant copy number.

The comparison between the plasmid and the genome clearly verified this statement. Identical synthetic sequence, comprising a reporter gene *gfp* (green fluorescence protein, GFP) and an antibiotic resistant gene *kan^R*, has been transformed into the host *Escherichia coli* cells, either in a low-copy plasmid (~ 20 copies per cell) or onto the genome. Once both cell types were induced to display similar green fluorescent intensities (*i.e.*, the same averaged concentration of GFP), the variance of the cellular GFP concentration obtained from plasmid carriers was much larger than that from genomic carriers (Ying *et al*, 2010). It demonstrates that the variety of gene expression levels among the cells carrying the synthetic sequence in plasmid format is much larger than that in the genomic format. Using a very low-copy plasmid (< 5 copies per cell) reduced the cell-to-cell variation but still showed larger variance than the genome format (unpublished data). It greatly suggests that it is noise in the gene copy number that increases the variation in a genetically identical population. In addition, less heterogeneous in the transformant phenotypes and longer generations were often observed in genome-integrated carriers than those of plasmid insertion (personal communication). Taken together, construction of a synthetic gene circuit on the genome will greatly improve the stability of the genetic structure itself and reduce the biological noise from copy number variation.

3. Applications: Synthetic gene circuits work as design principles

An applicable protocol for solid and efficient genome recombination is described above. An assortment of well-designed synthetic circuits built-in *Escherichia coli* cells are going to be introduced as successful applications. Precise reconstruction of synthetic gene circuits to mimic gene expression and regulation is generally employed in a plasmid with a chromosomal reporter gene (Blake *et al*). Using the modified method, we successful integrated a variety of gene circuits into the bacterial genome. First of all, a relatively long genetic loop of mutual inhibitory structure displayed a strong hysteric expression pattern was successfully constructed, which was assumed to be resulted from the decreased fluctuation due to the genome recombination. Applying the mutual inhibitory design, a bistable dual function genetic switch was built in the genome. Selective expression of the genome integrated genetic switch sensitive to the environmental transition was clearly observed, which suggested that the synthetic circuits could show physiological activity in living cells and play an important role in population adaptation.

3.1 Mutual inhibitory construct behaved as design principles

Mutual inhibitory structure has been successfully constructed in bacterial cells on the plasmids (Gardner *et al*, 2000; Kashiwagi *et al*, 2006), and the pretty work of this structure in genome format was firstly reported using the refined method (Ying *et al*, 2010). As shown in Figure 2A, the mutual inhibitory structure, which was integrated into the genome and showed the expected features, was finally acquired using the method described here. The *Escherichia coli* DH1 cells were used as hosts for the construct. To prevent disturbance of the expression from the promoter P_{lac} by the endogenous level of LacI protein, the native lacI gene and the related genes lacY and lacZ were deleted from the genome. The reporter genes, *gfp* and *rfp*, were employed as two different visible phenotypes and were used for quantitative evaluation of the protein levels of the two repressor proteins, TetR and LacI. The two antibiotic resistant genes were used as selection markers during genetic construction. According to the design principle, when the two expression units (*i.e.*, *gfp-lacI-kan^R* and *rfp-tetR-cat^R*), which are regulated by the promoters P_{lac} and P_{tet}, respectively, are expressed, they will inhibit each other, and the cells will show a "red" or "green" phenotype. The *gfp-lacI-kan^R* unit is highly induced ("green" phenotype) when the chemical inducer, doxycycline (Dox), is added. Similarly, the *rfp-tetR-cat^R* unit is greatly induced ("red" phenotype) by isopropyl β-D-1-thiogalactopyranoside (IPTG).

The engineered *Escherichia coli* cells carrying this mutual inhibitory structure (Figure 2A) displayed either green or red fluorescent intensity (Figure 2B), representing the two discrete phenotypes of highly induced expression of either unit. As known, the positive feedback structure could accelerate the expression of any of the two expression units occasionally showing slightly higher expression level, while the mutual inhibitory structure would suppress expression of the other unit, leading to a fixation effect of the expressed unit (Gardner *et al*, 2000; Kashiwagi *et al*, 2006; Ozbudak *et al*, 2004). The discrete expression of the two units verified the bistable nature, involved in the mutual inhibitory structure, as designed features.

A   B

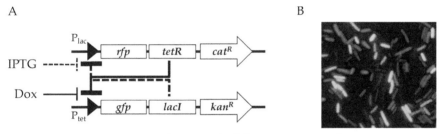

Fig. 2. *Escherichia coli* cells carrying the mutual inhibitory circuit. A. Genetic construction of the mutual inhibitory structure. B. Immerged image of the *Escherichia coli* cells observed under fluorescent microscope. Cells carrying the synthetic circuit as illustrated in A were grown in the minimal medium with induced conditions.

Following preliminary incubation in a doxycycline-free or doxycycline-supplemented (40 nM) medium, *Escherichia coli* cells carrying the mutual inhibitory structure were cultured in various concentrations of doxycycline (*i.e.*, 0, 5, 10, 15, 20, 30 and 40 nM) to induce the expression of *gfp-lacI-kan^R*, the "green" unit. Due to the different initial states, that is, the "green" or "red" induced expression level, the *gfp* expression levels varied even under the same condition, for instance, in the presence of 5 nM doxycycline, cells with high expression of the "red" unit at preincubation (doxycycline-free condition) showed an induced expression of *rfp* after incubation (Figure 3, upper panel), while those with high expression of "green" unit in preincubation (doxycycline-supplemented condition) showed an induced expression of *gfp* after incubation (Figure 3, bottom panel). The hysteresis in gene expression was clearly observed in the engineered *Escherichia coli* cells, under the inductions of 5 − 15 nM doxycycline, and kept for several days undergoing serial transfer, equal to approximate 30 to 50 generations. It indicated that the applicable synthetic approach of building genome integrated gene circuits successfully presenting the designed principles.

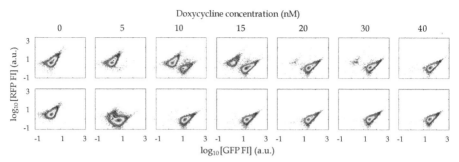

Fig. 3. Memory effect of the mutual inhibitory structure. Cells populations growing under the different induced conditions were measured using flow cytometry (FC500, Beckman). Each dot spot represents a single cell. Every 10 000 cells were collected and shown in the density maps here. GFP FI and RFP FI represent the green and red fluorescence intensity, respectively.

The success of the synthetic design was due to the easy genomic construction method. Construction of a simple toggle switch on a plasmid and transformation into bacterial cells was previously reported (Gardner *et al*, 2000). A genomic version of the similar toggle switch is introduced here. It could be steadily maintained along with propagation and cell division. Note that, to optimize this structure (*i.e.*, for strong hysteresis of gene expression), the "green" unit (*i.e.*, *gfp-lacI-kan^R*) was fixed and the regulatory region of the "red" one (*i.e.*, *rfp-tetR-cat^R*) was flexible, by using various operators or promoters. As differences in the sequence of promoters and the number of operators could strongly influence the binding affinity of the repressor proteins, regardless of the identical promoter, the operator sequences have been adjusted to amend the expression level of the repressors. The promoters and operators have been optimized for the enhanced binding affinity with the repressors, strong hysteresis (memory effect) was clearly observed in this construction. Owing to the greatly reduced copy number of the genome integrated synthetic construct, the changes in binding affinity of operator and repressor were clearly observed. The final construct of the satisfying promoter and operator sequences was decided by means of the "Plug and Play" strategy (Haseltine *et al*, 2007). Thus, optimization of the genetic construct in genome becomes practical, as the modification of the method reduces the steps and increases the efficiency of genome recombination.

3.2 Synthetic switch sensitive to environmental transitions

An assortment of synthetic circuits of varied genetic designs has been successfully constructed into the *Escherichia coli* genome (Kashiwagi *et al*, 2009; Ying *et al*, 2010). Experimental investigation demonstrated that these synthetic circuits were functional in living cells and could survive cells from starvation (Tsuru *et al*, 2011; Shimizu *et al*, in revision). Among these constructs, a relatively complex circuit of mutual inhibitory structure (Matsumoto *et al*, 2011) is introduced here.

3.2.1 Design principles

As described, bistability can be easily introduced into the genetic design of the synthetic circuit to produce two discrete stable states, "red" and "green", as shown in Figure 2B. Once the biological functions, for instance, which are crucial for cell growth, are introduced to the synthetic circuit, the two stable states will represent two phenotypes of physiological activities. For instance, two additional genes, *geneA* and *geneB*, can be inserted into the two expression units, "red" and "green", as shown in Figure 4. The *geneA* and *geneB* encode the proteins (*e.g.*, enzymes) that catalyze the biological reactions promoting the two independent physiological pathways (Figure 4, Function A and Function B), respectively. Switching between the two expression units is supposed to have selective activation of the two pathways, resulting in growth recovery of cell population and/or causing improved production of target proteins.

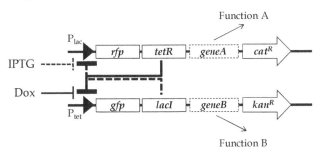

Fig. 4. Synthetic circuit designed for physiological functions. Dual functions are introduced in the mutual inhibitory structure. Induced expression of *geneA* and *geneB* represent the activation of Function A and Function B (*e.g.*, amino acid biosynthesis), respectively.

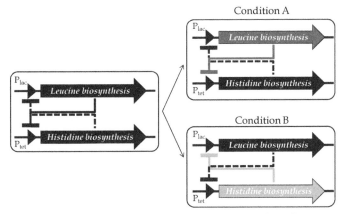

Fig. 5. Discrete phenotypes designed for the synthetic circuit. Induced expression of either "red" or "green" unit is supposed to be decided by the growing environment, either condition A (*e.g.*, leucine-free medium) or condition B (*e.g.*, histidine-free medium). Selectively induced expression of the two units is shown in red or green.

Such synthetic gene circuit was finally designed as described elsewhere (Matsumoto *et al*, 2011) and constructed as follows (Figure 4 and Figure 5). The two expression units, "red" and "green", representing a dual-function synthetic switch, were built in the *Escherichia coli* genome by homologous recombination as described in 2.1. The "red" unit contained three genes, *rfp*, *tetR* and *leuB*, encoding a red fluorescent protein (RFP), a repressor protein (*i.e.*, TetR) for blocking expression of the "green" unit (the promoter P_{tet}) and an enzyme contributing to leucine biosynthesis, respectively. The "green" unit consisted of three genes, *gfp*, *lacI* and *hisC*, encoding a green fluorescent protein (GFP), a repressor protein (*i.e.*, LacI) inhibiting expression of the "red" unit (the promoter P_{lac}) and an enzyme involved in histidine biosynthesis, respectively. That is, the *geneA* and *geneB* were replaced by *leuB* and *hisC*, and leucine and histidine biosynthesis represented the Function A and B, respectively (Figure 4). By the way, to oblige the cells to use the genes, *hisC* and *leuB*, within the synthetic switch, the native regulation of *Leu* and *His* operons was disturbed by removing *leuB* and *hisC* from their native chromosomal locations. Thus, the expression of *leuB* and *hisC* only inside the "red" and "green" units were reported by the red and green fluorescence, respectively. Thus, a synthetic switch based on a mutual inhibitory structure showing two discrete physiological states, the induced leucine (red) and histidine (green) productions, was constructed as designed (details in Matsumoto *et al*, 2011).

Rewiring of the stress-stringent genes (*i.e.*, *leuB* and *hisC*) to the synthetic circuit allows us not only to investigate the unknown survival strategy in living systems but also to search the possibility of metabolism reconstitution. As genome recombination promises a stable genetic carrier, this synthetic dual-function circuit can be applied to mimic cellular behaviour. According to the design principle, the *Escherichia coli* cells carrying this synthetic switch are able to show two different phenotypes, "red" and "green", representing the induced expression of *leuB* and *hisC*, and related to two physiological functions, *i.e.*, leucine and histidine biosynthesis, respectively (Figure 5). Bistability, resulting from the mutual inhibitory structure, was assumed to confer the "memory effect" on the cells carrying this structure, as shown in Figure 3, where the same promoter and repressor cassette is used. That is, the cells are thought to be able to show two distinct phenotypes under identical culture conditions due to the diverse histories (induction) of gene expression.

Here, two diverse biological functions leucine and histidine biosynthesis are designed, both of which result in a fitness recovery depending on the external conditions (Figure 5). For instance, it is the cells only showing an induced expression level of the "red" unit, which contacting the gene *leuB* essential for leucine biosynthesis that could survive under leucine-depleted conditions (*i.e.*, Condition A). In general, *Escherichia coli* cells use the *Leu* operon and *His* operon to respond to starvation (Keller and Calvo, 1979; Searles *et al*, 1983; Wessler and Calvo, 1981). Depletion of leucine will lead to the induced expression of structural genes in the *Leu* operon; similarly, histidine depletion will cause an increase in expression of proteins encoded within the *His* operon (Henkin and Yanofsky, 2002; Keller *et al*, 1979). *leuB* and *hisC*, which are located within the *Leu* and *His* operons (Gama-Castro *et al*, 2008), are responsible for leucine and histidine biosynthesis, respectively. Rewiring these stringent starvation genes to the synthetic circuit not only introduces physiological activities to the synthetic design but also disturbs the original native regulation.

3.2.2 Experimental investigation

Firstly, whether the constructed synthetic circuit was functional was examined. The addition of IPTG induced the expression of the "red" unit that comprising *leuB* greatly improved the cell growth under the leucine-depleted conditions; similarly, the addition of doxycycline (Dox) induced the expression of *hisC* within the "green" unit, and allowed the cells to grow in histidine-depleted conditions (Table 2). Obviously, the selective full induction of the two expression units enabled the cells to grow under starved conditions. These results verified the following points: 1) the rewired genes, *leuB* and *hisC*, were biologically active, regardless of the chromosomal replacement; 2) the repressor-promoter interactions involved in the mutual inhibitory structure were strong; 3) the synthetic circuit was genetically stable and functional in the living cells.

	Cell growth				Gene Expression		
	No add.	+IPTG	+Dox		No add.	+IPTG	+Dox
+Leu	0.24	0.07	0.25	**+Leu**	hisC	leuB	hisC
+His	0.10	0.35	0.10	**+His**	hisC	leuB	hisC
Both AA	0.37	0.40	0.40	**Both AA**	hisC	leuB	hisC

Table 2. Cell growth and gene expression under varied conditions. +Leu, +His and Both AA indicate the addition of leucine, histidine and both amino acids, respectively. +IPTG, +Dox and No add. indicate the addition of IPTG, doxycycline, and in the absence of inducers, respectively. Cell growth is shown in the growth rate (h^{-1}), and gene expression is marked as the induced gene name in red or green, indicating the fluorescence of the cell population.

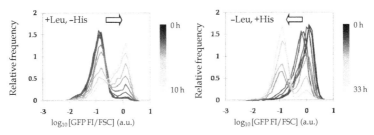

Fig. 6. Population shift in response to environmental transition. The distributions of newly formed populations (from 0 to 10 h or 33 h) are shown from light to dark grey, respectively. Green fluorescence intensity (GFP FI) and forward scattering (FSC) represent the abundances of GFP expressed in single cells and the relative cell size, respectively. GFP FI/FSC indicates the expression level of the "green" unit (*i.e.*, *hisC*) in cells. The figures are partially modified from the original paper (Matsumoto *et al*, 2011).

Subsequently, whether the synthetic circuit comprised of the rewired genes can be used by the *Escherichia coli* cells as a functional genetic switch in response to environmental transition was investigated. Under the hysteretic conditions, exponentially growing cells were transferred to the fresh media without the essential amino acid, leucine or histidine, for growth. Temporal changes in population dynamics were analyzed using flow cytometry, as described somewhere else (Matsumoto *et al*, in submission). When the "red" cell population of repressed expression of *hisC* was transferred to histidine-free conditions, the composition of the cell population changed gradually. More and more "green" cells of induced expression of *hisC* were born within the population, resulted in the population transition from "red" to "green" (Figure 6, left). Similarly, once the "green" cell population of repressed expression of *leuB* was transferred to leucine-free conditions, the "red" cells of induced expression of *leuB* were born and the fast growth allowed the "red" cells to take over the whole population, leading to the distribution shift from "green" to "red" (Figure 6, right). Note that in the rich medium, both "red" and "green" populations kept their distributions as initial expression level. Taken together, the *Escherichia coli* cells carrying this synthetic circuit formed the new population in accordance with the nutritional status.

Furthermore, the relation between cell growth and expression of the synthetic circuit was evaluated. The temporal trajectory of the relation between cell growth and relative expression level in the cell population showed that the growth rate decreased significantly under conditions of leucine or histidine depletion but recovered gradually, accompanied by a gradual increase in *leuB* ("red" unit) or *hisC* ("green" unit) expression level. The growth recovery and population transition due to histidine depletion were faster than those due to leucine depletion. The "green" unit (*hisC*) was much easier to induce than the "red" unit (*leuB*), possibly due to slight leakage of gene expression from P_{tet}, diverse essentiality in amino acid requirement, or as yet unknown synchronised expression changes in other related genes, *etc*. Further applications using other genes and amino acids are required to determine the universality of the capacity of the synthetic circuit to respond to external perturbations and to function as genetic switch sensitive to surroundings.

The experiments demonstrated that bacterial cells carrying this synthetic circuit formed diverse populations in response to the nutritional conditions and survived under conditions of nutrient depletion. A genome-integrated dual-function synthetic circuit sensitive to an environmental transition was successfully acquired. It strongly suggested that the synthetic design of proto-operons sensitive to external perturbations is practical for native cells. In summary, the method of genetic engineering and the application studies introduced here provides an efficient constructive approach for the studies or analysis in bacterial systems biology. Here, the genes involved in physiological functions responsive to external changes are introduced into the two expression units, the selective expression of the two units in cells can be considered a synthetic operon contributing to survival and/or adaptation.

4. Conclusion

A modified method for the integration of complicated genetic circuits into the *Escherichia coli* genome is introduced. The methodology provides an efficient synthetic approach for the dynamic and stochastic study of genetic networks. Linear artificial sequences as long as ~ 9 Kbps can be easily integrated into the bacterial genome at one time. The applications clearly showed accurate phenotypic behavior of the genome-integrated synthetic gene circuits corresponding to the design principle, which confirmed that the improved method allows the efficient construction of a single copy of a complicated genetic circuit in cells. As the genome recombination generally minimizes the copy number noise in the genetic circuit, it allows the precise design and interpretation of the cellular network. The availability of long-fragment insertions allows the easy reconstruction of complicated networks on the genome, and provides a powerful tool for synthetic and systems biology. Furthermore, the *Escherichia coli* cells carrying the synthetic circuit showed selective expression pattern in accordance with the environmental conditions, demonstrated the successful application of the genome-integrated synthetic circuit in cells. The applications owing to the simplified protocol demonstrated that the synthetic construct in the genome could be physiologically functional and sensitive to environmental transition. Synthetic approaches not only leads to the technical evolution for industrial use, but also can be employed to observe novel phenomena in living organisms. Further applications and improvement may bring us the completely synthetic genome functionally works in protocells.

5. Acknowledgments

We thank Natsuko Yamawaki, Junko Asada and Natsue Sakata for technical assistance, and the group members for fruitful discussion. This work was partially supported by Grants-in-Aid for Challenging Exploratory Research 22657059 (to BWY) and the "Global COE (Centers of Excellence) program" of the Ministry of Education, Culture, Sports, Science and Technology, Japan.

6. References

Baba T, Ara T, Hasegawa M, Takai Y, Okumura Y, Baba M, Datsenko KA, Tomita M, Wanner BL, Mori H (2006) Construction of Escherichia coli K-12 in-frame, single-gene knockout mutants: the Keio collection. *Mol Syst Biol* 2: 2006 0008.

Balagadde FK, Song H, Ozaki J, Collins CH, Barnet M, Arnold FH, Quake SR, You L (2008) A synthetic Escherichia coli predator-prey ecosystem. *Mol Syst Biol* 4: 187.

Blake WJ, M KA, Cantor CR, Collins JJ (2003) Noise in eukaryotic gene expression. *Nature* 422: 633-637.

Cox RS, 3rd, Surette MG, Elowitz MB (2007) Programming gene expression with combinatorial promoters. *Mol Syst Biol* 3: 145.

Dabert P, Smith GR (1997) Gene replacement with linear DNA fragments in wild-type Escherichia coli: enhancement by Chi sites. *Genetics* 145: 877-889.

Datsenko KA, Wanner BL (2000) One-step inactivation of chromosomal genes in Escherichia coli K-12 using PCR products. *Proc Natl Acad Sci U S A* 97: 6640-6645.

Ellermeier CD, Janakiraman A, Slauch JM (2002) Construction of targeted single copy lac fusions using lambda Red and FLP-mediated site-specific recombination in bacteria. *Gene* 290: 153-161.

Elowitz MB, Levine AJ, Siggia ED, Swain PS (2002) Stochastic gene expression in a single cell. *Science* 297: 1183-1186.

Gama-Castro S, Jimenez-Jacinto V, Peralta-Gil M, Santos-Zavaleta A, Penaloza-Spinola MI, Contreras-Moreira B, Segura-Salazar J, Muniz-Rascado L, Martinez-Flores I, Salgado H, Bonavides-Martinez C, Abreu-Goodger C, Rodriguez-Penagos C, Miranda-Rios J, Morett E, Merino E, Huerta AM, Trevino-Quintanilla L, Collado-Vides J (2008) RegulonDB (version 6.0): gene regulation model of Escherichia coli K-12 beyond transcription, active (experimental) annotated promoters and Textpresso navigation. *Nucleic Acids Res* 36: D120-124.

Gardner TS, Cantor CR, Collins JJ (2000) Construction of a genetic toggle switch in Escherichia coli. *Nature* 403: 339-342.

Guet CC, Elowitz MB, Hsing W, Leibler S (2002) Combinatorial synthesis of genetic networks. *Science* 296: 1466-1470.

Guido NJ, Wang X, Adalsteinsson D, McMillen D, Hasty J, Cantor CR, Elston TC, Collins JJ (2006) A bottom-up approach to gene regulation. *Nature* 439: 856-860.

Haseltine EL, Arnold FH (2007) Synthetic gene circuits: design with directed evolution. *Annu Rev Biophys Biomol Struct* 36: 1-19.

Hasty J, McMillen D, Collins JJ (2002) Engineered gene circuits. *Nature* 420: 224-230.

Henkin TM, Yanofsky C (2002) Regulation by transcription attenuation in bacteria: how RNA provides instructions for transcription termination/antitermination decisions. *Bioessays* 24: 700-707.

Ito Y, Toyota H, Kaneko K, Yomo T (2009) How selection affects phenotypic fluctuation. *Mol Syst Biol* 5: 264.

Kang Y, Durfee T, Glasner JD, Qiu Y, Frisch D, Winterberg KM, Blattner FR (2004) Systematic mutagenesis of the Escherichia coli genome. *J Bacteriol* 186: 4921-4930.

Kashiwagi A, Sakurai T, Tsuru S, Ying BW, Mori K, Yomo T (2009) Construction of Escherichia coli gene expression level perturbation collection. *Metab Eng* 11: 56-63.

Kashiwagi A, Urabe I, Kaneko K, Yomo T (2006) Adaptive response of a gene network to environmental changes by fitness-induced attractor selection. *PLoS ONE* 1: e49.

Kato C, Ohmiya R, Mizuno T (1998) A rapid method for disrupting genes in the Escherichia coli genome. *Biosci Biotechnol Biochem* 62: 1826-1829.

Keller EB, Calvo JM (1979) Alternative secondary structures of leader RNAs and the regulation of the trp, phe, his, thr, and leu operons. *Proc Natl Acad Sci U S A* 76: 6186-6190.

Klappenbach JA, Dunbar JM, Schmidt TM (2000) rRNA operon copy number reflects ecological strategies of bacteria. *Appl Environ Microbiol* 66: 1328-1333.

Link AJ, Phillips D, Church GM (1997) Methods for generating precise deletions and insertions in the genome of wild-type Escherichia coli: application to open reading frame characterization. *J Bacteriol* 179: 6228-6237.

Maeda YT, Sano M (2006) Regulatory dynamics of synthetic gene networks with positive feedback. *J Mol Biol* 359: 1107-1124.

Matsumoto Y, Ito Y, Tsuru S, Ying BW, Yomo T (2011) Bacterial cells carrying synthetic dual-function operon survived starvation. *J Biomed Biotechnol*, in press.

McDaniel R, Weiss R (2005) Advances in synthetic biology: on the path from prototypes to applications. *Curr Opin Biotechnol* 16: 476-483.

Mileyko Y, Joh RI, Weitz JS (2008) Small-scale copy number variation and large-scale changes in gene expression. *Proc Natl Acad Sci U S A*.

Ozbudak EM, Thattai M, Lim HN, Shraiman BI, Van Oudenaarden A (2004) Multistability in the lactose utilization network of Escherichia coli. *Nature* 427: 737-740.

Pedraza JM, van Oudenaarden A (2005) Noise propagation in gene networks. *Science* 307: 1965-1969.

Posfai G, Kolisnychenko V, Bereczki Z, Blattner FR (1999) Markerless gene replacement in Escherichia coli stimulated by a double-strand break in the chromosome. *Nucleic Acids Res* 27: 4409-4415.

Rosenfeld N, Elowitz MB, Alon U (2002) Negative autoregulation speeds the response times of transcription networks. *J Mol Biol* 323: 785-793.

Rosenfeld N, Young JW, Alon U, Swain PS, Elowitz MB (2005) Gene regulation at the single-cell level. *Science* 307: 1962-1965.

Rosenfeld N, Young JW, Alon U, Swain PS, Elowitz MB (2007) Accurate prediction of gene feedback circuit behavior from component properties. *Mol Syst Biol* 3: 143.

Searles LL, Wessler SR, Calvo JM (1983) Transcription attenuation is the major mechanism by which the leu operon of Salmonella typhimurium is controlled. *J Mol Biol* 163: 377-394.

Serra-Moreno R, Acosta S, Hernalsteens JP, Jofre J, Muniesa M (2006) Use of the lambda Red recombinase system to produce recombinant prophages carrying antibiotic resistance genes. *BMC Mol Biol* 7: 31.

Sprinzak D, Elowitz MB (2005) Reconstruction of genetic circuits. *Nature* 438: 443-448.

Stevenson BS, Schmidt TM (2004) Life history implications of rRNA gene copy number in Escherichia coli. *Appl Environ Microbiol* 70: 6670-6677.

Tsuru S, Yasuda N, Murakami Y, Ushioda J, Kashiwagi A, Suzuki S, Mori K, Ying BW, Yomo T (2011) Adaptation by stochastic switching of a monostable genetic circuit in Escherichia coli. *Mol Syst Biol* 7: 493.

Wessler SR, Calvo JM (1981) Control of leu operon expression in Escherichia coli by a transcription attenuation mechanism. *J Mol Biol* 149: 579-597.

Ying BW, Ito Y, Shimizu Y, Yomo T (2010) Refined method for the genomic integration of complex synthetic circuits. *J Biosci Bioeng* 110: 529-536.

12

The Thermostable Enzyme Genes of the dTDP-L-Rhamnose Synthesis Pathway (*rmlBCD*) from a Thermophilic Archaeon

Maki Teramoto[1], Zilian Zhang[1], Motohiro Shizuma[2], Takashi Kawasaki[1], Yutaka Kawarabayasi[1] and Noriyuki Nakamura[1]

[1]*National Institute of Advanced Industrial Science and Technology (AIST),*
Amagasaki, Hyogo
[2]*Osaka Municipal Technical Research Institute (OMTRI),*
Joto-ku, Osaka,
Japan

1. Introduction

The biosynthesis of saccharides is important because these diverse molecules mediate various functions from structure and storage to signaling. The biosynthesis of oligosaccharides, polysaccharides and glycoconjugates involves glycosyltransferases (EC 2.4), which transfer a sugar moiety from a nucleotide-activated donor sugar onto acceptors (Breton et al., 2006).

L-rhamnose is found widely in bacteria and plants (Giraud & Naismith, 2000). It is a common component of the cell wall and the capsule of many pathogenic bacteria, and has been indicated to play an essential role in many pathogenic bacteria (Giraud & Naismith, 2000). L-rhamnose is also found in the cytoplasmic membrane of archaea (Sprott et al., 1983), while pathogenic archaea have not been identified (Eckburg et al., 2003).

The nucleotide-activated L-rhamnose, dTDP-L-rhamnose, is synthesized from dTTP and glucose-1-phosphate (G-1-P) by a conserved four-step reaction. In the first reaction, RmlA (G-1-P thymidylyltransferase, EC 2.7.7.24) transfers the thymidylmonophosphate nucleotide to G-1-P. RmlB (dTDP-D-glucose 4,6-dehydratase, EC 4.2.1.46) then catalyzes oxidation of the C4 hydroxyl group of the sugar, followed by dehydration. Third, RmlC (dTDP-4-dehydrorhamnose 3,5-epimerase, EC 5.1.3.13) catalyzes an unusual double epimerization reaction at positions C3 and C5. Finally, RmlD (dTDP-4-dehydrorhamnose reductase, EC 1.1.1.133) reduces the C4 keto group to generate the final product, dTDP-L-rhamnose (Giraud & Naismith, 2000).

Thermal instability of the Rml enzymes has been raised as an issue (Graninger et al., 1999, 2002). Thus far, the highest reported temperature for the dTDP-L-rhamnose synthesis reaction is around 50 °C using thermophilic bacterial enzymes (Graninger et al., 2002; Novotny et al., 2004). The presence of dTDP-L-rhamnose has not been reported in archaea.

However, putative *rmlABCD* genes have been identified in the genomes of thermophilic archaea, including *Pyrococcus horikoshii* (Kawarabayasi et al., 1998), *Archaeoglobus fulgidus* (Klenk et al., 1997), *Sulfolobus solfataricus* (She et al., 2001) and *Sulfolobus tokodaii* (Kawarabayasi et al., 2001). Their *rml* genes possibly encode thermostable enzymes, and could thus greatly contribute to the thermostability of the Rml enzymes by *in vitro* protein evolution techniques such as family shuffling (Kikuchi et al., 2000). *S. tokodaii* 7 grows optimally at 80°C (Suzuki et al., 2002), and thus its putative *rmlABCD* genes were functionally and biochemically analyzed in this study.

2. Materials and methods

2.1 Sequence analysis

Sequences were compared to those in the SWISS-PROT protein sequence database (Bairoch & Apweiler, 2000) using BLAST (Altschul et al., 1990).

2.2 Vector construction

The four genes *rmlCDAB* (genes ST1969 to ST1972; DDBJ accession number BA000023) were PCR-amplified with each of the following sets of primers designed to introduce an *Nde*I site (overlapping the initiating ATG codon) and a downstream *Hind*III or *Xho*I site (the restriction sites are underlined): 5'-GGAATTC<u>CATATG</u>CCTTTTGAATTCGAAAATCTGGG-3' (forward; contains the *Nde*I site) and 5'-GCGGCCGC<u>AAGCTT</u>TTAATCAAAGACTTCAGCCTTTTC-3' (reverse; contains the *Hind*III site) for *rmlC* (gene ST1969); 5'-GGAATTC<u>CATATG</u>CGAACACTAATAACTGGTGC-3' (forward; contains the *Nde*I site) and 5'-GGTGC<u>TCGAG</u>CACCACCATACCGTCTAGATCC-3' (reverse; contains the *Xho*I site) for *rmlD* (gene ST1970) (His tag at the C terminal); 5'-GGAATTC<u>CATATG</u>GGAGGCGGTAATTTTACAC-3' (forward; contains the *Nde*I site) and 5'-GGTGC<u>TCGAG</u>TCATAATATCACCGAAGAATTCTC-3' (reverse; contains the *Xho*I site) for *rmlA* (gene ST1971); 5'-GGAATTC<u>CATATG</u>ATAATTATTGGTGGTGCTGG-3' (forward; contains the *Nde*I site) and 5'-GGTGC<u>TCGAG</u>TTTACTAACTTTTACTTTCCAAGGC-3' (reverse; contains the *Xho*I site) for *rmlB* (gene ST1972) (His tag at the C terminal). PCR was performed in 50 µl of standard PCR mixture with 1 unit of KOD-plus DNA polymerase (Toyobo, Japan), 10 ng of *S. tokodaii* 7 genomic DNA (NITE Biological Resource Center, Chiba, Japan) and 100 pmol of each primer. The amplification program was as follows: (i) 94°C for 2 min; (ii) 25 cycles of 94°C for 30 sec, X°C for 30 sec (X = 55 for *rmlCDB*; X = 58 for *rmlA*) and 68°C for 1 min. PCR products were digested at the restriction sites in the primers described above and ligated into the same restriction sites of the pET21(b) expression vector (Novagen). Nucleotide sequences of the resultant plasmids were confirmed after transformation of *Escherichia coli* DH5α (Takara, Japan) with the resultant plasmids and purification of the plasmids from the *E. coli* (QIAprep Spin Miniprep kit; Qiagen). The resultant plasmids to express *rmlC*, *rmlD*, *rmlA* and *rmlB* were designated pSTC, pSTD, pSTA and pSTB, respectively.

2.3 Preparation of supernatants containing the expressed proteins

E. coli BL21-Codon Plus (DE3)-RIL cells (Stratagene) were transformed with either pSTC, pSTD, pSTA or pSTB. The transformed *E. coli* was cultured at 37°C in 600 ml LB medium supplemented with 100 µg ml^{-1} ampicillin and 20 µg ml^{-1} chloramphenicol to an OD$_{600}$ of X

(X = 0.49 for RmlC; X = 0.56 for RmlD; X = 0.12 for RmlA; X = 0.60 for RmlB). Isopropyl β-D-thiogalactopyranoside was then added to the culture at a final concentration of 0.1 mM, and the culture was grown at 16°C overnight. The culture was then harvested by centrifugation, washed twice with 50 mM Tris-HCl buffer (pH 7.5), and resuspended in 2.5 ml of the 50 mM Tris-HCl buffer. The suspension was sonicated on ice for 4 min with a UD-201 ultrasonic disruptor (output, 6; duty, 50%; Tomy Seiko, Japan), heated at 80°C for 20 min, and centrifuged at 20,000 g for 10 min at 4°C. The resultant supernatant (2.5 ml) was desalted through a PD10 Sephadex G-25 column (GE Healthcare) with the 50 mM Tris-HCl buffer as the elution buffer. The resultant supernatant (3.5 ml) containing the expressed protein was stored at 4°C.

2.4 SDS-PAGE

Eight μl of each supernatant was analyzed on an E-R12.5L polyacrylamide gel (ATTO, Japan) at 30 mA constant current together with 10 μl of the broad range molecular weight standards (Bio-rad), and visualized with Coomassie Brilliant Blue R-250. The gel was photographed with a luminescent image analyzer Las-4000 mini (Ver 2.1; Fuji film, Japan). The amount of expressed protein was determined using MultiGauge software (Ver 3.2; Fuji film) in comparison to the molecular weight standards.

2.5 Reaction of the expressed protein

The reaction to detect G-1-P thymidylyltransferase activity of RmlA was performed at 80°C for 3 h in 50 mM Tris-HCl buffer (pH 7.5), 2 mM MgCl$_2$, 10 mM dTTP, 10 mM UTP and 10 mM G-1-P with the RmlA supernatant (40 μl of the supernatant, which contained 4 μg RmlA, was added to the reaction mixture for a total volume of 100 μl). UTP was also used as a substrate candidate because a product of a *rmlA* homolog (ST0452) from *S. tokodaii* 7 shows the G-1-P thymidylyltransferase activity (RmlA activity) as well as G-1-P uridylyltransferase activity (Zhang et al., 2005).

The reaction to determine substrate specificity of RmlB was performed at 80°C for 3 h in 50 mM Tris-HCl buffer (pH 7.5), 2 mM MgCl$_2$, 10 mM dTDP-D-glucose, and 10 mM UDP-D-glucose with the RmlB supernatant (40 μl of the supernatant, which contained 4 μg RmlB, was added to the reaction mixture for a total volume of 100 μl). To serve as controls, this reaction was performed with the RmlC supernatant or the RmlD supernatant (40 μl of the supernatant, which contained 20 μg of RmlC or RmlD, was added to the reaction mixture for a total volume of 100 μl) instead of the RmlB supernatant. Optimal reaction temperature for RmlB was determined in the same way as just described above except that UDP-D-glucose was not included in the reaction mixture and the reaction was preformed at the indicated temperature for 2 h. The reaction to characterize the thermostability of RmlB was carried out in the same way as described above for determining the optimal temperature except that the reaction was performed at 80°C for the indicated time period.

The dTDP-L-rhamnose synthetic reaction was performed at 80°C for 3 h in 50 mM Tris-HCl buffer (pH 7.5), 2 mM MgCl$_2$, 10 mM dTDP-D-glucose, 5 mM NADPH, and 5 mM NADH with a combination of the supernatants (4 μgs each of RmlB, RmlC and/or RmlD were added to the reaction mixture for a total volume of 100 μl). Each reaction was stopped by adding ten times the reaction volume of 500 mM KH$_2$PO$_4$.

2.6 HPLC of the reaction

A 50-µl aliquot of the reaction solution mixed with 500 mM KH_2PO_4 was subjected to HPLC using a LaChrom ELITE system with a L-2420 UV-VIS detector and a L-2130 pump (Hitachi) equipped with a Wakosil 5C18-200 column (4.6 x 250 mm; Wako, Japan). The 500 mM KH_2PO_4 was run as the elution buffer at a constant flow rate of 1 ml min⁻¹. The substrates and products were detected at 254 nm, and the peak area was used for calculation of the amount.

2.7 Mass spectrometry (MS) of the reaction products

The reaction samples for MS were prepared with a smaller amount of the substrate dTDP-D-glucose (2 mM) and thus with 2 mM NADPH and 2 mM NADH to completely consume the dTDP-D-glucose (this reaction was otherwise performed in the same way as described above), as a standard sample dTDP-L-rhamnose (the predicted product; Genechem Inc., Taejon, Korea) was unable to be separated from dTDP-D-glucose by HPLC used for MS (described below).

Reaction samples, to which 500 mM KH_2PO_4 was not added, were separated by HPLC (LC-10AT, Shimadzu, Japan) performed on a Wakosil 5C18-200 column using a water/methanol/acetic acid mixture (98.9/1.0/0.1, v/v/v) as the elution buffer at a flow rate of 1 ml min⁻¹. Mass spectra of the peaks detected by the HPLC were then measured by an electrospray ionization/ion trapped mass spectrometer (LCQ[DECA], ThermoQuest, USA) connected to the HPLC under the following instrumental conditions: detection mode, negative; mass range, m/z 100–1500; spray voltage, 5 kV; capillary temperature, 350°C; capillary voltage, -10 V; sheath gas flow, 100 units; auxiliary gas flow, 20 units (1 unit is roughly 0.025 L min⁻¹).

3. Results and discussion

3.1 Putative dTDP-L-rhamnose synthesis gene cluster (*rmlABCD*) from the thermophilic archaeon *S. tokodaii* 7

The *rml* genes from *S. tokodaii* 7 are clustered in the order of *rmlCDAB*, and the *rmlC* is oriented in a different direction from the others *rmlDAB* (Kawarabayasi et al., 2001). The RmlBCD were most similar to the dTDP-L-rhamnose synthesis enzymes (Table 1). However, the most similar homolog of the RmlA was UDP-N-acetylglucosamine pyrophosphorylase/glucosamine-1-phosphate N-acetyltransferase from *Methanocaldococcus jannaschii* (39% identity and 79% similarity; Q58501; Namboori & Graham, 2008), and its second closest homolog was RmlA from *Streptococcus mutans* (29% identity and 71% similarity; P95778; Tsukioka et al., 1997).

The RmlA from *S. tokodaii* possessed a signature sequence for recognition of nucleoside triphosphate, $GXGTRX_8PK$, and that of G-1-P, LVEKP (Thorson et al., 1994). RmlB contained motifs characteristic of dTDP-D-glucose 4,6-dehydratase, PSSPYSASKA and GGAGFIG (Allard et al., 2001). A region covering approximately 150 residues and conserved in dTDP-4-dehydrorhamnose 3,5-epimerase for substrate binding (Giraud et al., 2000) was found in RmlC. RmlD possessed the $PX_3YX_3KX_3E$ motif, which is characteristic of a reductase/epimerase/dehydrogenase/dehydratase superfamily for L-rhamnose synthesis from G-1-P, and another motif, STDYVF, unique in RmlD sequences (Graninger et al., 1999). However, the conserved NAD(P) binding motif (Wierenga et al., 1985), $GXGX_2G$, was not found in RmlD.

Product	Deduced MW (kDa)	Homolog	% Identity /% similarity	Accession no.
RmlB	37.0	dTDP-D-glucose 4,6-dehydratase (RmlB)		
		from *Salmonella typhimurium*	41/73	P26391
		from *Escherichia coli*	41/79	P55293
RmlC	20.5	dTDP-4-dehydrorhamnose 3,5-epimerase (RmlC)		
		from *Shigella flexneri*	45/75	P37780
		from *Salmonella typhimurium*	40/77	P26394
RmlD	31.0	dTDP-4-dehydrorhamnose reductase (RmlD)		
		from *Salmonella typhimurium*	28/63	P26392
		from *Shigella flexneri*	27/62	P37778

Table 1. The *rmlBCD* products from *Sulfolobus tokodaii* 7 and their homologs

3.2 Activity of RmlA

RmlA was expressed as a 35-kDa protein (data not shown), in agreement with the molecular weight (MW) of 38 kDa deduced from its nucleotide sequence. The predicted G-1-P thymidylyltransferase (RmlA) activity was assayed at 80°C with dTTP and G-1-P, but was not detected. RmlA may thus function as the UDP-N-acetylglucosamine pyrophosphorylase/glucosamine-1-phosphate N-acetyltransferase as deduced from its sequence similarity. This deserves further investigation. A product from another *rmlA* homolog, located away from the *rmlABCD* cluster in *S. tokodaii* 7, shows RmlA activity (Zhang et al., 2005). This is consistent with the reports that *rmlABCD* genes are not always found together (Cole et al., 1998; Giraud & Naismith, 2000).

3.3 RmlB as a thermophilic dTDP-D-glucose 4,6-dehydratase

RmlB was expressed as a 35-kDa protein (Fig. 1), which corresponded to the deduced MW (Table 1). The predicted activity (Table 1) was assayed at 80°C with dTDP-D-glucose and UDP-D-glucose (Fig. 2). UDP-D-glucose was also used as a substrate candidate because UDP-D-glucose as well as dTDP-D-glucose can be produced by the product from the another *rmlA* homolog (Zhang et al., 2005). As shown in Fig. 2, dTDP-D-glucose was used as a substrate by RmlB, whereas UDP-D-glucose was not. As controls, RmlC (Fig. 2) and RmlD (data not shown) did not use dTDP-D-glucose and UDP-D-glucose as the substrate. RmlB (+B) produced a broad peak in the mass chromatogram by the selected ion monitoring of the m/z 545 peak (data not shown; this broad peak was not obvious in Fig. 2, +B, 3h), which corresponded to the deprotonated molecule $(M-H)^-$ of the dTDP-D-glucose 4,6-dehydratase (RmlB) product, dTDP-4-dehydro-6-deoxy-D-glucose (MW = 546). Consistent with this result, dTDP-4-dehydro-6-deoxy-D-glucose has previously been eluted as a broad peak from a C18 column (Nakano et al., 2000; Watt et al., 2004). Together with the sequence homology, these data indicate that RmlB from *S. tokodaii* was thermostable dTDP-D-glucose 4,6-dehydratase (RmlB). Peaks 1 and 2, which became prominent by the addition of RmlB (Fig. 2), were indicated by MS to be from TMP and TDP, respectively (data not shown). It is unclear if RmlB degraded dTDP-D-glucose to TDP and TMP.

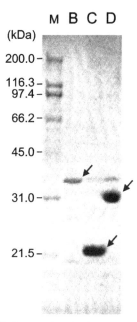

Fig. 1. SDS-PAGE analysis of RmlBCD. Eight µl of each cell-free supernatant from *E. coli* cells expressing *rmlB* (lane B), *rmlC* (lane C) or *rmlD* (lane D) was analyzed, and contained 0.8, 4.0 or 4.0 µg, respectively, of the deduced product (indicated by arrows). M, molecular weight markers.

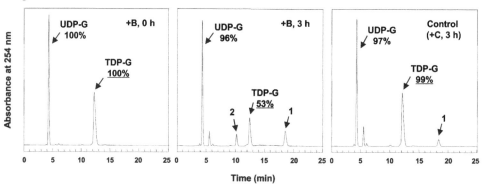

Fig. 2. Substrate specificity of RmlB shown by HPLC. RmlB-containing supernatant was incubated with dTDP-D-glucose (TDP-G) and UDP-D-glucose (UDP-G) at 80°C for the indicated period (+B). RmlC-containing supernatant was also incubated in the same way instead of the RmlB supernatant as a control (+C). Relative amounts of TDP-G and UDP-G are indicated compared to the amounts of TDP-G and UDP-G in the +B sample at 0 h, respectively. Peaks 1 and 2 of the reaction products were indicated by MS to be from TMP and TDP, respectively.

Temperature range for the dTDP-D-glucose-utilizing activity of RmlB was measured from 60 to 99°C, and the optimal temperature was shown to be 80°C (Fig. 3A). Therefore, RmlB from *S. tokodaii* was thermophilic; its optimal temperature for the activity coincided with the optimal growth temperature of 80°C for its host *S. tokodaii* 7 (Suzuki et al., 2002). The activity of RmlB gradually diminished at 80°C over hours (Fig. 3B). Specific dTDP-D-glucose-utilizing activity of RmlB was calculated to be 4.2 U/mg protein based on the data from the first 1 h of Fig. 3B. On the other hand, *E. coli* RmlB shows a high activity of approx. 3700 U/mg protein (Marolda & Valvano, 1995).

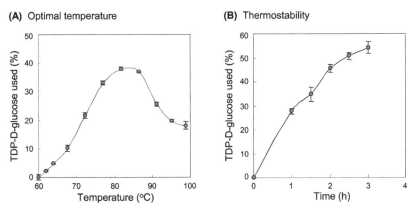

Fig. 3. Thermophilic TDP-D-glucose-utilizing activity of RmlB. (A) Optimal temperature of the activity. RmlB supernatant was incubated with TDP-D-glucose for 2 h at the indicated temperature. The amount of TDP-D-glucose used was shown as a percentage of the amount of TDP-D-glucose in the control sample incubated for 2 h at 80°C without supernatant (100% remained). (B) Thermostability of the activity at the optimal temperature of 80°C. RmlB supernatant was incubated with TDP-D-glucose at 80°C for the indicated period. The amount of TDP-D-glucose used was expressed as a percentage of the amount of TDP-D-glucose in the sample at 0 h (100% remained). Each value is the mean ± standard error from two independent experiments.

3.4 The dTDP-L-rhamnose synthesis reaction from dTDP-D-glucose at 80°C catalyzed by RmlBCD

RmlC and RmlD were expressed as a 22-kDa protein and a 30-kDa protein, respectively (Fig. 1), which corresponded to their respective deduced MWs (Table 1). The dTDP-L-rhamnose synthesis reaction catalyzed by RmlBCD, suspected based on their homology (Table 1), was analyzed at 80°C using dTDP-D-glucose and NAD(P)H as substrates (Fig. 4). A combination of RmlB plus RmlD produced peak 3 (+BCD and +BD in Fig. 4); the retention time and MS spectrum of this peak were identical to those of the standard sample dTDP-L-rhamnose. RmlB(C)D from *S. tokodaii* were thus shown to synthesize dTDP-L-rhamnose from dTDP-D-glucose at 80°C (discussed in the next paragraph). Without NAD(P)H, peak 3 was not produced in the reaction (data not shown). RmlB plus RmlC (+BC in Fig. 4) did not yield peak 3. The broad *m/z* 545 peak produced by RmlB disappeared with the addition of RmlD (+BD and +BCD in Fig. 4; MS data not shown). Together with the results indicating that RmlB from *S. tokodaii* was dTDP-D-glucose 4,6-dehydratase (RmlB) and with the sequence homology, the results strongly suggest that RmlD from *S. tokodaii* was thermostable dTDP-4-dehydrorhamnose reductase (RmlD).

It is possible that peak 3 produced by the combination of RmlB plus RmlD (+BD) could have been an epimer of dTDP-L-rhamnose produced without the possible epimerase RmlC (Table 1). Addition of RmlC showed no effect on the broad m/z 545 peak produced by RmlB (+BC in Fig. 4; MS data not shown), which is consistent with the previous observation using a C18 column (Watt et al., 2004). Therefore, unfortunately, the dTDP-4-dehydrorhamnose 3,5-epimerase (RmlC) activity, predicted activity of RmlC from *S. tokodaii*, was unable to be detected with the system used.

The concentrations of peak 3 (indicated from dTDP-L-rhamnose) and dTDP-D-glucose in the +BCD sample (Fig. 4) were determined to be 2.4 and 4.8 mM, respectively, showing that 52% of the added dTDP-D-glucose was used and that 46% of the dTDP-D-glucose used was converted to dTDP-L-rhamnose in the reaction. Consequently, RmlB and RmlD were estimated to show their respective activities of at least 0.33 U/mg protein.

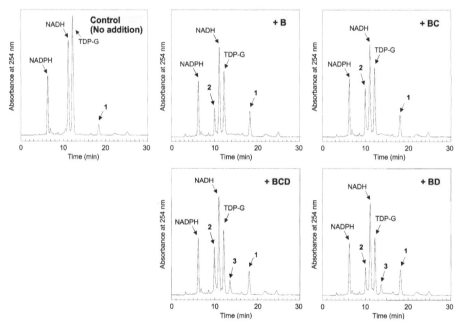

Fig. 4. The dTDP-L-rhamnose synthesis reaction from TDP-D-glucose by RmlBCD shown by HPLC. Combinations of RmlB (B), RmlC (C) and RmlD (D) supernatants were incubated with TDP-D-glucose (TDP-G), NADPH and NADH for 3 h at 80°C. The sample treated in the same way without the supernatants is also shown as a control. Peaks 1, 2 and 3 of the reaction products were indicated to be from TMP, TDP and dTDP-L-rhamnose, respectively, by MS.

4. Conclusions

Genes for thermostable RmlB and RmlD of the dTDP-L-rhamnose synthesis pathway were functionally identified from a thermophilic archaeon *S. tokodaii* 7. *S. tokodaii* Rml enzymes were suggested to be functionally identical to the bacterial counterparts, and exhibited superior thermostability. The temperature level of 80°C that was tested in this study is the highest value yet reported for the dTDP-L-rhamnose synthesis reaction from dTDP-D-glucose. Therefore, *S. tokodaii rml* genes could confer thermostability on the high-activity Rml enzymes, including the *E. coli* RmlB (Marolda & Valvano, 1995), by *in vitro* protein evolution techniques such as family shuffling (Kikuchi et al., 2000), and are useful for a broad field of potential applications requiring Rml enzyme including production of rhamnose-containing antigens as vaccines (Hsu et al., 2006; Prakobphol & Linzer, 1980).

5. Acknowledgment

We thank Hirofumi Sato of OMTRI for kind advice about MS and Hiromi Murakami of OMTRI for useful discussion about sugar metabolism.

6. References

Allard, S. T., Giraud, M. F., Whitfield, C., Graninger, M., Messner, P. & Naismith, J. H. (2001). The crystal structure of dTDP-D-Glucose 4,6-dehydratase (RmlB) from *Salmonella enterica* serovar Typhimurium, the second enzyme in the dTDP-l-rhamnose pathway. *J Mol Biol*, 307, 283–295.

Altschul, S. F., Gish, W., Miller, W., Myers, E. W. & Lipman, D. J. (1990). Basic local alignment search tool. *J Mol Biol*, 215, 403–410.

Bairoch, A. & Apweiler, R. (2000). The SWISS-PROT protein sequence database and its supplement TrEMBL in 2000. *Nucleic Acids Res*, 28, 45–48.

Breton, C., Snajdrová, L., Jeanneau, C., Koca, J. & Imberty, A. (2006). Structures and mechanisms of glycosyltransferases. *Glycobiology*, 16, 29R–37R.

Cole, S. T., Brosch, R., Parkhill, J., Garnier, T., Churcher, C., Harris, D., Gordon, S. V., Eiglmeier, K., Gas, S. & other authors (1998). Deciphering the biology of Mycobacterium tuberculosis from the complete genome sequence. *Nature*, 393, 537–544.

Eckburg, P. B., Lepp, P. W. & Relman, D. A. (2003). Archaea and their potential role in human disease. *Infect Immun*, 71, 591–596.

Giraud, M. F., Leonard, G. A., Field, R. A., Berlind, C. & Naismith, J. H. (2000). RmlC, the third enzyme of dTDP-L-rhamnose pathway, is a new class of epimerase. *Nat Struct Biol*, 7, 398–402.

Giraud, M. F. & Naismith, J.H. (2000). The rhamnose pathway. *Curr Opin Struct Biol*, 10, 687–696.

Graninger, M., Kneidinger, B., Bruno, K., Scheberl, A. & Messner, P. (2002). Homologs of the Rml enzymes from *Salmonella enterica* are responsible for dTDP-β-L-rhamnose biosynthesis in the gram-positive thermophile *Aneurinibacillus thermoaerophilus* DSM 10155. *Appl Environ Microbiol*, 68, 3708–3715.

Graninger, M., Nidetzky, B., Heinrichs, D. E., Whitfield, C. & Messner, P. (1999). Characterization of dTDP-4-dehydrorhamnose 3,5-epimerase and dTDP-4-dehydrorhamnose reductase, required for dTDP-L-rhamnose biosynthesis in *Salmonella enterica* serovar Typhimurium LT2. *J Biol Chem*, 274, 25069-25077.

Hsu, C. T., Ganong, A. L., Reinap, B., Mourelatos, Z., Huebner, J. & Wang, J. Y. (2006). Immunochemical characterization of polysaccharide antigens from six clinical strains of Enterococci. *BMC Microbiol*, 6, 62–70.

Kawarabayasi, Y., Hino, Y., Horikawa, H., Jin-no, K., Takahashi, M., Sekine, M., Baba, S., Ankai, A., Kosugi, H. & other authors (2001). Complete genome sequence of an aerobic thermoacidophilic crenarchaeon, *Sulfolobus tokodaii* strain7. *DNA Res*, 8, 123–140.

Kawarabayasi, Y., Sawada, M., Horikawa, H., Haikawa, Y., Hino, Y., Yamamoto, S., Sekine, M., Baba, S., Kosugi, H. & other authors (1998). Complete sequence and gene organization of the genome of a hyper-thermophilic archaebacterium, *Pyrococcus horikoshii* OT3 (supplement). *DNA Res*, 5, 147–155.

Kikuchi, M., Ohnishi, K. & Harayama, S. (2000). An effective family shuffling method using single-stranded DNA. *Gene*, 243, 133–137.

Klenk, H. P., Clayton, R. A., Tomb, J. F., White, O., Nelson, K. E., Ketchum, K. A., Dodson, R. J., Gwinn, M., Hickey, E. K. & other authors (1997). The complete genome sequence of the hyperthermophilic, sulphate-reducing archaeon *Archaeoglobus fulgidus*. *Nature*, 390, 364–370.

Marolda, C. L. & Valvano, M. A. (1995). Genetic analysis of the dTDP-rhamnose biosynthesis region of the *Escherichia coli* VW187 (O7:K1) *rfb* gene cluster: identification of functional homologs of *rfbB* and *rfbA* in the *rff* cluster and correct location of the *rffE* gene. *J Bacteriol*, 177, 5539–5546.

Nakano, Y., Suzuki, N., Yoshida, Y., Nezu, T., Yamashita, Y. & Koga, T. (2000). Thymidine diphosphate-6-deoxy-L-lyxo-4-hexulose reductase synthesizing dTDP-6-deoxy-L-talose from *Actinobacillus actinomycetemcomitans*. *J Biol Chem*, 275, 6806–6812.

Namboori, S. C. & Graham, D. E. (2008). Acetamido sugar biosynthesis in the Euryarchaea. *J Bacteriol*, 190, 2987–2996.

Novotny, R., Schäffer, C., Strauss, J. & Messner, P. (2004). S-layer glycan-specific loci on the chromosome of *Geobacillus stearothermophilus* NRS 2004/3a and dTDP-L-rhamnose biosynthesis potential of *G. stearothermophilus* strains. *Microbiology*, 150, 953–965.

Prakobphol, A. & Linzer, R. (1980). Purification and immunological characterization of a rhamnose-glucose antigen from *Streptococcus* mutans 6517-T2 (serotype g). *Infect Immun*, 30, 140–146.

She, Q., Singh, R. K., Confalonieri, F., Zivanovic, Y., Allard, G., Awayez, M. J., Chan-Weiher, C. C., Clausen, I. G., Curtis, B. A. & other authors (2001). The complete genome of the crenarchaeon *Sulfolobus solfataricus* P2. *Proc Natl Acad Sci USA*, 98, 7835–7840.

Sprott, G. D., Shaw, K. M. & Jarrell, K. F. (1983). Isolation and chemical composition of the cytoplasmic membrane of the archaebacterium *Methanospirillum hungatei*. *J Biol Chem*, 258, 4026–4031.

Suzuki, T., Iwasaki, T., Uzawa, T., Hara, K., Nemoto, N., Kon, T., Ueki, T., Yamagishi, A. & Oshima, T. (2002). *Sulfolobus tokodaii* sp. nov. (f. *Sulfolobus* sp. strain 7), a new member of the genus *Sulfolobus* isolated from Beppu Hot Springs, Japan. *Extremophiles*, 6, 39–44.

Thorson, J. S., Kelly, T. M. & Liu, H. W. (1994). Cloning, sequencing, and overexpression in *Escherichia coli* of the α-D-glucose-1-phosphate cytidylyltransferase gene isolated from *Yersinia pseudotuberculosis*. *J Bacteriol*, 176, 1840–1849.

Tsukioka, Y., Yamashita, Y., Oho, T., Nakano, Y. & Koga, T. (1997). Biological function of the dTDP-rhamnose synthesis pathway in *Streptococcus mutans*. *J Bacteriol*, 179, 1126–1134.

Watt, G., Leoff, C., Harper, A. D. & Bar-Peled, M. (2004). A bifunctional 3,5-epimerase/4-keto reductase for nucleotide-rhamnose synthesis in *Arabidopsis*. *Plant Physiol*, 134, 1337–1346.

Wierenga, R., De Maeyer, M. & Hol, W. (1985). Interaction of pyrophosphate moieties with .alpha.-helixes in dinucleotide-binding proteins. *Biochemistry*, 24, 1346–1357.

Zhang, Z., Tsujimura, M., Akutsu, J., Sasaki, M., Tajima, H. & Kawarabayasi, Y. (2005). Identification of an extremely thermostable enzyme with dual sugar-1-phosphate nucleotidylyltransferase activities from an acidothermophilic archaeon, *Sulfolobus tokodaii* strain 7. *J Biol Chem*, 280, 9698–9705.

Part 4

Biotechnological Applications of Tissue Engineering

13

Experimental Lichenology

Elena S. Lobakova and Ivan A. Smirnov
Moscow State University,
M.V. Lomonosov,
Russia

1. Introduction

The late 19th and, especially, the early 20th century were marked by the introduction of experimental approaches in various biological disciplines. The methods of accumulative and axenic microorganism cultures were already widely used in microbiology of that period; in animal and plant sciences, attempts were made to grow whole organisms, individual organs, tissues and/or individual cells under controlled laboratory conditions (Vochting, 1892; Harrison, 1907). By the early 20th century, some results had already been achieved in cultivating animal tissues (Krontovsky, 1917 cited in Butenko, 1999), and, in the 1920s, plant and animal cells and tissues (Czech, 1927; Prat, 1927; Gautheret, 1932; White, 1932). An important step in plant tissue cultivation was the discovery of phytohormones and development of specialized cultivating media that allowed inducing, on the one hand, dedifferentiation and callus formation, or, on the other hand, cell differentiation. These achievements helped to solve a number of problems, both theoretical and applied (Street, 1977; Butenko, 1999). With time, the spectrum of organisms introduced in cultures was widening, the principal methods of growing plant cells *in vitro* were developed, and the foundations were laid for microclonal propagation.

The said period was also marked by the formation and development of the notion of symbiosis. The revolutionary works of A.S. Famintsyn (1865) and S. Schwendener (1867) (as cited in Famintsyn, 1907) discovered the dual nature of lichens. The notion of symbiosis was formulated in 1879 by A. de Bary. In the early 20th century, K.S. Mereschkowski established the theory of symbiogenetic origin for the eukaryotic cell and formulated the notion of two "plasms" (Mereschkowski, 1907, 1909).

Symbiosis is currently studied by a special scientific discipline, symbiology, and regarded as a stable super-organism system undergoing balanced growth and characterized by specific interrelations of components, and by unique biochemistry and physiology (Ahmadjian & Paracer, 1986; Paracer & Ahmadjian, 2000).

It is noteworthy that the development of each of the above-mentioned fields of study has not been independent. Constantly intervening with each other, works in all these fields were conductive to the formation of a new branch, already within the new science of symbiology. In the 1990s, this new branch was termed experimental symbiology.

2. Specifics of lichens as experimental systems. Peculiarities of the terminology

Lichens are a classic example of symbiotic associations with multicomponent composition as their principal feature. According to the number of partners forming the thallus, two- and three-component lichens are recognized. The former consist of a fungal component (the mycobiont) and a photosynthetic component (the photobiont). In two-component lichens, the photobiont is represented either with a green alga or a cyanobacterium; in three-component lichens, with both: a green alga in the basal part of the thallus and a cyanobacterium in specialized formations, cephalodia (Rai, 1990, Paracer & Ahmadjian, 2000).

According to the type of localization in the lichen, internal (intra-thallus) and external (surface) cephalodia are recognized. In nature, lichens with internal cephalodia are probably prevalent. Some investigators, e.g., P.A. Genkel and L.A. Yuzhakova (1936) (the history of the question is described in: A.N. Oksner, 1974) suggested that nitrogen-fixing bacteria (such as *Azotobacter* spp.) also constitute an obligatory symbiotic component of lichens. Experimental evidence did not support this view (Krasilnikov, 1949). On the other hand, it is currently believed that bacteria are associated, minor symbionts in the lichen system, participating in the morphogenesis of the thallus (Ahmadjian, 1989).

In addition to morphology, lichens as symbiotic systems demonstrate a number of peculiar biochemical and ecological features. Only occasional findings of the so-called lichen compounds in monocultures of lichen symbionts (in most cases, mycobionts) have been reported (Ahmadjian, 1961, 1967). At the same time, large amounts of phenolic compounds (mainly depsides and depsidones), found almost nowhere else, are present in lichens (Culberson, 1969; Vainshtein, 1982a, 1982b, 1982c). The functions of these compounds are not yet fully known. Various compounds probably play different roles in the vital functions of lichens: some participate in the initiation of symbiotic interactions (Ahmadjian, 1989), some provide for the exchange of nutrients between the symbionts (Vainshtein, 1988), and some are used for adaptation to environmental conditions (e.g., in substrate destruction or in competition: Tolpysheva, 1984a, 1984b, 1985; Vainshtein & Tolpysheva 1992; Manojlovic et. al., 2002). Symbiosis helped lichens to become extremely widespread, but they are prevalent in extreme or simply oligotrophic habitats. This probably reflects the fact that lichens are capable of surviving considerable changes of temperature, drying, poor substrates, but at the same time, due to slow growth, it is hard for them to survive competition with higher plants (Paracer & Ahmadjian, 2000).

The multicomponent composition of lichens makes it difficult to use them in biotechnology. Lichens are super-organism multicomponent systems, and we believe that it is necessary to discuss here the terminology used for growing lichens in culture. In English-language literature, the word "culture" is used for laboratory manipulations with lichen thalli and their fragments, but different authors understand this term differently. Taking into account the fact that experimental lichenology developed largely on the basis of approaches borrowed from plant physiology, we believe that it is advisable to define the notion of "culture" more accurately, in the light of the sense this term has in plant physiology, where it means the growing of dedifferentiated parts of an organism on growth media under controlled laboratory conditions (Street, 1977; Butenko, 1999). Lichens have no true tissues, only plectenchymas, and thus the term "tissues" can be used only in quotation marks. We believe that it is possible, by analogy, to use also a number of other terms, such as explants, "callus cultures", etc.

3. History and classification of the experimental approaches in lichenology

Two principal groups of experimental approaches can be recognized in lichenology (Fig. 1, 2): first, resynthesis of lichens and producing of model systems (developed by V. Ahmadjian, M. Galun), and second, obtaining lichen "tissue cultures" (developed by Y. Yamamoto). The former groups of approaches is based on the dissociation of the initial thallus into components (Fig. 1) and subsequent attempt to resynthesize the initial sample or model it based on similar systems (Ahmadjian, 1973a, 1973b). In the latter case, dedifferentiated biomasses of the lichen are obtained (Fig. 2), using thallus fragments, soredia, or isidia (Yoshimura & Yamamoto, 1991; Yoshimura et al., 1993; Yamamoto et al., 1993).

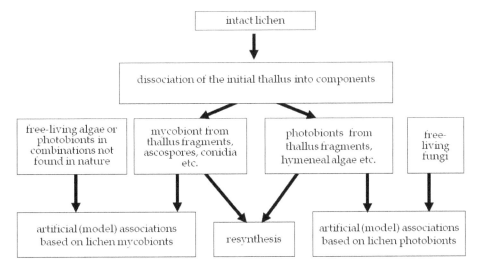

Fig. 1. Main stages of the method of lichen resynthesis.

The former group of approaches reconstructs the general situation emerging in lichens in the course of sexual reproduction, and the latter resembles rather the processes that take place in nature in the course of their vegetative reproduction. For instance, soredia are dedifferentiated parts of the thallus consisting of cells of both symbionts, which makes them very convenient sources for obtaining "tissue cultures". Isidia, morphologically structured thallus fragments, are natural explants of microclonal propagation (Fig. 2). But soredia and isidia are found only in some lichens, which makes obtaining "lichen tissue cultures" from thallus fragments an important approach (Yamamoto et al., 1993; Yoshimura et al., 1993). Interestingly, lichen thalli develop similarly in all cases, independently of the approach used (Ahmadjian, 1973a, 1973b; Yoshimura et al., 1993; Stocker-Worgotter & Turk, 1989).

Attempts to synthesize lichens from separate components (fungi and green algae or cyanobacteria) were made simultaneously with the discovery of the dual nature of lichens by S. Schwendener in 1867. In the same year, A.S. Faminsyn and O.V. Baranetsky, cultivating fragments of lichen thalli, succeeded in describing for the first time the growth of the photobiont outside the thallus. However, almost all attempts to resynthesize a lichen from separate components in the late 19th century were unsuccessful (Bornet, 1873; Bonnier, 1889; critique of these works: Ahmadjian & Jacobs, 1983). The cause of these failures was the lack of both theoretical and experimental foundations laid for such experiments. Successful experiments on lichen construction were performed only three decades later, due to the development of theoretical data on the biology of lichens and laying of methodological foundations for cultivating lichen component monocultures on artificial media.

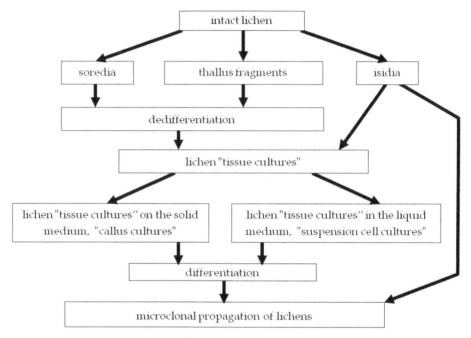

Fig. 2. Main stages of the method of "lichen tissue cultures".

In 1939, E. Thomas was the first to successfully resynthesize the lichen *Cladonia pyxidata* (Paracer, Ahmadjian V., 2000). However, he was unable to reproduce his experiment: none of the 800 subsequent attempts to resynthesize the lichen was successful. The 1950s mark the start of the epoch of experimental lichenology. In different countries, several scientific schools emerged, each of them developing approaches of its own and following its own general direction of studies: in the United States, V. Ahmadjian (resynthesis and construction of artificial associations); in Israel, M. Galun (sensory signal interactions); in Germany, S. Ott and his colleagues (lichen morphogenesis); in Japan, Y. Yamamoto (obtaining and biotechnological application of lichen "tissue cultures"). In the former Soviet Union, methods for isolating photobionts from thalli were developed by E.A. Vainshtein and I.A. Shapiro. During this period, methods for extracting mycobionts and photobionts from thalli were perfected, appropriate media for cultivating symbiont monocultures were developed, and new substrates for reconstructing lichen thalli were proposed (Ahmadjian, 1973a).

Lichenologists became considerably more interested in resynthesizing the thalli of two-component lichens thanks to the classic studies of Ahmadjian (1961, 1967, 1973a, 1973b, 1990). His contribution to this area is hard to overestimate: over 40% of publications on lichen thalli resynthesis were produced by him alone or in collaboration with others. He developed the method for extracting and cultivating the mycobiont from apothecia. Thanks to Ahmadjian's work, thallus resynthesis of some lichen species became common practice. The principal advantages and disadvantages of the methods he proposed are now clear, but while the advantages are currently widely used in experimental studies, the disadvantages have not yet been overcome, restricting the area where this approach is applied (Ahmadjian, 1973b, 1990).

Integrating the first group of approaches and methods used for obtaining plant tissue cultures, Y. Yamamoto (1985) founded another branch of experimental lichenology: cultivation of ground lichen thallus fragments on artificial growth media (Fig. 2). One of the most important advantages of this approach was using vegetative parts of the thallus, which made the approach equally applicable to lichens with different modes of reproduction. In addition, Yamamoto's method allowed obtaining, under controlled conditions and on a tight timetable, a biological system consisting of both components and displaying a number of properties typical of intact lichens (Yoshimura et al., 1993; Yamamoto et al., 1993). Taking into account the fact that under particular conditions for cultivating such systems, thalli morphologically and anatomically similar to the natural ones are formed (Yoshimura & Yamamoto, 1991; Yoshimura et al., 1993), this approach can be used for reproducing lichens *in vitro*.

The interest towards these studies can be illustrated by the fact that the number of lichen species cultivated *in vitro* increased from two in 1985 to 193 in 1993, and the collection of mycobiont and photobiont cultures and "tissue cultures" by that time comprised over 400 lines (Yamamoto et al., 1993) and 400 species (Yamamoto et al., 1995). These included lichens from different climatic and geographical zones, growing in Canada, England, Finland, Israel, Japan, Antarctica, and Malaysia.

4. Lichen resynthesis and producing artificial (model) associations

The method of lichen resynthesis includes the following stages: (1) dissociation of the natural lichen into components; (2) obtaining monocultures of the mycobiont and photobiont; (3) mixed cultivation of the mycobiont and photobiont; and (4) producing a stable morphogenetic association (Fig. 1).

4.1 Lichens used in attempts of resynthesis

Although works on the experimental resynthesis of lichens are in progress since over 70 years ago, the number of species of lichens resynthesized to date remains not too high (Table 1). Experiments were successful with far from all systems, and those lichen species that produced more or less developed thalli in resynthesis experiments, such as *Cladonia cristatella* (Ahmadjian & Jacobs, 1983) and *Xanthoria parietina* (Bubrick et al., 1985; Galun, 1989), were repeatedly used subsequently for solving particular lichenological problems (Fig. 3). Most resynthesized species are crustose two-component forms, and only one, *Stereocaulon vulcani*, is three-component, which probably reflects the complexity of such systems (Table 1).

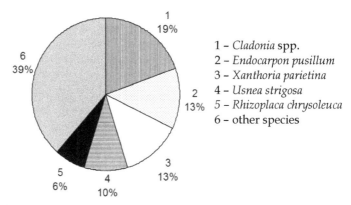

1 – *Cladonia* spp.
2 – *Endocarpon pusillum*
3 – *Xanthoria parietina*
4 – *Usnea strigosa*
5 – *Rhizoplaca chrysoleuca*
6 – other species

Fig. 3. Lichen species used in attempts of resynthesis

Most attempts to reconstruct thalli were made on lichens with green algal photobiont (Fig. 4). Among lichens with cyanobacterial photobiont, successful resynthesis was achieved only in members of the genus *Peltigera*; attempts to resynthesize *Leptogium issatschenkovi* were unsuccessful (Ahmadjian, 1989). Among the almost 30 genera of green algae occurring as photobionts in lichens (Oksner, 1974; Ahmadjian & Paracer, 1986), members of only 10 were used in resynthesis experiments; in 60% of the studies, species of the genera *Trebouxia*, *Desmococcus* (= *Protococcus*) and *Chlorella* were used (Fig. 4).

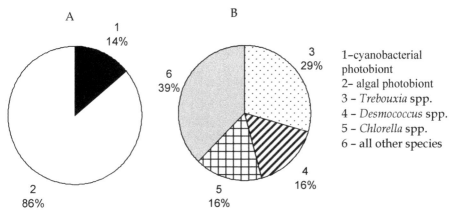

1–cyanobacterial photobiont
2– algal photobiont
3 – *Trebouxia* spp.
4 – *Desmococcus* spp.
5 – *Chlorella* spp.
6 – all other species

Fig. 4. Proportion of resynthesized lichens with different photobionts: A, proportion of cyanolichens (1) and phycolichens (2); B, different genera of green algae

Lichen name	Photobiont genera	Source
Acarospora fuscata	*Trebouxia*	Cited by: Oksner, 1974
Buelia spp.	*Trebouxia*	Cited by: Oksner, 1974
Bacidia bagliettoana	*Chlorella*	Bornet, 1873
Caloplaca decipiens	*Chlorella*	Cited by: Oksner, 1974
Cladonia cristatella	*Trebouxia* or *Chlorella*	Remmer, Ahmadjian, Livdahl, 1986 — cited by: Ahmadjian & Paracer, 1986 ; Ahmadjian, 1967
C. furcata	*Trebouxia* or *Chlorella*	Jahns, 1978 — cited by: Ahmadjian & Paracer, 1986
C. pyxidata	*Trebouxia* or *Chlorella*	Thomas, 1939 — cited by: Oksner, 1974
Collema limosum	*Nostoc*	Rees, 1871 — cited by: Oksner, 1974
Dermatocarpon miniatum	*Hyalococcus* or *Protococcus*	Stocker-Worgotter & Turk, 1989
Endocarpon pusillum	*Myrmecia*	Bertsch, Butin, 1967; Stahl, 1877 — цит по Stocker-Worgotter & Turk, 1988
Graphidaceae g. sp.	*Trentepohlia*	Herriset, 1946 — cited by: Oksner, 1974
Heppia echinulata	*Scytonema,* cyanobacterium	Marton, Galun, 1976 — cited by: Ahmadjian & Paracer, 1986
Huilia albocaerolescens	green alga	Cited by: Ahmadjian & Paracer, 1986
Lecanora spp.	*Protococcus*	Ahmadjian, Russell, Hildreth, 1980; Culberson, Ahmadjian, 1980 — cited by: Ahmadjian & Paracer, 1986
Lecidia spp.	*Trebouxia, Pseudochlorella, Coccobotrys* or *Chlorosarcinopsis*	Cited by: Ahmadjian & Paracer, 1986
Lepraria spp.	*Stichococcus* or *Chlorella*	Cited by: Ahmadjian & Paracer, 1986
Leptogium issatschenkovi	*Nostoc*	Danilov, 1929 — cited by: Oksner, 1974
Peltigeria canina	*Nostoc*	Ahmadjian, 1989
Physia stellaris	*Trebouxia*	Bonnier, 1888
Rhizoplaca chrysoleuca	*Protococcus*	Cited by: Ahmadjian & Paracer, 1986
Rinodina sophodes	green alga	Bonnier, 1888
Sarcogyna spp.	*Myrmecia*	Cited by: Ahmadjian & Paracer, 1986
Stereocaulon vulcani	*Trebouxia* + cyanobacterium	Cited by: Ahmadjian & Paracer, 1986
Staurothele rugulosa	*Protococcus*	Stahl, 1877 — cited by: Stocker-Worgotter & Turk, 1988
Usnea strigosa	green alga	Ahmadjian & Jacobs, 1983, Ahmadjian & Jacobs, 1985
Verrucaria macrostoma	*Coccobotrys* or *Protococcus*	Stocker-Worgotter & Turk, 1989
Xanthoria parietina	*Trebouxia*	Bornet, 1873; Bonnier, 1889 — cited by: Oksner, 1974

Table 1. Lichen species used in attempts of synthesis

The proportion of such studies performed on photobionts from the genus *Trebouxia* is especially high. This is explained by the fact that 90% of lichen species have photobionts belonging to this genus (Oksner, 1974). However, using photobionts of the genus *Tribouxia* in experiments on lichen thallus resynthesis (Ahmadjian, 1973a) did not guarantee success; this is why in 75% of experiments photobionts of other genera were used. In their natural environment, all resynthesized lichen species form sexual reproductive organs (apothecia or perithecia), and most form asexual reproductive organs as well: conidia, pycnoconidia, etc. (Fig. 5A).

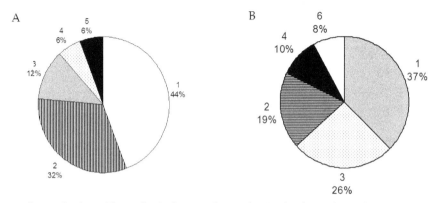

A: 1, sexual reproduction with apothecia; 2, asexual reproduction (with conidia etc.); 3, vegetative reproduction with soredia; 4, vegetative reproduction with isidia; 5, sexual reproduction with perithecia.

B: 1, sexual reproduction with apothecia; 2, vegetative reproduction with soredia; 3, asexual reproduction (with conidia etc.); 4, vegetative reproduction with isidia; 5, sterile species (reproduction lacking or not observed).

Fig. 5. Proportion of different modes of reproduction in resynthesized (A) and cultivated (B) lichens.

Vegetative reproduction structures, soredia and isidia, are found only in three of the reconstructed lichen species. Experimental resynthesis requires a culture of the mycobiont, and Ahmajian (1973a) proposed using spores for extracting it from the naturally growing lichen. This is probably why lichens with mainly vegetative mode of reproduction were seldom used in such *in vitro* experiments.

It has been found that successful resynthesis of a lichen thallus from mycobiont and photobiont monocultures requires experimental conditions imitating the natural environment: the substrate and medium should be poor, so that the lichen does not dissociate into components, and neither the mycobiont or photobiont get any advantages for growth at any stage of development. To induce thallus differentiation in a mixed culture, alternating drying and moistening periods should be imposed, and the drying of biomass should be gradual (Ahmadjian, 1973b).

4.2 Producing artificial (model) associations

Work on the resynthesis of lichens and producing model association has laid the foundation for solving a number of basic problems of the current lichenology. One of these problems is the study of selectivity and specificity of symbiont interaction in lichens. For solving this problem in the course of lichen thallus resynthesis, it is possible to cultivate the mycobiont in pairs with free-living algae or photobionts in combinations not found in nature. In 50% of such experiments, the model systems revealed the lichenization of photobiont cells: the initial contact between the partners of the association took place, and pre-thalli were formed (Tables 2, 3). The same mycobiont showed both positive and negative results with algae of different species of the same genus or with photobionts extracted from the same lichen. The cause of this selective interaction of partners in pairs has not yet been revealed; however, several factors affecting the success of this process were determined, the most important of them being the source of the isolated photobiont and its symbiotrophics (Ahmadjian & Jacobs, 1983).

Lichen from which mycobiont was extracted	Natural photobiont	Potential photobiont	Source of photobiont	Result
Cladonia cristatella	Trebouxia ereci	Friedmannia israeliensis	free-living	+
		Myrmecia sp.	not specified	-
		Nostoc sp.	cycads	-
		Pleurastrum terrestre	free-living	-
		Pseudochlorella sp.	not specified	-
		Pseudotrebouxia sp.	Xanthoria parietina	-
		Trebouxia sp.	Pilophoron sp.	+
		Trebouxia sp.	Stereocaulon sp.	+
		Trebouxia sp.	Gymnoderma sp.	+
		Trebouxia sp.	Lecidia sp.	+
		Trebouxia sp.	Lepraria sp.	+
		Trebouxia sp.	Parmelia sp.	-
		Trebouxia sp.	Xanthoria aureola	-
		Trebouxia italiana	Xanthoria parietina	+
Endocarpon pusillum	Myrmecia biatorellae	Protococcus staurothelis	Staurothele regulosa	-
Glyphis lepida	Trentepohlia	Trentepohlia sp.	Pyrenula nitidula	+
Phaeographina fulganata	Trentepohlia	Trentepohlia sp.	Pyrenula nitidula	+
Staurothele regulosa	Protococcus staurothelis	Myrmecia biatorellae	Endocarpon pusillum	-

Table 2. Model associations based on the mycobiont. Note: +, synthesis continued to prethallus; -, attempts of synthesis were unsuccessful

For instance, species of the genus Trebouxia competent for the mycobiont C. cristatella were, with one exception, derived from lichens taxonomically close to the family Cladoniaceae, while attempts to synthesize thalli with photobionts of the same genus extracted from lichens of distantly related families were unsuccessful. Interestingly, the mycobiont

C. cristatella formed prethallus also with the free-living alga *Friedmannia israeliensis* (Ahmadjian & Jacobs, 1983). The synthesis, however, did not go any further, and, as in all the other cases with constructing lichen-based model systems, the mycobiont killed the algal cells. Unfortunately, these experiments were not continued. It is very likely that some changes in media and/or cultivating conditions would produce a new lichen species: after all, in experiments with compatible cultures of natural symbionts parasitism of the mycobiont on photobiont is also observed under certain conditions.

Potential photobiont	Source of photobiont	Potential mycobiont	Result
Friedmannia israeliensis	free-living	*Cladonia cristatella*	+
Myrmecia sp.	not specified	*Cladonia cristatella*	-
Myrmecia biatorellae	*Endocarpon pusillum*	*Staurothele regulosa*	-
Nostoc sp.	cycads	*Cladonia cristatella*	-
Pleurastrum terrestre	free-living	*Cladonia cristatella*	-
Protococcus staurothelis	*Staurothele regulosa*	*Endocarpon pusillum*	-
Pseudochlorella sp.	not specified	*Cladonia cristatella*	-
Pseudotrebouxia sp.	*Xanthoria parietina*	*Cladonia cristatella*	-
Trebouxia sp.	*Pilophoron* sp.	*Cladonia cristatella*	+
	Stereocaulon sp.	*Cladonia cristatella*	+
	Gymnoderma sp.	*Cladonia cristatella*	+
	Lecidia sp.	*Cladonia cristatella*	+
	Lepraria sp.	*Cladonia cristatella*	+
	Parmelia sp.	*Cladonia cristatella*	-
	Xanthoria aureola	*Cladonia cristatella*	-
Trebouxia italiana	*Xanthoria parietina*	*Cladonia cristatella*	+
Trentepohlia sp.	*Pyrenula nitidula*	*Glyphis lepida*	+
Trentepohlia sp.	*Pyrenula nitidula*	*Phaeographina fulganata*	+

Table 3. Model associations based on the photobiont. Note: +, synthesis continued to prethallus; -, attempts of synthesis were unsuccessful

On the other hand, photobionts isolated from lichen *X. parietina* — *Pseudotrebouxia* sp. and *Trebouxia* sp. — interacted differently with the mycobiont *C. cristatella*: the mycobiont quickly *Pseudotrebouxia* sp., but formed prethallus with *Trebouxia* sp. (Table 2). It has been found that some photobionts (*Trebouxia* spp., *Trentepohlia* spp.) and mycobionts (*Graphidaceae* g. spp.) are better than others (e.g., *Myrmecia* spp.) at forming artificial associations. All the attempts to synthesize model associations using the photobiont *Trentepohlia* sp. and mycobionts extracted from lichens of the family Graphidaceae were successful. This was probably determined, on the one hand, by the greater or smaller resistance of the alga to parasitism and by the degree of mycobiont aggression, and, on the other hand, by the proximity of the potential photobiont in a number of characters (type of vegetative reproduction, etc.) to the natural photobiont. In addition, there are more and less specialized and symbiotrophic genera or species of photobionts and mycobionts.

Unfortunately, there are no data in the literature on model associations based on lichen photobionts and free-living fungi (Table 3). Meanwhile, we believe that this problem is very interesting, since the photobiont, intensely influenced by the mycobiont in the lichen, has developed a number of protective adaptations. These adaptations may be conductive to success in producing model associations of the photobiont and free-living symbiotic fungi.

Another problem is the construction of artificial three-component associations. For solving this problem, a third component, a cyanobiont, can be added to the resynthesized lichen to study the process of cephalodia formation (Ahmadjian, 1989; Ahmadjian & Jacobs, 1983; Ahmadjian & Paracer, 1986). Attempts to model three-component lichens have been made by adding strains of the symbiotrophic and free-living *Nostoc* sp. to two-component lichens with a green alga (e.g., to *C. cristatella*). In all such experiments, the mycobiont parasitized the cyanobiont, and the formation of cephalodia was not observed.

A large field for work in experimental lichenology is opened by the opportunity to obtain protoplasts of the symbionts. The methodology of this procedure for a lichen mycobiont grown from a spore was developed by V. Ahmadjian (1991). Kinoshito (Yamamoto et al., 1993; Kinoshito et al., 2001) modified this procedure to develop a method independent of the way of mycobiont extraction, making it possible to obtain mycobiont protoplasts from vegetative parts of the thallus (experiments on obtaining mycobiont protoplasts were performed only with members of the genus *Cladonia*). Using isolated protoplasts makes it possible to model intracellular interactions of the symbionts. Interestingly, a similar enigmatic organism is known in nature, the fungus *Geosiphon pyriforme*, which hosts inside the cell the cyanobacterium *Nostoc punctatum* (Wolf, Schüßler, 2005).

5. "Tissue cultures" of lichen

The field of experimental lichenology has been developed especially to obtain "callus tissue cultures" of lichens. Currently, lichens cultivated on solid growth media include 52 genera from 22 families, represented mainly by fruticose forms (52%), somewhat fewer foliose forms (36%), and only 12% crustose forms. The vast majority of the species maintained in "callus cultures" are two-component lichens with green algal photobionts. Attempts to introduce three-component lichens to a "callus culture" result in their dissociation into components and formation of "chimeric" forms (see below).

The ratio of different modes of reproduction among cultivated lichens differs from that among reconstructed ones, and there is no pronounced prevalence of forms with sexual or asexual reproduction (necessary for extracting the mycobiont according to the first group of experimental approaches). On the contrary, there is quite a high proportion of lichens with vegetative reproductive structures, soredia and isidia: 36% together, the same as the proportion of lichens with apothecia (37%). Furthermore, the universality of the approach allows cultivating fragments of lichens that form apothecia extremely rarely (Fig. 5B).

5.1 Problems emerging in the course of lichen cultivation

Among difficulties of this experimental approach, the high proportion of explants infected with accompanying microorganisms and the low proportion of thallus fragment germination can be named. In the case of *Usnea rubescens*, growth was observed only in 27%

of the inoculated thallus fragments; in *Peltigera praetextata*, it was observed in 0.4–1.2% cases (maximum and minimum values are given). These parameters are influenced by a number of factors, such as species, habitat, area of the lichen's thallus, number of symbionts (Yoshimura et al., 1993; Yoshimura & Yamamoto, 1991; Smirnov & Lobakova, 2007). The best results are obtained by using soredia of two-component epiphytic fruticose lichens. One possible explanation of this is the fact that 1 g of wet lichen weight contains up to 10^{10} microorganism cells, while in soredia their number is considerably lower (Krasilnikov, 1949); thus, using specialized vegetative reproduction structures (both isidia and soredia) is probably quite promising.

Yoshimura et al. (1993) note that attempts to cultivate fragments of the cyanolichen *P. praetextata* on growth media result in higher levels of infection by contaminant bacteria and myxomycetes, compared to attempts to cultivate explants of lichens with green algal symbionts. This is usually explained in the literature by the fact that symbiotic cyanobacteria have strong, complexly organized surface structures (polysaccharide sheaths), serving as habitats for associated bacteria, complicating the process of obtaining sterile cyanolichen explants (Yoshimura & Yamamoto, 1991; Gusev & Mineeva, 1992). However, in our experiments, in spite of the higher affection level actually observed in fragments of cyanolichens cultivated on growth media, compared to other kinds of lichens, cyanobiont isolates extracted from cyanolichen thallus fragments contained fewer accompanying bacteria than green algal isolates from the same lichens (Smirnov & Lobakova, 2007).

One drawback of the method for obtaining lichen "tissue cultures" described in the literature is the total lack of any primary sterilization of explants. Although some studies refer to unsuccessful attempts to sterilize thallus fragments with mercuric chloride (Ahmadjian, 1989), our data (Smirnov & Lobakova, 2007) demonstrate the efficiency of consistent complex usage of "mild" sterilizing agents (such as hydrogen peroxide, alcohol, chlorhexidine).

In our opinion, works aimed by obtaining "suspension cell cultures" of lichens are especially promising. Experiments in this branch of lichenology are rare and fragmentary, while maintaining symbiont monocultures in liquid media are currently a common practice (Ahmadjian & Jacobs, 1983). In "suspension cultures" of lichens, thalli are not formed; this is why in most studies aimed at obtaining structured lichen thalli, even if thallus fragments were inoculated into liquid growth media, the experiments were never completed, because provisional results did not comply with the initial aims. On the other hand, obtaining "suspension cultures" of lichens is of considerable interest for biotechnology, since methods based on "suspension cultures" are easier to introduce into the industry. Thus, this mode of growth will probably allow obtaining large amounts of dedifferentiated cell biomass and using it in biotechnology as sources of lichen compounds.

A promising approach for obtaining both "suspension cultures" and "callus cultures" of lichens involves using "nurse cultures" (Smirnov & Lobakova, 2008) and conditioning (processing) the cultivating media with metabolites of associated organisms. Data available in the literature give evidence that the initiation of symbiont growth (especially of the mycobiont) requires compounds secreted by the photobiont or associated bacteria. It has been found that the development of the mycobiont was quicker and more intense if (1) non-sterile tree bark was used as the substrate; (2) the cultivating medium was conditioned by metabolites of the photobiont or bacteria; (3) the spores were infected by bacteria (Ahmadjian, 1989; Yolando et al., 2002; Smirnov, 2006). Our results show that conditioning the media with simple metabolites (after sterilization) is inefficient, compared to using native metabolites (dialysis cultivation).

5.2 Study of lichen morphogenesis

A special place among experimental works in lichenology is occupied by the branch involving the study of lichen thallus morphogenesis and revealing the factors influencing this process. A number of studies have addressed the problem of inducing morphogenesis in "callus cultures" of soredia, both in the laboratory, and in the natural environment (Stocker-Worgotter & Turk, 1988; Stocker-Worgotter & Turk, 1989; Yoshimura & Yamamoto, 1991; Armaleo, 1991; Yoshimura et al., 1993). Comparison of natural lichen thalli with those obtained by inducing morphogenesis in *in vitro* systems demonstrates their anatomical and morphological similarity; the same layers are formed: upper cortex (in some species, lower cortex), photobiont layer, medulla. One drawback of this approach is the fact that non-homogeneous material, e.g. in the shape and size of scales, is often formed in the laboratory, probably because of the heterogeneity of various thallus parts caused by the parasexual process (Stocker-Worgotter & Turk, 1988) or by somatic variation (Street, 1977; Butenko, 1999).

Lichens with different modes of reproduction (sexual, asexual and vegetative) under laboratory conditions with morphogenesis induction undergo the same stages of development (lag phase, arachnoid phase, prethallus and thallus: Ahmadjian, 1973a, 1973b; Ahmadjian & Jacobs, 1983; Stocker-Worgotter & Turk, 1989; Yoshimura et al., 1993) as in nature (Ott, 1988). The only difference is the duration of particular stages, depending on the type of the explant and conditions of its cultivation. In *P. didactyla*, thallus develops quicker than in other species (especially at the final stages). It has been found (Stocker-Worgotter & Turk, 1988), that using soredia as explants is conductive to the quick (2–4 times quicker) formation of thallus *in vitro*; however, rates of morphogenesis as high as in nature have not yet been achieved under laboratory conditions.

The experimental approaches in lichenology described here are currently used for solving a number of basic problems like those persistent in biotechnology. The former approaches include studying the ecological and morphological plasticity of lichens and revealing differentiation factors of thalli and the share of each partner in the formation of the unique super-organism system.

In this respect it is especially interesting to study the development of the "tissue cultures" of three-component lichens, such as *Peltigera aphthosa*, with a green alga as the photobiont and a cyanobiont in cephalodia. In "tissue cultures" of *P. aphthosa*, explants often formed a homoiomerous cyanolichen, and the green alga was expelled from the association and remained in the culture as free-living colonies. The drying of the system increased the number of green algal colonies, and they were included into the composition of non-differentiated mixed aggregates. The slightly raised and drying areas of the homoiomerous cyanothallus became colourless (the cyanobiont disappeared) and were gradually colonized by the green alga, while in other areas, which preserved contact with the substrate, cyanobacteria were preserved. These areas resembled primordia of new green lobes; however, no further development was observed. Due to the difficulties of moisture control, normal thalli did not form in the experiments; the formation of cephalodial primordia was, nevertheless, observed.

The phenomenon described provides an experimental confirmation of the idea that mycobionts can include several morphotypes (as analysis of their DNA has also shown), and/or the formation of chimeric lichens is possible. While the existence of such chimeras was earlier considered unproven, now the reality of this phenomenon has been confirmed both by some field studies (for review, see: Plyusnin, 2002) and by laboratory experiments.

Morphogenetic "tissue cultures" of lichens are convenient experimental models for the study of this phenomenon. The results of using them allow us to state that the formation of a particular morphotype or chimeric lichen depends on moisture. For instance, these results allow suggesting that the cyanobacterial morphotype is more widespread than has been believed earlier and unidentified species of the genus *Peltigera* with cyanobacteria often represent one of the morphotypes of three-component lichens (Yoshimura et al., 1993).

It can be assumed that experimental approaches will also play an important role in the molecular biology of cyanolichens: they will allow studying the exchange of genes, inferred by some authors, between the symbionts by means of plasmids in the course of morphogenesis (Ahmadjian, 1991). Reviewing the data available in the literature has shown that for studying the early stages of lichen thallus morphogenesis, it is better to use methods of resynthesis, while for the study of specificity and selectivity of interactions between components of this symbiosis, as well as of different stages of thallus differentiation, the "tissue culture" and morphogenesis induction methods are more suitable.

Dedifferentiated mixed cellular aggregates of a "callus culture" of lichens can be used in the study of the genetic control over symbionts in the course of the formation of a balanced super-organism system (Yamamoto et al., 1993; Yoshimura et al., 1993).

5.3 The biotechnological potential of lichen "tissue cultures"

Using experimental approaches is promising also for producing from lichens their unique secondary metabolites, the lichen compounds. The biosynthesis of lichen compounds in "tissue cultures" is usually no different from that in the natural thallus in the composition of depsides, tridepsides, and depsidones; triterpenoid compounds are, however, a more labile class of substances, and in "callus cultures" of lichens they often disappear (Table 4).

In most cases, the concentration of lichen compounds in a "culture" is considerably lower than in a natural thallus: the content of the usnic acid in *Usnea rubescens* is 0.9% in the natural state and 0.162% in a "callus culture", i.e., five times higher; in *Ramalina yasudae*, it is even 100 times higher (Yamamoto et al., 1985). But since "tissue cultures" of some lichen species grow considerably quicker (their biomass increases at least by a factor of 5 over 14 weeks), using the Yamamoto method for industrial production of lichen compounds (Yamamoto et al., 1985; Yamamoto et al., 1993) is very promising. Importantly, using "tissue cultures" of lichens, we can decrease the number of lichens that are removed from their natural environment, and extremely slowly regenerating in nature.

Class of compounds	Compounds	*Usnea strigosa*		*Usnea rubescens*		*Ramalina yasudae*		*Peltigera pruinosa*		*Peltigera aphthosa*	
		t	r	t	c	t	c	t	c	t	c
depsides and depsidones	globin acid	+	-								
	connorsticic acid	-	+								
	cryptostictic acid	-	+								
	methyl lecanorate			-	-	-	-	-	-	+	+
	norstictic acid	+	+								
	protocetraric acid			+	+	+	+	-	-	-	-
	salazanic acid			+	-	+	-	-	-	-	-
	usnic acid	+	+	+	+	+	+	-	-	-	-
	fumaroprotocetraric acid	+	-								
	evernic acid			+	-	+	-	-	-	-	-
tridepsides	methyl gyrophorate			-	-	-	-	+	+	+	+
	tenuiorin			-	-	-	-	+	+	+	+
triterpenoids	dolichorrhizin			-	-	-	-	+	-	-	-
	zeorin			-	-	-	-	+	-	-	-
	phlebeic acid			-	-	-	-	-	-	+	-

Table 4. Comparison of lichen compound production by "tissue cultures", resynthesized thalli, and natural thalli, from: Ahmadjian & Jacobs, 1983; Yamamoto et al., 1985; Yoshimura & Yamamoto, 1991. Note: +, compound present; -, compound not found; t, compound extract from natural thallus; c, from resynthesized thallus; c, from "tissue culture".

The expediency of using lichen "tissue cultures" for obtaining biologically active compounds is also supported by the fact that their methanol and acetone extracts demonstrate a levels of superoxide dismutase activity, and have antibacterial (against Gram-positive bacteria: Fig. 6) and antiviral (when EBV test system is used: Fig. 7) effects (Yamamoto et al., 1993; Yamamoto et al., 1995).

The degrees of antibacterial and antiviral activities strongly vary between different lichens, even among species of the same genus (Fig. 7). In most cases, the inhibitory action of extracts of natural thalli is higher than that of "tissue culture" extracts; there are, however, some exceptions: laboratory extracts of *Cladia aggregata* and *Evernia prunastri* displayed higher levels of activity than extracts of their natural thalli. Interestingly, "tissue cultures" of lichens of the genera *Cetraria*, *Evernia* and *Cladonia*, the extracts of which demonstrated considerable levels of antiviral activity, had no antibacterial effect.

RI

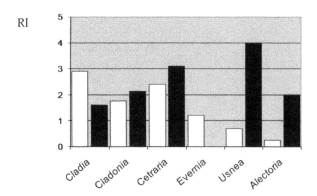

Fig. 6. Antiviral activity of extracts from thalli and "tissue cultures" of lichens (on the base: Yamamoto et al., 1993, 1995). EBV test system was used. RI, ratio of CV in experiments with particular lichen extract and CV in control samples; CV(cell viabilility), percentage of surviving cells 48 hours after the start of the experiment.

On the other hand, "tissue cultures" of lichens of the genera *Usnea*, *Umbilicaria* and *Ramalina*, which strongly inhibited the growth of Gram-positive bacteria, poorly inhibited viral growth in a EBV test system (Fig. 7). One exception was the "tissue culture" of *Cladia aggregata*, which demonstrated considerable activity in both cases.

AA

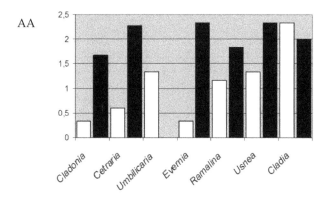

Fig. 7. Antibacterial effect of extracts from thalli and "tissue cultures" of lichens (on the base: Yamamoto et al., 1993). Antibacterial activity (AA) is given in relative units. Tests were performed on the species *Propionibacterium acnes*, *Staphylococcus aureus*, *Bacillus subtilis*.

Interestingly, the concentration of lichen compounds in reconstructed lichen thalli is often higher than in nature; Ahmadjian and Jacobs (1985) explain this by the more favourable conditions for lichen development formed in the course of resynthesis. It is noteworthy that producing artificial associations, with symbiont combinations not found in nature, can be used as a promising source of new antibiotic compounds. The possibility of this application is demonstrated by the two novel compounds, not typical of this species in nature, found in the thallus of *Usnea strigosa* in the course of resynthesis (Table 4). The biotechnological application of this approach for producing lichen compounds is currently restricted by the low rate of the system's growth, surmountable in the future by optimizing cultivation methods.

A special place among the problems of current lichenology is occupied by the conservation of rare lichen species and their re-introduction into the natural environment. The above-described experimental approaches can be used, among other purposes, for solving these problems. Methods of rare species gene pool conservation in collections and cryobanks are well-developed for higher plants (Street, 1977; Butenko, 1999). Some authors (Tolpysheva, 1998) believe that it would be useful to apply this experience to lichens as well.

6. Conclusion

Among experimental approaches in lichenology, two groups of methods can be recognized: lichen resynthesis and cultivation. The former approach helped to find the answers to many questions of lichen biology, but currently it faces a number of insoluble problems (e.g., the failure of attempts to produce mature spores in sporocarps), due to which the number of studies on lichen reconstruction has considerably decreased (Ahmadjian, 1990). The latter approach is promising for introducing lichens into the field of biotechnological developments. However, this is largely hindered by the low yield of lichen biomass in the course of cultivation. Two principal causes of this can be named: the considerable level of infection with fungi and bacteria (Yamamoto et. al, 2004) and the insufficiently quick growth of the culture of the lichen itself. The solution to the problem of "explant" infection with contaminant species may be found in surface sterilization of lichens, similar to that used in plant physiology (Smirnov & Lobakova, 2007). The solution to the problem of culture growth acceleration may be found in conditioning the media with secondary metabolites of various origins. The analysed literature contained no mentions of using "nurse cultures", a method widely used in plant physiology, considerably increasing the rate of growth in cultures (Street, 1977; Butenko, 1999; Butenko et al., 1987). At the same time, a number of authors have shown that secondary metabolites, both of associated fungi and algae, extracted from lichens (Vainshtein, 1988), and of accompanying fungi and algae (Ahmadjian, 1989), can accelerate growth in cultures of isolated symbionts, both mycobionts and phycobionts. Another way of accelerating the growth of cultures, both of the symbionts and of the lichen as a whole, may be found in using suspension cultures. Conditioning of media and suspension cultures can also be useful in the first group of experimental approaches, especially in producing model associations based on lichen photobionts (according to the literature, in most cases it was the mycobiont that served as the basis for novel associations).

7. Acknowledgments

The authors are grateful to Yu.T. Dyakov for the idea to write a paper on this subject, to A.K. Eskova for useful discussions and to P.N. Petrov for his invaluable help in the English text of the manuscript.

8. References

Ahmadjian V. & Jacobs J. B. (1985). Artificial reestablishment of lichens IV. Comparison between natural and synthetic thalli of *Usnea strigosa*. *Lichenologist* 17: 149 – 165.

Ahmadjian V. & Jacobs, J. B. (1983). *Algal-fungal relationships in lichens: recognition, synthesis, and development.* — In Goff. L. J., (Ed.): Algal symbiosis, pp. 147 – 172. — Cambridge: Cambridge University Press.

Ahmadjian V. & Paracer S. (1986). *Symbiosis in introduction in biological association.* Clark University Press. pp. 14 – 36.

Ahmadjian V. (1961). Studies on lichenized fungi. *The Bryologist* 64: 168 – 179.

Ahmadjian V. (1967). *The Lichen Symbiosis.* Blaisdell, Waltham, MA. 152 pp.

Ahmadjian V. (1973a). Methods of isolation and culturing lichen symbionts and thalli (pp. 653 – 660). In: Ahmadjian V, Hale M. E. (eds) *The Lichens.* Academic Press, New York.

Ahmadjian V. (1973b). Resynthesis of lichens, pp. 565 – 579. In V. Ahmadjian & M. E. Hale (eds.), *The Lichens.* New York and London.

Ahmadjian V. (1989). Studies on the isolation and synthesis of bionts of the cyanolichen *Peltigeria canina* (Peltigeraceae). *Pl. Syst. Evol.* 165: 29 – 38.

Ahmadjian V. (1990). What have synthetic lichens told us about real lichens? *Bibl. Lichenol.* 38: 3 – 12.

Ahmadjian V. (1991). Molecular biology of lichens: a look to the future. *Symbiosis.* 11: P. 249 – 254.

Armaleo D. (1991). Experimental microbiology of lichen. *Symbiosis.* 11: P. 163 – 178.

Bertsch & Butin (1967). Die Kultur der Erdflechte *Endocarpon pusillum* im Labor. *Planta* 72: 29-42.

Bonnier, G (1888). Germination des spores des lichens sur les protonemas des mousses et sur des algues differentes des gonidies du lichen. *Compt. Rend. Soc. Biol.* Paris 40: 541-543.

Bornet J.-B.-E. (1873). Recherches sur les gonidies des Lichens. *Annal, d. se. nat.,* 5 c, V. XVII.

Bubrick, P.; Frensdorff A. & Galun M. (1985): Proteins from the lichen *Xanthoria parietina* (L.) Th. Fr. which bind to phycobiont cell walls: isolation and partial purification of an algal-binding protein. - *Symbiosis* 1: 85 - 95.

Butenko R.G. (1999). [*Biology of plant cells in vitro and biotechnologies based on them*]. Moscow: FBK-press, 159 p.

Butenko R.G.; Gusev M.V.; Kirkin A.F.; Korzhenevskaya T.G. & Makarova E.N. (1987). [*Biotechnology*]. Book 3. Moscow: Vyshaya Shkola, 127 p.

Culberson C.F. (1969). *Chemical and botanical guide to lichen products.* Chapel Hill: University of North Carolina, 628 p.

Czech H. 1927. Kultur von pflanziichen Gevebezellen. *Arch. Expti. Zeilforsch.* Bd. 3. S. 176 – 200.

Famintsyn A.S. (1907). [On the role of symbiosis in the evolution of organisms] *Zap. Imperatorskoy Akademii Nauk.* Ser. 8. V. 20. No. 3. P. 15-39.

Galun M. (1989). *CRC Handbook of Lichenology.* M. Galun (ed.). - Vol. 1. CRC Press Inc., Boca Raton, Florida, 1989. - 297 p.

Gautheret R. (1932). Sur la culture d'extremites de racines. Compt. *Rend Soc. Biol.* t. 109, P. 1236 – 1238.

Gusev M.V. & Mineeva L.A. (1992). [*Microbiology.*] Moscow: MSU, 448 p.

Harrison R. (1907). Observation on the living developing nerve fiber. *Proc. Soc. Expit. Biol. Mag.* v. 4, P. 140 – 143.

Kinoshito Y.; Yamamoto Y.; Kurokawa T. & Yoshimura I., 2001 Influence of nitrogen source on usnic acid production in a cultured mycobiont of lichen *Usnea hirta* (L.) Wigg. *Bioscience, biotechnology, biochemistry.* 65 (8): 1900 – 1902.

Komine, M.; Iwasaki, Y.; Yamamoto, Y. & Hara, K. (2004). Developing a suitable growth substrate for lichen forced cultivation under an artificial environment. *Lichens in focus – IAL 6.*

Krasilnikov N.A. (1949). [Lichen microflora.] *Mikrobiologia.* V. 18. No. 3. P. 3 – 24.

Manojlovic N. T.; Solojuc S. & Sukdolak S. (2002). Antimicrobial activity of an extract and antraquinones from *Caloplaca shaeveri Lichenologist.* Vol. 34. N. 1. P. 83 – 85.

Mereschkowski K.S. (1907). [*The laws of endochrome.*] Doct. Sci. Dissertation. Kazan: Kazan Imperial University. 402 pp.

Mereschkowski K.S. (1909). [Theory of two plasms as the foundation of the symbiogenesis theory, a new doctrine on the origin of organisms] *Uch. zap. Kazanskogo un-ta.* V. 76. 97 p.

Oksner A.N. (1974). [*Guide to the lichens of the USSR. Issue 2. Morphology, systematics and geographical distribution.*] Leningrad: Nauka, 283 p.

Ott S. (1988). Photosymbiodemes and their development in *Peltigera venosa.* Lichenologist 20: 361-368. *Pl. Syst. Evol.* 165: 29-38.

Paracer S. & Ahmadjian V. (2000). *Symbiosis: An Introduction to Biological Associations* (Oxford Univ. Press, Oxford, 2nd ed.). ISBN 0-195-11806-5, 261 p.

Plyusnin S.N. (2002). [Intrathallic variation of lichens] *Vestnik Instituta biologii Komi NTs UrO RAN.* No. 53. P. 15–16.

Prat S. (1927). The toxity of tissue juices for cells of the tissue. *Amer. J. Bot.* 14: 121.

Rai A.N. (1990). Cyanobacterial-fungal symbioses: the cyanolichens. – In Handbook of Symbiotic Cyanobacteria. Rai A.N. ed. pp. 9 – 41. – CRS Press, Boca Raton: Florida. USA.

Smirnov I.A. & Lobakova E.S. (2007). [Peculiar features of lichen photobiont cultivation] *Fundamentalnye i prikladnye aspedky issledovaniya simbioticheskikh system. Materials of All-Russia Conference with International Participation.* Saratov: Nauchnaya Kniga. P. 32.

Smirnov I.A. & Lobakova E.S. (2008). [Morphophysiological description of a mixed cultures of *Pleurotus ostreatus* and the nitrogen-fixing cyanobacteria *Anabaena variabilis*] *Vyshshie bazidialnye griby: individuumy, populyatsii, soobshchestva. Materials of Conference on the Centenary of M.V. Gorlenko.* Moscow: Vostok-Zapad. p. 198–199.

Smirnov I.A. (2006). [Micromycetes associated with *Cetraria islandica*] *Materials of XIII International Conference "Lomonosov – 2006"* Moscow: MAKS press. P. 211.

Stocker-Worgotter E. & Turk R. (1988). Culture of the cyanobacterial lichen from soredia under laboratory conditions. *Lichenologist* 20: 369-375.

Stocker-Worgotter E. & Turk R. (1989). Artificial cultures of lichen *Peltigera didactyla* in natural environment. *Plant Systematics and Evolution* 165, 39-48.

Street H.E. (ed.) (1977). *Plant Tissue Culture.* Botanical Monographs. 11. Blackwell Scientific Publications, Oxford, London, Edinburg, Melbourne.

Tolpysheva T.Yu. (1984a). [Effect of extracts from lichens on fungi. 1. Effect of water extracts of *Cladina stellaris* and *C. rangiferina* on the growth of soil fungi] *Mikologiya i Phitopatologiya.* T. 18. No. 4. P. 287–293.

Tolpysheva T.Yu. (1984b). [Effect of extracts from lichens on fungi. 2. Effect of integrated preparations from *Cladina stellaris* and *C. rangiferina* on the growth of soil fungi] *Mikologiya i Phitopatologiya*. V. 18. No. 5. P. 384–388.

Tolpysheva T.Yu. (1985). [Effect of extracts from lichens on fungi. 3. Effect of usnic acid and atranorin on the growth of soil fungi] *Mikologiya i Phitopatologiya..* V. 19. No. 6. P. 482–489.

Tolpysheva T.Yu. (1998). [*Red data book of Moscow Oblast (lichens).*] Moscow: Argus; Russky Universitet. P. 501–514.

Vainshtein E.A. & Tolpysheva T.Yu. (1992) [Effect of extract from the lichen *Hypogymnia physodes* (L.) Nyl. and of pure lichen acids on wood-rotting fungi] *Botanichesky Zhurnal*. V. 26. No. 6. P. 448–455.

Vainshtein E.A. (1982a). *Lichen compounds of secondary origin*. P. 1. Leningrad: Deposited in VINITI, nos. 210–83. p. 1–238.

Vainshtein E.A. (1982b). *Lichen compounds of secondary origin*. P. 2. Leningrad: Deposited in VINITI, nos. 210–83. p. 239–485.

Vainshtein E.A. (1982c). *Lichen compounds of secondary origin*. P. 3. Leningrad: Deposited in VINITI, nos. 210–83. p. 486–717.

Vainshtein E.A. (1988). [*Lichen symbiosis and physiological and biochemical regulation of the interactions between the fungal and algal components.*] Extended Abstract of Doct. Sci. Dissertation. Leningrad., 45 p.

Vochting H. (1892). *Uber transplantation am Pflanzenkorper. Untersuchungen zur Physiologic und Pathologie*. Tubingen.

White Ph. (1932). Influence of some environmental condition on the growth of excised root tips of wheat seedlings of liquid media. *Plant Physiolol*. v. 4, P. 613 – 628.

Wolf E. & Schii.ler A. (2005). Phycobiliprotein fluorescence of *Nostoc punctiforme* changes during the cycle and chromatic adaptation: characterization by spectral CLSM and spectral unmixing *Plant, Cell and Erivironment*. V. 2. P. 480-491.

Yamamoto Y.; Miura Y.; Higuchi M.; Kinoshita Y. & Yoshimura I. (1993). Using lichen tissue cultures in modem biology. *The Bryologist* 96: 384 393.

Yamamoto Y.; Miura Y.; Kinoshita Y.; Higuchi M.; Yamada Y.; Muracami A.; Ohigashi H. & Koshimizu K. (1995). Screening of tissue cultures and thalli of their active constituents for inhibition of tumor promoter-induced Epstein-bar virus activation. *Chem. Pharm. Bull.* 43 (8), 1388 – 1390.

Yamamoto Y.; Mizuguchi R. & Yamada Y. (1985). Tissue cultures of *Usnea rubescens* and Ramalina yasudae and production of usnic acid in their cultures. Agricultural Biological Chemistry 49: 3347 – 3348.

Yamamoto Y.; Takeda M.; Hara K.; Komine M.; Inamoto T.; Kawakatsu, M. & Miyagawa H., (2004). Screening for antibacterial activities and isolation of antibiotics from mycobiont cultures. *Lichens in focus – IAL 6*.

Yolando et al., (2002). Bioprodaction of lichens phenolics by immobilized lichen cels with emphasis on the role of epiphytic bacteria. *J. Hattori Bot. Lab.* №92, 245 – 260

Yoshimura I. & Yamamoto Y. (1991). Development of *Peltigera praetextata* lichen thalli in culture. *Symbiosis*. 11: P. 109 – 117.

Yoshimura I.; Kurokawa T.; Yamamoto Y. & Kinoshita Y. (1993). Development of lichen thalli *in vitro*. *Bryologist* 96: 412 – 421.

14

Magnetic Particles in Biotechnology: From Drug Targeting to Tissue Engineering

Amanda Silva[1*], Érica Silva-Freitas[2*], Juliana Carvalho[2], Thales Pontes[2],
Rafael Araújo-Neto[2], Kátia Silva[2], Artur Carriço[2] and Eryvaldo Egito[2]

[1]*Université Paris 7,*
[2]*Universidade Federal do Rio Grande do Norte,*
[1]*France*
[2]*Brazil*

1. Introduction

Iron oxide nanoparticles are responsible to magnetic field allowing them to be manipulated, tracked, imaged and remotely heated. Such key features open up a wide field of applications in medicine which includes cell separation, magnetic force-based tissue engineering, MRI tracking of transplanted cells, magnetic drug targeting and hyperthermia.

In most applications reported in the literature, magnetic systems are typically composed of an inorganic core and an organic coating. Although cores have been made from different materials, iron oxide nanoparticles constituted of magnetite (Fe_3O_4) and maghemite (γ-Fe_2O_3) are used at a great extent. While the core provide nanocontainers with magnetic properties, the shell functions to (i) protect against core agglomeration, (ii) provide chemical handles for the conjugation of drug molecules, and (iii) limit opsonization. Additionally, shell coatings have been engineered to enhance pharmacokinetics and tailor in vivo fate. Organic shells main comprise phospholipid bilayered membranes or polymeric coating of dextran, for instance. Magnetic system design with such different materials can be achieved via a number of approaches, including in situ coating, post-synthesis adsorption and end-grafting. In fact, several methods have been proposed for their synthesis, coating, and stabilization, mainly comprising the precipitation route together with a surface functionalization step by means of polymers or surfactants. This point will be the focus of the next chapter section – "Producing magnetic particles."

Once produced, these magnetic carriers must meet certain criteria for use in the human body. For therapeutic purposes, magnetic carriers must be water-based, biocompatible, biodegradable, and nonimmunogenic. Besides, special care should be focused on the particle size, surface properties, magnetic properties, and administration route, as will be discussed in the third chapter section, entitled "Magnetic particles: concerns towards *in vivo* use."

The fourth chapter section comprises the applications of magnetic particles in the field of biotechnology. They can be divided into therapeutic and diagnostic ones. Chapter subsections will focus on both. Also discussed is a novel application of magnetic

---

* These authors contributed equally to this work

nanoparticles – the use of magnetic force for tissue engineering, termed "magnetic force-based Tissue Engineering (Mag-TE)." Since cells labeled with magnetic nanoparticles can be manipulated using magnets, this novel tissue engineering methodology using magnetic force and functionalized magnetic nanoparticles may hold great promise in reproducing *in vitro* patterned tissues for organ regeneration.

The fifth and last section of this chapter provides concluding remarks while addressing future perspectives in regard to magnetic particles in biotechnology.

2. Producing magnetic particles

2.1 Synthesis of magnetic carriers

In most applications reported in the literature, iron oxides, such as magnetite and maghemite, are the magnetic material of choice. The synthesis, coating, and stabilization of such particles will be discussed below. The most common synthetic route to produce magnetite (Fe_3O_4) is the coprecipitation of hydrated divalent and trivalent iron salts in an alkaline medium (A. K. Silva et al., 2008).

Nanoreactors can be employed for the precipitation reaction. They provide a constrained domain, which limits the growth of the particles. This method offers numerous advantages over the previous ones when higher homogeneity of size and shape are concerned. A discussion of these follows.

Microemulsions are colloidal nano-dispersions of water in oil (or oil in water) stabilized by a surfactant film. The synthesis of magnetic particles by this means is carried out when water droplets interact and exchange their contents. Experimental results have confirmed that the microemulsion method allows good control of the particles by preventing their growth and providing particles small enough to get stable magnetic fluids. On the other hand, magnetic particles prepared by coprecipitation may undergo aggregation. Microemulsions, which are thermodynamically stable dispersions, can be considered as truly nanoreactors that can be used to carry out chemical reactions and, in particular, to synthesize nanomaterials. The main idea behind this technique is that by appropriate control of the synthesis parameters, these nanoreactors can produce smaller and more uniform particles than the ones produced by other standard methods. Particle size was found to depend on the molar ratios of water and surfactant (Lopez-Quintela, 2003).

Liposomes are also used as nanoreactors for the precipitation as they provide a constrained domain, which limits the growth of the particles. Alternatively, encapsulation of magnetic particles into liposomes may be performed after synthesis (A. A. Kuznetsov et al., 2001). Magnetoliposomes have been found to be a promising approach that offers some unique advantages when the magnetic nanoparticles are applied in biological systems. Lipid systems present the advantage of low toxicity due to their composition, mainly physiological lipids, compared to the polymeric particles. In fact, encapsulation of the magnetic nanoparticles in liposomes increases their biocompatibility under physiological conditions, making them suitable for a large variety of biological applications. Furthermore, it is known that magnetic particles tend to agglomerate, and are chemically unstable with respect to oxidation in air. Encapsulation of the magnetic nanoparticles in liposomes protects them from aggregation and oxidation (Heurtault et al., 2003).

Concerning the production of polymer-based magnetic carriers, three different methods may be used. The emulsification/polymerization method has been successfully employed to produce magnetic microcapsules. In this process, particles are synthesized in the internal aqueous phase of an inverse emulsion/microemulsion. Afterwards, polymerization by a cross-linking agent takes place. In such microcapsules, the drug and the magnetic particles are in the inner compartment (Saravanan et al., 2004). Alternatively, polymer-covered magnetic particles can be produced by in situ precipitation of magnetic materials in the presence of a polymer that acts as a stabilizer.

Magnetic polymer nanoparticles have been produced in the presence of water-soluble dextran, poly (vinyl alcohol), sodium poly (oxyalkylene di phosphonates), carboxymethyl starch, and amylose starch, just to name a few. In all cases, magnetic particles are surrounded by a hydrophilic polymer shell. Such systems are functionalized by the introduction of chemical groups so that they are able to bind active molecules. For instance, dextran-coated magnetic particles, which are highly hydrophilic, uniform, and nontoxic magnetic carriers, may be activated by the periodate oxidation method. Thus, magnetic polyaldehyde dextran is formed and may be conjugated to different molecules (Hong et al., 2004).

Another method for producing magnetic polymer particles consists of separately synthesizing magnetic particles and polymer particles and then mixing them together to enable either physical or chemical adsorption of the polymer onto the magnetic material to be achieved (E. L. Silva et al., 2009).

3. Magnetic particles: Concerns towards *in vivo* use

For therapeutic purposes, magnetic carriers must be water-based, biocompatible, biodegradable, and nonimmunogenic (Häfeli, 2004; A. K. Silva et al., 2007a). Concerning the *in vivo* use, the following parameters are critical: (a) particle size, (b) surface characteristics of the particle, (c) concentration of the fluid, (d) volume of the fluid, (e) administration route, (f) duration/rate of the injection/infusion, (g) geometry and strength of the magnetic field, (h) duration of magnetic field application, (i) particle stability, and (j) magnetic properties. Physiological parameters of the patient organism are also important. They comprise: a) size, weight, and body surface, (b) blood volume, (c) cardiac output and systemic vascular resistance, d) circulation time, (e) tumor volume and location, (f) vascular content of target area, and (g) blood flow in that area (A. A. Kuznetsov et al., 1999; Lübbe et al., 1999).

Markedly, size is a crucial factor. Large microspheres can physically irritate the surrounding tissue or even embolize small blood vessels and capillaries. Besides, stable suspensions of dense particles larger than 2µm are rarely prepared, and it is difficult to inject suspensions of such particles through a catheter. On the other hand, very small particles (less than 0.1 µm in diameter) have a small magnetic moment. In such a case, magnetic forces may not be high enough to counteract the linear blood-flow rates in the tissue. As a consequence, the magnetic field may fail in successfully concentrating particles at the target organ, with also the possibility of a significant fraction of them accumulating in the liver (Häfeli, 2004; Lübbe et al., 2001).

Surface charge is known to play an important role in blood half-lives of particles. It is generally agreed that highly positively and negatively charged particles present a decreased circulation time. In such a case, particles undergo phagocytosis, resulting in distribution mainly in the liver or spleen. The clearance from circulation is mediated by interaction with cells, especially those of the reticulo-endothelial system. Functional groups on cell surfaces alter the circulation time. A usual approach consists of grafting magnetic systems with PEG (polyethylene glycol), which may be achieved by the precipitation of the particles in the presence of this polymer. By such a technique, sterically stabilized carriers are produced due to the induced sterical hindrance, which avoids protein binding and macrophage recognition (Bulte et al., 1999).

Since the particles must be effectively controlled by the applied magnetic field, their magnetic properties, their dispersion, and their degree of agglomeration are important. It has been observed that an increase in stability of the particles leads to a decrease in toxicity. Low coercive force will prevent aggregation of the particles prior to superimposition of the field. As a result, superparamagnetic particles seem to be ideal. Besides superparamagnetism, high magnetic susceptibility and high saturation magnetization allow the particles to be effectively controlled by a relatively weak field (A. A. Kuznetsov et al., 1999).

The fate of magnetic particles also depends strongly on the administration route:

3.1 Intravenous administration

After intravenous administration, smaller particles are subject to rapid renal elimination or are removed by cells capable of endocytosis (i.e., by B and T lymphocytes), while larger ones undergo uptake by the liver, spleen, and bone marrow. The blood half-lives of many iron oxide nanoparticles administered in patients vary from 1 h to 24–36 h (Corot et al., 2006). Iron oxide particles present low toxicity and are well tolerated in the human body. Inside the cells, such systems are expected to be degraded relatively fast. In fact, degradation into iron (Fe) and oxygen is presumed to occur in intracellular lysosomes of macrophages under the influence of a variety of hydrolytic enzymes, low pH, and protein mobilization and utilization according to natural Fe pathways. The human body contains around 3–4 g of Fe, for example, in the proteins ferritin, hemosiderin, transferritin, and hemoglobin. As the magnetic nanoparticles start to break down, any soluble Fe becomes part of this normal Fe pool, which is then regulated by the body. A clinical dose would likely include just a few milligrams of Fe per kilogram of body weight, which is low compared to the total participating in Fe metabolism. Iron oxides have been shown to degrade *in vivo* and integrate Fe stores in the human body. Therefore, they are not expected to be toxic to the organism (Mornet et al., 2004). This assumption concerning the low toxicity was confirmed by the lethal dose LD50 of magnetic systems. For instance, the LD50 of dextran–iron oxide complex was found to be 2000–6000 mg of Fe kg–1 of body mass. Besides, the systemic safety of several iron oxide nanoparticles has been evaluated after injection in humans, indicating that these products have a satisfactory safety profile according to standard toxicological and pharmacological tests (Corot et al., 2006).

3.2 Subcutaneous and intratumoral administration

Small particles injected locally infiltrate into the interstitial space around the injection site and are gradually absorbed by the lymphatic capillaries into the lymphatic system. For this reason, subcutaneously or locally injected (intratumoral administration) nanoparticles can be used for lymphatic targeting, i.e., as a tool for chemotherapy against lymphatic tumors or metastases. In order to achieve a good uptake in regional lymph nodes following subcutaneous injection, colloidal carriers should be small (60 nm or smaller) and the surface of the particles should be neither too hydrophilic nor too hydrophobic. Concerning intratumoral administration, different studies indicate its feasibility (Hilger et al., 2002, 2005).

3.3 Oral administration

Magnetic particles exhibit strong potential as externally modulated oral systems for both *in vivo* imaging and targeted drug delivery. In this approach, imaging agents or drugs can be localized to specific sites through the application of an external magnetic field. Similarly to the other routes of administration, the fate of the particles in the gastrointestinal tract is closely related to the particle size. Particles under 5 μm can be removed via lymphatic drainage, particles up to 500 nm can cross the membrane of epithelial cells through endocytosis, and particles less than 50 nm can achieve the paracellular passage between intestinal epithelial cells. The use of the magnetic force to delay the transit of orally administered drugs may become an attractive strategy for enhancing the efficacy of orally delivered systems. Despite the promising properties, magnetic particles may dissolve in acid media (Silva-Freitas et al., 2011). Such possible particle loss could reduce the efficiency of the magnetic system used as a drug carrier. In order to avoid it, magnetic particles may be coated using polymers to protect them from the gastric environment. Our recent studies have shown that xylan and Eudragit®S100 coatings are able to protect magnetite from the gastric pH, successfully preventing particle degradation (A. K. Silva et al., 2007b; É. L. Silva et al., 2009).

4. Applications of magnetic particles in the field of biotechnology

4.1 The therapeutic applications

4.1.1 Magnetic drug targeting

One of the major problems related to drug administration is the difficulty in targeting a tissue or an area of the body. After administration, drugs tend to distribute to various organs in a process that occurs depending on the physicochemical properties of the molecule. In order to reach an acceptable therapeutic level at the desired site, large amounts of the drug must be administered. However, only a part of the dose will actually reach the intended tissue or disease site, while the other fraction can cause toxic side effects at non-target organs (Häfeli, 2004; A. K. Silva et al., 2010).

Magnetic drug targeting (MDT) presents a solution to this problem by using magnetic particles as controllable carriers of therapeutic agents which are either encapsulated or attached to the surface of these particles (Mangual et al., 2011). Due to its non-invasiveness, high efficiency, quick-impact, and reduced toxicity in the non-target regions, many researchers are engaging in this area (Cao et al., 2011).

MDT typically uses an external magnetic field source to capture and retain magnetic drug carrier particles at a specific site after being administered in the body. The accumulation of the magnetic particles in the desired area depends on the interplay of the magnetic forces, fluid resistance, and diffusive motions (Cao et al., 2011).

There are some significant limitations of MDT. One limitation is the gradient problem, once the retention of the magnetic particles is quite low due to the relatively weak nature of the magnetic force, which must overcome the hydrodynamic force (A. K. Silva et al., 2006). It can be difficult to use external magnets to target areas deep within the body by the fact that the strength of the magnetic field generated from a permanent magnet decreases sharply with distance (Mangual et al., 2011).

An approach to overcome this limitation uses implant assisted-MDT that takes advantage of a magnetic implant placed inside the body that can become magnetized in the presence of an external magnet creating a localized magnetic field around the implant and increasing the magnetic force on nearby magnetic particles (Mangual et al., 2011).

In addition, magnetic properties and the internalization of particles depend strongly on the size of the magnetic particles. Some hydrodynamic parameters, such as blood flow rate, particle concentration and infusion route play significant roles. Also, there are several forces acting on magnetic particles in a viscous environment and magnetic field, such as magnetic force due to all field sources, viscous drag force, inertia, gravity, thermal kinetics, particle fluid interactions and inter-particle effects such as magnetic dipole interactions, electric double layer interactions, and van der Waals force (Babincova & Babinec, 2009).

The possibilities of MDT applications have drastically increased in recent years. In the clinical area of human medicine, these particles are being used as delivery systems for several drugs, mainly chemotherapy drugs. Such approach may represent a new method to treat cancer. Also, MDT has been used in radiotherapy and immunotherapy. Fig. 1 demonstrates a schematic MDT model for cancer treatment (Alexiou et al., 2011).

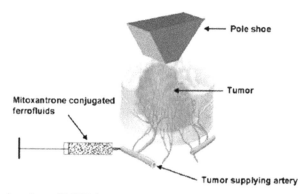

Fig. 1. Schematic drawing of MDT for tumor treatment. Reprinted from the *Journal of Magnetism and Magnetic Materials*, Vol 323, C Alexiou, R Tietze, E Schreiber, R Jurgons, H Richter, L Trahms, H Rahn, S Odenbach, S Lyer, Cancer therapy with drug loaded magnetic nanoparticles--magnetic drug targeting, 1404-1407, Copyright (2011), with permission from Elsevier. License number 2690150177916.

Our research group has developed a new technology for oral delivery of drugs for the treatment of *Helicobacter pylori* infections. The focus of this work was the development of a stomach-specific formulation of amoxicillin based on drug-containing Eudragit®S100 microparticles with a magnetite core. This could be the first attempt to prepare a magnetic system for local delivery of antibiotics, in particular for the treatment of *H. pylori* infections (É. L. Silva et al., 2009).

4.1.2 Magnetofection

Several systems, including viral and non-viral carriers, have been developed to transfer foreign genetic material into cells with the aim of enhancing gene transfer *in vitro* and *in vivo*. Viral vectors can provide a high transfection rate and a rapid transcription of the genetic material inserted in the viral genome. However, viral carriers present potential problems for patients, such as the immunogenicity of the viral proteins, lack of desired tissue selectivity, potential for oncogenesis due to chromosomal integration, and generation of infectious viruses due to recombination.

Non-viral vectors are less immunogenic, are easy to produce in large scale and capable of delivering large genetic material, exhibit enhanced biosafety, and can be associated with tissue targeting. However, unlike viral analogues that have evolved to overcome cellular barriers and immune defense mechanisms, non-viral gene carriers exhibit significantly lower transfection efficiency compared with the viral ones. Among the non-viral vectors, the most used compounds are cationic and biodegradable polymers, lipids, liposomes, and niosomes (He et al., 2010).

In order to overcome those disadvantages, a new transfection method called magnetically guided gene transfection or magnetofection has been developed. Magnetofection employs magnetic nanoparticles combined with transfection agents to form magnetic gene vectors so that the vectors can be rapidly concentrated on the surface of target cells under the attraction of a magnetic field (Yunfeng, 2010).

Since the first reports on magnetically enhanced nucleic acid delivery, it has become a well-established method and has been predominantly used to potentiate viral and non-viral gene delivery. The nucleic acids can be directly associated with magnetic nanoparticles in naked form or can be incorporated into a complex composed of magnetic particles and other components, e.g., cationic lipids and polymers. Fig. 2 shows the principle of magnetofection *in vitro* (Schillinger et al., 2005).

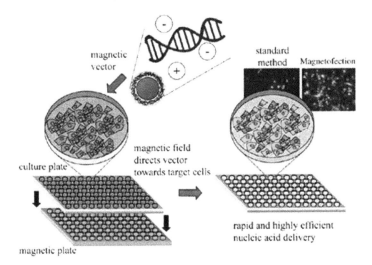

Fig. 2. Illustration of magnetofection in cell culture. The magnetic force pulls magnetic nanoparticles with surface-bound gene vectors toward the cells and results in rapid kinetics and high efficiency nucleic acid delivery. Reprinted from the *Journal of Magnetism and Magnetic Materials*, Vol 293, U Schillinger, T Brill, C Rudolph, S Huth, S Gersting, F Krötz, J Hirschberger, C Bergemann, C Plank, Advances in magnetofection--magnetically guided nucleic acid delivery, 501-508, Copyright (2005), with permission from Elsevier. License number 2690150501010.

The magnetofection process has substantial advantages over traditional transfection methods. The process time is substantially reduced; high transfection rates can be obtained with lower vector doses; an increase in the gene transfer efficiency can be reached; and gene delivery can be achieved with non-permissive cells. The association of gene vectors with superparamagnetic nanoparticles under a magnetic field can boost the efficiency of many vector types up to several hundred-fold and this technique is available for most gene vectors. The accumulation of the vectors in the target area and the dose–response relationship with a smaller amount of DNA required for sufficient gene expression can be enhanced by magnetofection methods (Holzbach et al., 2010).

4.1.3 Magnetic embolization

The conventional cancer treatment is based on chemotherapy and radiotherapy, which are effective treatment, but these systemic approaches typically have an effect on both healthy and disease tissues (Yang et al., 2011). Therefore, it is necessary to study novel strategies to eradicate cancer cells. An alternative approach to overcome these disadvantages is magnetic embolization, which is a new technique in cancer therapy presenting less toxicity than chemotherapy and less invasiveness than surgery.

Magnetic embolization consists of injecting (in a blood vessel) a magnetorheological (MR) fluid, which is a suspension of micron-sized magnetizable particles such as Fe or iron oxide particles (e.g., magnetite). The microscopic structures of these fluids change in the presence of a magnetic field, which leads to a phase transition from a liquid to a solid. MR fluids solidify only under a magnetic field. A seal is formed, which mechanically blocks the tumor blood vessels, causing its death. Once the field is removed, thermal energy makes the fluid return to its original liquid state (J. Liu et al., 2001).

Flores and Liu showed that blocking the fluid flow is possible within a single-tube system, simulating one blood vessel. They found that the seal remains stable even at pressures exceeding those found inside human arterioles and capillaries (Flores et al., 1999; J. Liu et al., 2001). Despite some promising results, it is difficult to obtain selective embolization of small blood vessels when these are positioned at a large distance from the magnetic field source (e.g., approximately >3 cm inside the body of the patient). In this case, a suitable magnetic field intensity and gradient in the vicinity of the vessel is required (A. K. Silva et al., 2010).

A possible solution to this problem may be obtained by a new approach referred to as magnetic resonance navigation (MRN). This has been proposed to steer and track in real time endovascular magnetic carriers in deep tissues to target areas of interest (Pouponneau et al., 2010), and restrain the systemic carrier distribution. MRN is achieved with a clinical magnetic resonance imaging (MRI) scanner upgraded with an insert of steering coils. Therefore, MRN can overcome the problem of a weaker magnetic field in deep tissues observed with an external magnet. By controlling chemoembolic material distribution, MRN could improve embolization and drug concentration in the tumor area while limiting chemoembolization of healthy blood vessels and the hepatic complications (Kennedy et al., 2010).

MRN significantly controlled the distribution of these therapeutic particles as compared to the control. More importantly, a decrease in the TMMC (Therapeutic Magnetic Microcarrier) levels in the untargeted lobe was obtained. Steering efficiency was higher with the left steering compared to the right steering. Therefore, MRN was more efficient in preserving the right liver lobe from the chemoembolization than left lobes. Only one-third of the TMMC dose reached the right lobe without steering because of the natural difference in blood supply between the lobes (the right lobes have less weight than left lobes) (Pouponneau et al., 2011) (Fig. 3). Thus, further work is necessary to confirm the feasibility of MRN applied to magnetic embolization.

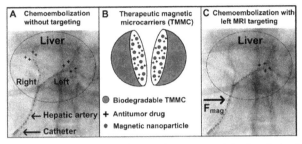

Fig. 3. Representation of MRI targeting with PLGA microparticles loaded with the doxorubicin and magnetic FeCo nanoparticles for liver chemoembolization. Images A and C are fluoroscopy images of the rabbit hepatic artery with superposed images of the microparticle distribution without (A) and with (C) the MRI targeting. On image A, the microparticles are released from the catheter in the artery and distributed to both lobes. Image B illustrates a schematic representation of a cut of the microparticles loaded with an antitumor drug and magnetic nanoparticles embedded into a biodegradable matrix. Image C displays the MRI targeting of the left bifurcation using the magnetic force (Fmag) to preserve the right lobe from the chemoembolization. Reprinted from Biomaterials, Vol 32, P Pouponneau, J-C Leroux, G Soulez, L Gaboury, S Martel, Co-encapsulation of magnetic nanoparticles and doxorubicin into biodegradable microcarriers for deep tissue targeting by vascular MRI navigation, 3481-3486, Copyright (2011), with permission from Elsevier. License number 2690150825494.

4.1.4 Tissue engineering

Advances in cell therapy research gave rise to a fast-growing multidisciplinary field that integrates knowledge of engineering, biology and medicine. Tissue engineering (TE) is a promising technology for overcoming the organ transplantation limitations related to organ donor shortage. It consists of appropriately using cells, materials, and physics/biochemical processes to restore, maintain, or improve tissue function (Fig. 4).

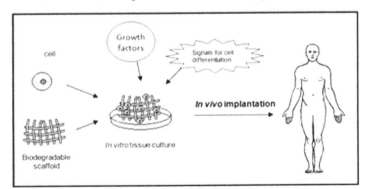

Fig. 4. Schematic picture of tissue engineering. The combination of a scaffold, cells and soluble factors facilitate the formation of structural and functional tissue units.

Despite successful efforts and results, tissue-engineered constructs lack structural complexity. Well-defined spatial cell organization is required in the attempt to reproduce living tissue complexity and succeed in creating functional tissue constructs. In this regard, several cell patterning methods such as microcontact printing and lithography have been developed. However, these methods demand specialized surfaces to be used as substrates and fabrication is time consuming. In order to bypass these shortcomings, innovative active patterning approaches have been based on the use of an external force. This constraint may be generated by an electric field, as in the dielectrophoresis technique, or by an optical trap. Alternatively, Ino et al. (Ino et al., 2007, 2008) suggested the use of magnetic force to induce two-dimensional patterning of magnetically-labeled cells on submillimetric scales.

This group developed a novel methodology for cell patterning using magnetic force and magnetite cationic liposomes (MCLs), which contain 10 nm magnetite nanoparticles. MCLs were used as carriers to introduce magnetite nanoparticles into target cells since the positively charged particle surface interacts with the negatively charged cell surface. Given that cells labeled with magnetic nanoparticles can be manipulated by using a magnet, they were able to seed labeled cells onto a low-adhesive culture surface through the use of magnetic force to form a tissue construct (Ino et al., 2007). This technique was designated magnetic force-based tissue engineering (Mag-TE). By using it, complex cell patterns (curved, parallel, or crossing motifs) were successfully fabricated from several cell types (Ino et al., 2007). This group also showed that magnetically labeled keratinocytes were accumulated using a magnet, and stratification was promoted by a magnetic force to form a sheet-like 3D construct. Transmission electron microscopy revealed the presence of intercellular adhesion proteins, desmosomes, within keratinocyte sheets constructed by Mag-TE. This result indicated that the proposed method enabled intercellular contact and preserved adhesion proteins. For this reason, patterns were found to resist after magnetic force removal and manipulation was also feasible.

Following a related approach, Frasca et al. applied magnetic forces to create a 3D cell assembly with tuneable size and controlled geometry. Cells were magnetically labeled using anionic citrate-coated iron oxide nanoparticles. Focalized magnetic force ensured an efficient entrapment of the cells at the magnet vicinity. This technology could be applied with no restriction regarding the physicochemical nature of the substrate, the cell type, or the geometry of the imposed magnetic constraint (Frasca et al., 2009). The same group demonstrated that magnetic force-assisted cell seeding provided effective cell seeding into 3-D porous scaffolds. Moreover, precise spatial cellular organization inside the scaffold could be achieved by means of magnetic microtips that develop high magnetic forces.

4.1.5 Magnetic hyperthermia

Hyperthermia is a technique that increases the temperature of the local environment of a tumor, resulting in changing the physiology of diseased cells and finally leading to apoptosis. Depending on the degree of temperature rise, hyperthermia treatment can be classified into different types. In thermo ablation, a tumor is submitted to high temperatures of heat >46 °C (up to 56 °C), causing cells to undergo direct tissue necrosis, coagulation, or carbonization. Moderate hyperthermia (41 °C <T< 46 °C) has various effects both at the cellular and tissue levels. Diathermia uses lower temperatures (T<41 °C) for the treatment of rheumatic diseases in physiotherapy.

During moderate hyperthermia, which is traditionally termed hyperthermia treatment, cells undergo heat stress resulting in activation and/or initiation of many intra- and extracellular degradation mechanisms, including induction and regulation of apoptosis, signal transduction, multidrug resistance, and heat shock protein expression. The tissue-level effects include pH changes, perfusion, and oxygenation of the tumor microenvironment (Santos-Marques et al., 2006).

Nevertheless, traditional hyperthermia treatment presents several challenges: 1) unavoidable heating of healthy tissue resulting in burns, blisters, and discomfort; 2) limited penetration of heat into body tissues; and 3) thermal under-dosage in the target region, a nearly unsolved problem in the case of bone of the pelvis or scull, which shields deep tissues, often yielding recurrent tumor growth (Tanaka et al., 2005).

Improvements in this technique yielded the magnetic hyperthermia treatment, which has a number of advantages compared to conventional hyperthermia treatment: 1) cancer cells absorb magnetic nanoparticles (MNPs), therefore increasing the effectiveness of hyperthermia by delivering therapeutic heat directly to them; 2) MNPs can be targeted through cancer-specific binding agents, making the treatment much more selective and effective; 3) the frequencies of oscillating magnetic fields generally utilized pass harmlessly through the body and generate heat only in tissues containing MNPs; 4) MNPs can also cross the blood-brain barrier, and thus can be used for treating brain tumors; 5) effective and externally stimulated heating can be delivered at cellular levels through alternating magnetic fields; 6) stable MNPs can be administered through a number of drug delivery routes; 7) MNPs used for hyperthermia are only a few tens of nanometer in size and therefore allow easy passage into several tumors whose pore sizes are in the 380–780 nm range; 8) compared to macroscopic implants, MNP-based heat generation is much more efficient and homogeneous; 9) MNP-based hyperthermia treatment may induce antitumoral immunity (Ito et al., 2005); and 10) the last but most important aspect is that MNP-based hyperthermia can also be utilized for controlled delivery of drugs (Lu et al., 2005). This additional feature opens up possibilities for the development of multifunctional and multi-therapeutic approaches for treating a number of diseases.

For MNP-based hyperthermia, a general procedure involves the distribution of particles throughout the targeted tumor site, followed by generation of heat to the tumor using an external AMF. The absorption efficiency of any material to generate heat due to AMF is measured in terms of a specific absorption rate or specific loss of power. These terms are generally used to define the transformation of magnetic energy into heat. For a majority of applications, it is desirable to have higher temperature enhancement rates.

Heat generation can be attributed to two different phenomena: relaxation and hysteresis loss. The relaxation is of two types: Brownian and Néel. Brownian relaxation is due to the physical rotation of particles within the medium in which they are placed (external dynamics) and is hindered by the viscosity that tends to counter the movement of particles in the medium. Heat generation through Néel relaxation is due to rapidly occurring changes in the direction of magnetic moments relative to crystal lattice (internal dynamics). This is hindered by the energy of anisotropy that tends to orient the magnetic domain in a given direction relative to crystal lattice. For intracellular magnetic fluid hyperthermia, Néel relaxation is the major contributor for heat release (Hergt & Dutz, 2007).

A number of types of magnetic nanomaterials have been investigated for magnetic hyperthermia. Some of the well-known hyperthermic agents based on iron oxide are magnetite and maghemite nanoparticles. Recently, a high heating performance of 1300–1600W/g was reported based on FeCo metallic nanoparticles (Nojima et al., 2010). However, iron oxide-based MNPs continue to attract attention due to their lack of toxicity, excellent biocompatibility, and metabolization, which is carried out by heme oxygenase-1 to form blood hemoglobin and hence maintain Fe cell homeostasis in cells. In addition, magnetite was found to be superior to cobalt nanoparticles with respect to its high Curie temperature, saturation magnetization (90–98 emu/g), and lower toxicity in preclinical tests (Martina et al., 2008).

In vivo studies using magnetic hyperthermia have been conducted by Matsuoka et al. (Matsuoka, 2004). They have developed magnetic cationic liposomes based on superparamagnetic iron oxide nanoparticles and investigated their *in vivo* efficacy for hyperthermia treatment of hamster osteosarcoma. Magnetoliposomes were injected directly into the osteosarcoma and then subjected to an alternating magnetic field. The tumor was heated above 42°C, and complete regression was observed in 100% of the treated hamsters. Therefore, these results demonstrate the feasibility of magnetic hyperthermia.

4.2 The diagnostic uses

4.2.1 MRI contrast agent

MRI is a noninvasive and sensitive technique to obtain images by non-ionizing radiation. As it associates rapid *in vivo* acquisition of images, long effective imaging window, fine signal intensity contrast, high temporal and spatial resolution, and simultaneous information about physiology and anatomy of the desired area, MRI quickly became one of the major tomographic imaging modalities (G. Liu et al., 2011).

The images formed are the result of several parameters such as proton density, relaxation times (T1, T2, T2*), water diffusion, nuclear alignment, radio frequency excitation, spatial encoding, etc., providing a digital representation of tissue characteristics. The relative difference between the signal intensity of two adjacent regions is called contrast, and this difference is translated using a color scale (normally the grey scale for MRI). The spin–lattice ($1/T1 = R1$) and spin–spin ($1/T2 = R2$) relaxation rates of the water protons in tissues, which are the most important intrinsic factors for contrast, are dependent upon the local environment of the protons, so that different tissues will relax at different rates (Yurt & Kazanci, 2008). However, for specific studies of evaluation at the molecular and cellular level, the MRI sensibility is lower, being necessary the use of a contrast agent or even a selective binding attached to a contrast agent, in order to differentiate the target area (Lalatonne, 2010).

Contrast agents are normally defined based on their relaxation properties, their magnetic properties, and their biodistribution. When defining a contrast agent based on relaxation properties, the efficiency is described by the longitudinal and transverse relaxivity R1 and R2, respectively. The relaxivity reflects the change in the relaxation rate as a function of the contrast agent concentration. The relaxivities are affected by the size and the composition of these particles (Yurt & Kazanci, 2008).

To design a contrast agent, the choice of core and monolayer material is a critical step because this composition determines the primary physical and chemical properties besides reactivity, solubility, and interfacial interactions. Most common core among the MRI contrast agents are paramagnetic lanthanide metals (gadolinium, manganese and dysprosium ion complexes) and superparamagnetic magnetite particles (iron oxides) (Yurt & Kazanci, 2008). Iron oxide particles are widely investigated in MRI applications as they alter the relaxation times of tissues in which they are present and due to the low toxicity when compared to gadolinium chelates (Lalatonne, 2010). In this context, the superparamagnetic particles, which can be superparamagnetic iron oxide (SPIO) particles, ultrasmall superparamagnetic iron oxide (USPIO) and oral magnetic particles (OMPs), appear as preferred materials because (a) they have magnetic characteristics, (b) they are composed of biodegradable Fe, (c) their coating can be functionalized with various ligands, (d) they provide the greatest signal changes per unit of metal, and (e) they are easily detectable by light and electron microscopy (Bulte & Kraitchman, 2004).

Superparamagnetic iron oxides have substantially larger T2 relaxivity compared with gadolinium chelates in current clinical use, typically by an order of magnitude or more. This increase is confirmed by a superior magnetization. The T1-relaxivity can also be much higher for iron oxides than for gadolinium chelates. In addition, iron oxide nanoparticles may offer several advantages over existing agents due to their accumulation in macrophages combined with an intravascular distribution and higher relaxivity values (Bulte & Kraitchman, 2004).

Another field of research in development aims to use superparamagnetic contrast agents in drug delivery applications for real-time monitoring of drug distribution to the target tissue, as well as to follow the effect of therapeutics on the progression of disease.

4.2.2 Magnetic cell tracking

There is a great need to develop improved means of monitoring transplanted cells *in vivo*. A recent methodology involves the use of magnetic particles for intracellular magnetic labeling of cells. This technique, called magnetic cell tracking, allows *in vivo* tracking of implanted cells via MRI. Magnetic cell tracking can be used as a non-invasive tool to provide unique information on the dynamics of cell movements within and away from tissues *in vivo*. Alternatively, magnetic cell tracking could be applied in the future to monitor cell therapy in patients. Both approaches require magnetic labeling of cells as well as methods for analysis and evaluation of cell labeling (Vuu et al., 2005).

The magnetic cell tracking technique may overcome the limitations of individual *in vivo* imaging methods including low sensitivity, low resolution, or low soft tissue contrast. MRI provides excellent soft tissue contrast and due to its high resolution, MRI can be used for the visualization of single cells against a homogeneous background (Himmelreich & Dresselaers, 2009).

Several methods have been developed to incorporate sufficient quantities of iron oxide nanoparticles into cells. These methods mainly concern the prolonged incubation of the cells with the particles resulting in their passive internalization. Another possibility is the introduction of functional ligands chemically linked to the particles, in order to increase the uptake by cells. Besides, the transient increase in the membrane permeability using a magnetic field (magneto-electroporation) may result in a quick cytoplasmic accumulation of the magnetic particles (Dousset et al., 2008).

Some other examples of magnetic cell tracking applications include labeling mesenchymal stem cells, haematopoietic progenitor cells, Schwann cell transplants, neural stem cells, and NK cells.

4.2.3 Monitoring the gastrointestinal motility

The evaluation of the large intestine motility is usually made by intraluminal manometry, radiology, or scintigraphy. Most of the current knowledge about motility of the large intestine was generated by intraluminal manometry. Despite its providing quantitative assessment, intraluminal manometry is obviously invasive and uncomfortable for patients. Radiology offers qualitative or, at best, semi-quantitative information, and carries the risk of significant radiation exposure. Gamma-scintigraphy also imposes radiation exposure and depends on the availability of expensive equipment (Ferreira et al., 2004).

Among other methods, the investigation of intestinal movements by Magnetic Marker monitoring is considered to be a useful diagnostic tool. The colon exhibits complex motor patterns with variations of frequency and amplitude yielding compaction and movement of its contents along its extension. The arrival of a meal into the stomach is consistently associated with the unleashment of contractions of the large intestine, which causes movements of the colonic content, called gastrocolic reflex, and can be observed by an increase in the motor activity of the colon (Ferreira et al., 2004).

The oral route is still by far the most common way used for the administration of pharmacologically active substances. This is mainly due to the ease of administration and the general acceptance by the patients. Knowledge about the performance of dosage forms in the gastrointestinal tract is essential for the choice of the optimal formulation technology (Weitschies et al., 2010). In order to overcome restrictions that are associated with the use of radioisotopes, an alternative method for the investigation of the behavior of solid dosage forms in the gastrointestinal tract was developed. It is based on the labeling of the dosage as a magnetic dipole by means of incorporation of trace amounts of ferromagnetic particles, recording of the magnetic dipole field using biomagnetic measurement equipment, and data evaluation applying techniques established in magnetic source imaging (MSI). This method is known as Magnetic Marker Monitoring (MMM) or Magnetic Moment Imaging (MMI). (Goodman et al., 2010; Weitschies et al., 1994).

MMM is a new technique for the investigation of the gastrointestinal transit of magnetically marked solid drug dosage forms (Weitschies et al., 1999). The magnetic labeling of the dosage forms is achieved by the incorporation of small amounts of remanent ferromagnetic particles and their subsequent magnetization tracking. After ingestion of one magnetically marked dosage form, its magnetic dipole field is recorded during its gastrointestinal transit. Multichannel superconducting quantum interference devices (SQUID), developed for the detection of extremely weak biomagnetic fields, are employed for the measurement of the magnetic field (Drung, 1995). Finally, the parameters describing the magnetic dipole, i.e., its location $r = (x, y, z)$ and its magnetic moment $m = (m_x, m_y, m_z)$, are estimated from the recorded data by means of fitting procedures. After ingestion, their magnetic dipole field is recorded, and by means of fitting procedures, the location of the marked dosage form is estimated from the recorded data. The disintegration behavior is also assessed by this technique. The induction generated by the magnetic dipole moment of the oral dosage form during disintegration is used for the investigation of its mechanism and quantitative determination of the process (Weitschies et al., 2001a, 2001b).

Additionally, MMM has been applied for the determination of the performance of disintegrating and non-disintegrating solid dosage forms such as tablets, capsules, and pellets in the gastrointestinal tract, as well as for the determination of the *in vivo* drug release from modified release products such as enteric-coated tablets and enhanced release tablets (Weitschies et al., 2005a).

The combination of MMM with the pharmacokinetic measurements (pharmacomagnetography) enables the determination of *in vitro–in vivo* correlations and the delineation of absorption sites in the gastrointestinal tract (Weitschies et al., 2005b). The results obtained with MMM can also serve as a data base for the development of improved pharmacokinetic models.

5. Conclusion and perspectives

The use of magnetic particles in the medical field opens new prospect of selective treatment of local tissues where efficiency is increased through local concentrations while, at the same time, general side effects can be avoided. However, the use of magnetic carriers in the human body imposes several requirements on the magnetic carriers. Magnetic carriers must be water-based, biocompatible, biodegradable, and nonimmunogenic. Besides, special care should be focused on the particle size, surface properties, magnetic properties, and administration route, for example. In most of the reports in the literature, iron oxides are the material of choice for the development of magnetic systems for therapeutic purposes.

Several methods have been proposed for their synthesis, coating, and stabilization. Magnetic systems produced by different methods have found many applications in biotechnology. The safety aspect, the non-invasiveness, and the high targeting efficiency are promising advantages for the use of magnetic particles in therapeutics. The current challenge still consists of totally controlling the biocompatibility, stability, biokinetics, and properties of the particles. By incorporating advances in surface engineering, molecular imaging, and biotechnology, magnetic systems have great potential to enable physicians to diagnose and treat diseases with greater effectiveness than ever before.

6. Acknowledgment

This work was supported by CNPq and Capes-Brazil.

7. References

Alexiou, C.; Tietze, R.; Schreiber, E.; Jurgons, R.; Richter, H.; Trahms, L.; Rahn, H.; Odenbach, S. & Lyer, S. (2011). Cancer therapy with drug loaded magnetic nanoparticles--magnetic drug targeting. *Journal of Magnetism and Magnetic Materials*, Vol. 323, No. 10, pp. 1404-1407, ISSN 0304-8853

Babincova, M. & Babinec, P. (2009). Magnetic drug delivery and targeting: principles and applications. *Biomed Pap Med Fac Univ Palacky Olomouc Czech Repub*, Vol. 153, No. 4, (March 2010), pp. 243-50, ISSN 1213-8118

Bulte, J. W. M.; Cuyper, M. D.; Despres, D. & Frank, J. A. (1999). Preparation, relaxometry, and biokinetics of PEGylated magnetoliposomes as MR contrast agent. *Journal of Magnetism and Magnetic Materials*, Vol. 194, No. 1-3, pp. 204-209, ISSN 0304-8853

Bulte, J. W. M. & Kraitchman, D. L. (2004). Iron oxide MR contrast agents for molecular and cellular imaging. *NMR in Biomedicine*, Vol. 17, No. 7, pp. 484-499, ISSN 1099-1492

Cao, Q.; Han, X. & Li, L. (2011). Enhancement of the efficiency of magnetic targeting for drug delivery: Development and evaluation of magnet system. *Journal of Magnetism and Magnetic Materials*, Vol. 323, No. 15, pp. 1919-1924, ISSN 0304-8853

Corot, C.; Robert, P.; Idée, J.-M. & Port, M. (2006). Recent advances in iron oxide nanocrystal technology for medical imaging. *Advanced Drug Delivery Reviews*, Vol. 58, No. 14, pp. 1471-1504, ISSN 0169-409X

Dousset, V.; Tourdias, T.; Brochet, B.; Boiziau, C. & Petry, K. G. (2008). How to trace stem cells for MRI evaluation? *Journal of the Neurological Sciences*, Vol. 265, No. 1-2, pp. 122-126, ISSN 0022-510X

Drung, D. (1995). The Ptb 83-Squid System for Biomagnetic Applications in a Clinic. *Ieee Transactions on Applied Superconductivity*, Vol. 5, No. 2, pp. 2112-2117, ISSN 1051-8223

Ferreira, A.; Carneiro, A. A. O.; Moraes, E. R.; Oliveira, R. B. & Baffa, O. (2004). Study of the magnetic content movement present, in the large intestine. *Journal of Magnetism and Magnetic Materials*, Vol. 283, No. 1, pp. 16-21, ISSN 0304-8853

Flores, G. A.; Sheng, R. & Liu, J. (1999). Medical applications of magnetorheological fluids - A possible new cancer therapy. *Journal of Intelligent Material Systems and Structures*, Vol. 10, No. 9, pp. 708-713, ISSN 1045-389X

Frasca, G.;Gazeau, F., Wilhelm, C. (2009). Formation of a three-dimensional multicellular assembly using magnetic patterning. *Langmuir*, Vol. 25, No. 4, pp. 2348-2354, ISSN 1520-5827

Goodman, K.; Hodges, L. A.; Band, J.; Stevens, H. N. E.; Weitschies, W. & Wilson, C. G. (2010). Assessing gastrointestinal motility and disintegration profiles of magnetic tablets by a novel magnetic imaging device and gamma scintigraphy. *European Journal of Pharmaceutics and Biopharmaceutics*, Vol. 74, No. 1, pp. 84-92, ISSN 0939-6411

Häfeli, U. O.; Sweeney, S. M.; Beresford, B. A.; Sim, E. H. & Mackilis, R. M. (2004). Magnetically directed poly(lactic acid) 90Y-microspheres: Novel agents for targeted intracavitary radiotherapy. *Journal of Biomedical Materials Research*, Vol. 28, No. 8, pp. 901-908, ISSN 1552-4965

He, C.-X.; Tabata, Y. & Gao, J.-Q. (2010). Non-viral gene delivery carrier and its three-dimensional transfection system. *International Journal of Pharmaceutics*, Vol. 386, No. 1-2, pp. 232-242, ISSN 0378-5173

Hergt, R. & Dutz, S. (2007). Magnetic particle hyperthermia-biophysical limitations of a visionary tumour therapy. *Journal of Magnetism and Magnetic Materials*, Vol. 311, No. 1, pp. 187-192, ISSN 0304-8853

Heurtault, B.; Saulnier, P.; Pech, B.; Proust, J. E. & Benoit, J. P. (2003). Physico-chemical stability of colloidal lipid particles. *Biomaterials*, Vol. 24, No. 23, pp. 4283-4300, ISSN 0142-9612

Hilger, I.; Hergt, R. & Kaiser, W. A. (2005). Towards breast cancer treatment by magnetic heating. *Journal of Magnetism and Magnetic Materials*, Vol. 293, No. 1, pp. 314-319, ISSN 0304-8853

Hilger, I.; Hiergeist, R.; Hergt, R.; Winnefeld, K.; Schubert, H. & Kaiser, W. A. (2002). Thermal ablation of tumors using magnetic nanoparticles: an in vivo feasibility study. *Investigative Radiology*, Vol. 37, No. 10, pp. 580, ISSN 1536-0210

Himmelreich, U. & Dresselaers, T. (2009). Cell labeling and tracking for experimental models using magnetic resonance imaging. *Methods*, Vol. 48, No. 2, pp. 112-124, ISSN 1046-2023

Holzbach, T.; Vlaskou, D.; Neshkova, I.; Konerding, M. A.; Wörtler, K.; Mykhaylyk, O.; Gänsbacher, B.; Machens, H. G.; Plank, C. & Giunta, R. E. (2010). Non-viral VEGF(165) gene therapy--magnetofection of acoustically active magnetic lipospheres ('magnetobubbles') increases tissue survival in an oversized skin flap model. *Journal of Cellular and Molecular Medicine*, Vol. 14, No. 3, pp. 587-599, ISSN 1582-4934

Hong, X.; Guo, W.; Yuang, H.; Li, J.; Liu, Y. M.; Ma, L.; Bai, Y. B. & Li, T. J. (2004). Periodate oxidation of nanoscaled magnetic dextran composites. *Journal of Magnetism and Magnetic Materials*, Vol. 269, No. 1, pp. 95-100, ISSN 0304-8853

Ino, K.;Ito, A., Honda, H. (2007). Cell patterning using magnetite nanoparticles and magnetic force. *Biotechnology and Bioengineering*, Vol. 97, No. 5, pp. 1309-1317, ISSN 1097-0290

Ino, K.;Okochi, M.;Konishi, N.;Nakatochi, M.;Imai, R.;Shikida, M.;Ito, A., Honda, H. (2008). Cell culture arrays using magnetic force-based cell patterning for dynamic single cell analysis. *Lab on a Chip*, Vol. 8, No. 1, pp. 134-142, ISSN 1473-0189

Ito, A.; Shinkai, M.; Honda, H. & Kobayashi, T. (2005). Medical application of functionalized magnetic nanoparticles. *Journal of Bioscience and Bioengineering*, Vol. 100, No. 1, pp. 1-11, ISSN 1389-1723

Kennedy, A. S.; Kleinstreuer, C.; Basciano, C. A. & Dezarn, W. A. (2010). Computer modeling of yttrium-90-microsphere transport in the hepatic arterial tree to improve clinical outcomes. *International Journal of Radiation Oncology*Biology*Physics*, Vol. 76, No. 2, pp. 631-637, ISSN 0360-3016

Kotani, H.; Iwasaka, M.; Ueno, S. & Curtis, A. (2000). Magnetic orientation of collagen and bone mixture. *Journal of Applied Physics*, Vol. 87, No. 9, pp. 6191- 6193, ISSN 1089-7550

Kuznetsov, A. A.; Filippov, V. I.; Alyautdin, R. N.; Torshina, N. L. & Kuznetsov, O. A. (2001). Application of magnetic liposomes for magnetically guided transport of muscle relaxants and anti-cancer photodynamic drugs. *Journal of Magnetism and Magnetic Materials*, Vol. 225, No. 1-2, pp. 95-100, ISSN 0304-8853

Kuznetsov, A. A.; Filippov, V. I.; Kuznetsov, O. A.; Gerlivanov, V. G.; Dobrinsky, E. K. & Malashin, S. I. (1999). New ferro-carbon adsorbents for magnetically guided transport of anti-cancer drugs. *Journal of Magnetism and Magnetic Materials*, Vol. 194, No. 1-3, pp. 22-30, ISSN 0304-8853

Lalatonne, Y.; Jouni, H. M. M.; Serfaty, J. M.; Sainte-Catherine, O.; Lièvre, N.; Kusmia, S.; Weinmann, P.; Lecouvey, M. & Motte, L. (2010). Superparamagnetic bifunctional bisphosphonates nanoparticles: a potential MRI contrast agent for osteoporosis therapy and diagnostic. *Journal of Osteoporosis*, Vol. 2010, pp. 1-7, ISSN 2042-0064

Liu, G.; Wang, Z.; Lu, J.; Xia, C.; Gao, F.; Gong, Q.; Song, B.; Zhao, X.; Shuai, X.; Chen, X.; Ai, H. & Gu, Z. (2011). Low molecular weight alkyl-polycation wrapped magnetite nanoparticle clusters as MRI probes for stem cell labeling and in vivo imaging. *Biomaterials*, Vol. 32, No. 2, pp. 528-537, ISSN 0142-9612

Liu, J.; Flores, G. A. & Sheng, R. (2001). In-vitro investigation of blood embolization in cancer treatment using magnetorheological fluids. *Journal of Magnetism and Magnetic Materials*, Vol. 225, No. 1-2, pp. 209-217, ISSN 0304-8853

Lopez-Quintela, M. A. (2003). Synthesis of nanomaterials in microemulsions: formation mechanisms and growth control. *Current Opinion in Colloid & Interface Science*, Vol. 8, No. 2, pp. 137-144, ISSN 1359-0294

Lu, Z. H.; Prouty, M. D.; Guo, Z. H.; Golub, V. O.; Kumar, C. & Lvov, Y. M. (2005). Magnetic switch of permeability for polyelectrolyte microcapsules embedded with Co@Au nanoparticles. *Langmuir*, Vol. 21, No. 5, pp. 2042-2050, ISSN 0743-7463

Lübbe, A. S.; Alexiou, C. & Bergemann, C. (2001). Clinical Applications of Magnetic Drug Targeting. *Journal of Surgical Research*, Vol. 95, No. 2, pp. 200-206, ISSN 0022-4804

Lübbe, A. S.; Bergemann, C.; Brock, J. & McClure, D. G. (1999). Physiological aspects in magnetic drug-targeting. *Journal of Magnetism and Magnetic Materials*, Vol. 194, No. 1-3, pp. 149-155, ISSN 0304-8853

Mangual, J. O.; Avilés, M. O.; Ebner, A. D. & Ritter, J. A. (2011). In vitro study of magnetic nanoparticles as the implant for implant assisted magnetic drug targeting. *Journal of Magnetism and Magnetic Materials*, Vol. 323, No. 14, pp. 1903-1908, ISSN 0304-8853

Martina, M. S.; Wilhelm, C. & Lesieur, S. (2008). The effect of magnetic targeting on the uptake of magnetic-fluid-loaded liposomes by human prostatic adenocarcinoma cells. *Biomaterials*, Vol. 29, No. 30, pp. 4137-4145, ISSN 0142-9612

Matsuoka, F., Shinkai,M., Honda,H.,Kubo,T., Sugita,T., Kobayashi,T. (2004). Hyperthermia using magnetite cationic liposomes for hamster osteosarcoma. *BioMagnetic Research and Technology*. Vol. 2, No. 3, pp. 1-6, ISSN 1477-044X

Mornet, S.; Vasseur, S.; Grasset, F. & Duguet, E. (2004). Magnetic nanoparticle design for medical diagnosis and therapy. *Journal of Materials Chemistry*, Vol. 14, No. 14, pp. 2161-2175, ISSN 0959-9428

Nojima, K.; Ge, S.; Katayama, Y.; Ueno, S. & Iramina, K. (2010). Effect of the stimulus frequency and pulse number of repetitive transcranial magnetic stimulation on the inter-reversal time of perceptual reversal on the right superior parietal lobule. *Journal of Applied Physics*, Vol. 107, No. 9, ISSN 0021-8979

Pouponneau, P.; Leroux, J. C.; Soulez, G.; Gaboury, L. & Martel, S. (2011). Co-encapsulation of magnetic nanoparticles and doxorubicin into biodegradable microcarriers for deep tissue targeting by vascular MRI navigation. *Biomaterials*, Vol. 32, No. 13, pp. 3481-3486, ISSN 0142-9612

Pouponneau, P.; Savadogo, O.; Napporn, T.; Yahia, L. & Martel, S. (2010). Corrosion study of iron-cobalt alloys for MRI-based propulsion embedded in untethered microdevices operating in the vascular network. *Journal of Biomedical Materials Research Part B-Applied Biomaterials*, Vol. 93B, No. 1, pp. 203-211, ISSN 1552-4973

Santos-Marques, M. J.; Carvalho, F.; Sousa, C.; Remião, F.; Vitorino, R.; Amado, F.; Ferreira, R.; Duarte, J. A. & de Lourdes Bastos, M. (2006). Cytotoxicity and cell signalling induced by continuous mild hyperthermia in freshly isolated mouse hepatocytes. *Toxicology*, Vol. 224, No. 3, pp. 210-218, ISSN 0300-483X

Saravanan, M.; Bhaskar, K.; Maharajan, G. & Pillai, K. S. (2004). Ultrasonically controlled release and targeted delivery of diclofenac sodium via gelatin magnetic microspheres. *International Journal of Pharmaceutics*, Vol. 283, No. 1-2, pp. 71-82, ISSN 0378-5173

Schillinger, U.; Brill, T.; Rudolph, C.; Huth, S.; Gersting, S.; Krötz, F.; Hirschberger, J.; Bergemann, C. & Plank, C. (2005). Advances in magnetofection--magnetically guided nucleic acid delivery. *Journal of Magnetism and Magnetic Materials*, Vol. 293, No. 1, pp. 501-508, ISSN 0304-8853

Silva, A. K.; Egito, E. S.; Nagashima-Júnior, T.; Araújo, I. B.; Silva, É. L.; Soares, L. A. L. & Carriço, A. S. (2008). Development of superparamagnetic microparticles for biotechnological purposes. *Drug Development and Industrial Pharmacy*, Vol. 34, pp. 1111-1116, ISSN 1520-5762

Silva, A. K.; Silva, É. L.; Carriço, A. S. & Egito, E. S. (2007a). Magnetic carriers: a promising device for targeting drugs into the human body. *Current Pharmaceutical Design*, Vol. 13, pp. 1179-1185, ISSN 1381-6128

Silva, A. K.; Silva, É. L.; Carvalho, J. F.; Pontes, T. R.; Neto, R. P.; Carriço, A. S. & Egito, E. S. (2010). Drug targeting and other recent applications of magnetic carriers in therapeutics. *Key Engineering Materials*, Vol. 441, pp. 357-378, ISSN 1013-9826

Silva, A. K.; Silva, É. L.; Egito, E. S. & Carriço, A. S. (2006). Safety concerns related to magnetic field exposure. *Radiation and Environmental Biophysics*, Vol. 45, pp. 245-252, ISSN 1432-2099

Silva, A. K.; Silva, É. L.; Oliveira, E. E.; Nagashima-Júnior, T.; Soares, L. A. L.; Medeiros, A. C.; Araújo, J. H.; Araújo, I. B.; Carriço, A. S. & Egito, E. S. (2007b). Synthesis and characterization of xylan-coated magnetite microparticles. *International Journal of Pharmaceutics*, Vol. 334, pp. 42-47, ISSN 0378-5173

Silva, É. L.; Carvalho, J. F.; Pontes, T. R.; Oliveira, E. E.; Francelino, B. L.; Medeiros, A. C.; Egito, E. S.; Araujo, J. H. & Carriço, A. S. (2009). Development of a magnetic system for the treatment of Helicobacter pylori infections. *Journal of Magnetism and Magnetic Materials*, Vol. 321, No. 10, pp. 1566-1570, ISSN 0304-8853

Silva-Freitas, É. L.; Carvalho, J. F.; Pontes, T. R.; Araujo-Neto, R.P.; Carriço, A. S. & Egito, E. S. (2011). Magnetite Content Evaluation on Magnetic Drug Delivery Systems by Spectrophotometry: A Technical Note. *AAPS PharmSciTech*, Vol. 12, No. 2, pp. 521-524, ISSN 1530-9932

Tanaka, K.; Ito, A.; Kobayashi, T.; Kawamura, T.; Shimada, S.; Matsumoto, K.; Saida, T. & Honda, H. (2005). Heat immunotherapy using magnetic nanoparticles and dendritic cells for T-lymphoma. *Journal of Bioscience and Bioengineering*, Vol. 100, No. 1, pp. 112-115, ISSN 1389-1723

Vuu, K.; Xie, J.; McDonald, M. A.; Bernardo, M.; Hunter, F.; Zhang, Y.; Li, K.; Bednarski, M. & Guccione, S. (2005). Gadolinium-Rhodamine Nanoparticles for Cell Labeling and Tracking via Magnetic Resonance and Optical Imaging. *Bioconjugate Chemistry*, Vol. 16, No. 4, pp. 995-999, ISSN 1043-1802

Weitschies, W.; Blume, H. & Mönnikes, H. (2010). Magnetic Marker Monitoring: High resolution real-time tracking of oral solid dosage forms in the gastrointestinal tract. *European Journal of Pharmaceutics and Biopharmaceutics*, Vol. 74, No. 1, pp. 93-101

Weitschies, W.; Cardini, D.; Karaus, M.; Trahms, L. & Semmler, W. (1999). Magnetic marker monitoring of esophageal, gastric and duodenal transit of non-disintegrating capsules. *Pharmazie*, Vol. 54, No. 6, pp. 426-430, ISSN 0031-7144

Weitschies, W.; Hartmann, V.; Grutzmann, R. & Breitkreutz, J. (2001a). Determination of the disintegration behavior of magnetically marked tablets. *European Journal of Pharmaceutics and Biopharmaceutics*, Vol. 52, No. 2, pp. 221-226, ISSN 0939-6411

Weitschies, W.; Karaus, M.; Cordini, D.; Trahms, L.; Breitkreutz, J. & Semmler, W. (2001b). Magnetic marker monitoring of disintegrating capsules. *European Journal of Pharmaceutical Sciences*, Vol. 13, No. 4, pp. 411-416, ISSN 0928-0987

Weitschies, W.; Kosch, O.; Mönnikes, H. & Trahms, L. (2005a). Magnetic Marker Monitoring: An application of biomagnetic measurement instrumentation and principles for the determination of the gastrointestinal behavior of magnetically marked solid dosage forms. *Advanced Drug Delivery Reviews*, Vol. 57, No. 8, pp. 1210-1222, ISSN 0169-409X

Weitschies, W.; Wedemeyer, J.; Stehr, R. & Trahms, L. (1994). Magnetic markers as a noninvasive tool to monitor gastrointestinal transit. *Biomedical Engineering, IEEE Transactions on*, Vol. 41, No. 2, pp. 192-195, ISSN 0018-9294

Weitschies, W.; Wedemeyer, R.-S.; Kosch, O.; Fach, K.; Nagel, S.; Söderlind, E.; Trahms, L.; Abrahamsson, B. & Mönnikes, H. (2005b). Impact of the intragastric location of extended release tablets on food interactions. *Journal of Controlled Release*, Vol. 108, No. 2-3, pp. 375-385, ISSN 0168-3659

Xu, H. H. K.; Smith, D. T. & Simon, C. G. (2004). Strong and bioactive composites containing nano-silica-fused whiskers for bone repair. *Biomaterials*, Vol. 25, No. 19, pp. 4615-4626, ISSN 0142-9612

Yang, F.; Jin, C.; Yang, D.; Jiang, Y.; Li, J.; Di, Y.; Hu, J.; Wang, C.; Ni, Q. & Fu, D. (2011). Magnetic functionalised carbon nanotubes as drug vehicles for cancer lymph node metastasis treatment. *European Journal of Cancer*, Vol. In Press, Corrected Proof, ISSN 0959-8049

Yunfeng, S. (2010). In situ preparation of magnetic nonviral gene vectors and magnetofection in vitro. *Nanotechnology*, Vol. 21, No. 11, pp. 115103, ISSN 0957-4484

Yurt, A. & Kazanci, N. (2008). Investigation of magnetic properties of various complexes prepared as contrast agents for MRI. *Journal of Molecular Structure*, Vol. 892, No. 1-3, pp. 392-397, ISSN 0022-2860

Permissions

The contributors of this book come from diverse backgrounds, making this book a truly international effort. This book will bring forth new frontiers with its revolutionizing research information and detailed analysis of the nascent developments around the world.

We would like to thank Prof. Marian Petre, for lending his expertise to make the book truly unique. He has played a crucial role in the development of this book. Without his invaluable contribution this book wouldn't have been possible. He has made vital efforts to compile up to date information on the varied aspects of this subject to make this book a valuable addition to the collection of many professionals and students.

This book was conceptualized with the vision of imparting up-to-date information and advanced data in this field. To ensure the same, a matchless editorial board was set up. Every individual on the board went through rigorous rounds of assessment to prove their worth. After which they invested a large part of their time researching and compiling the most relevant data for our readers. Conferences and sessions were held from time to time between the editorial board and the contributing authors to present the data in the most comprehensible form. The editorial team has worked tirelessly to provide valuable and valid information to help people across the globe.

Every chapter published in this book has been scrutinized by our experts. Their significance has been extensively debated. The topics covered herein carry significant findings which will fuel the growth of the discipline. They may even be implemented as practical applications or may be referred to as a beginning point for another development. Chapters in this book were first published by InTech; hereby published with permission under the Creative Commons Attribution License or equivalent.

The editorial board has been involved in producing this book since its inception. They have spent rigorous hours researching and exploring the diverse topics which have resulted in the successful publishing of this book. They have passed on their knowledge of decades through this book. To expedite this challenging task, the publisher supported the team at every step. A small team of assistant editors was also appointed to further simplify the editing procedure and attain best results for the readers.

Our editorial team has been hand-picked from every corner of the world. Their multi-ethnicity adds dynamic inputs to the discussions which result in innovative outcomes. These outcomes are then further discussed with the researchers and contributors who give their valuable feedback and opinion regarding the same. The feedback is then collaborated with the researches and they are edited in a comprehensive manner to aid the understanding of the subject.

Apart from the editorial board, the designing team has also invested a significant amount of their time in understanding the subject and creating the most relevant covers. They scrutinized every image to scout for the most suitable representation of the subject and create an appropriate cover for the book.

The publishing team has been involved in this book since its early stages. They were actively engaged in every process, be it collecting the data, connecting with the contributors or procuring relevant information. The team has been an ardent support to the editorial, designing and production team. Their endless efforts to recruit the best for this project, has resulted in the accomplishment of this book. They are a veteran in the field of academics and their pool of knowledge is as vast as their experience in printing. Their expertise and guidance has proved useful at every step. Their uncompromising quality standards have made this book an exceptional effort. Their encouragement from time to time has been an inspiration for everyone.

The publisher and the editorial board hope that this book will prove to be a valuable piece of knowledge for researchers, students, practitioners and scholars across the globe.

List of Contributors

Marian Petre and Alexandru Teodorescu
Department of Natural Sciences, Faculty of Sciences, University of Pitesti, Romania

Grazina Juodeikiene and Dalia Eidukonyte
Kaunas University of Technology, Lithuania

Elena Bartkiene
Veterinary Academy, Lithuanian University of Health Sciences, Lithuania

Pranas Viskelis, Dalia Urbonaviciene and Ceslovas Bobinas
Institute of Horticulture, Lithuanian Research Centre for Agriculture and Forestry, Lithuania

Kenji Sakai
Graduate School of Bioresource and Bioenvironmental Sciences, Faculty of Agriculture, Kyushu University, Fukuoka, Japan

Pramod Poudel
Graduate School of Bioresource and Bioenvironmental Sciences, Faculty of Agriculture, Kyushu University, Fukuoka, Japan
National College (NIST), Department of Microbiology, Tribhuvan University, Kathmandu, Nepal

Yoshihito Shirai
Graduate School of Life Science and Systems Engineering, Kyushu Institute of Technology, Kitakyushu, Fukuoka, Japan

José G. C. Gomez and Luiziana F. Silva
Institute of Biomedical Sciences, University of São Paulo, Brazil

Beatriz S. Méndez and M. Julia Pettinari
Department of Biological Chemistry, Faculty of Sciences, University of Buenos Aires and National Council for Research (CONICET), Argentina

Pablo I. Nikel
Department of Biological Chemistry, Faculty of Sciences, University of Buenos Aires and National Council for Research (CONICET), Argentina
Institute for Research in Biotechnology, University of San Martín, Argentina

María A. Prieto
Department of Environmental Biology, Centro de Investigaciones Biológicas, Spain

Ioannou Irina and Ghoul Mohamed
Nancy University – ENSAIA, France

Yasushi Kageyama, Katsuya Ozaki, and Katsutoshi Ara
Biological Science Laboratories, Kao Corporation, Japan

Kouji Nakamura
Graduate School of Life and Environmental Sciences, University of Tsukuba, Japan

Hiroshi Kakeshita
Biological Science Laboratories, Kao Corporation, Japan
Graduate School of Life and Environmental Sciences, University of Tsukuba, Japan

Hendrik Waegeman and Marjan De Mey
Ghent University, Centre of Expertise-Industrial Biotechnology and Biocatalysis, Belgium

Takeko Kodama, Kenji Manabe and Shenghao Liu
Biological Science Laboratories, Kao Corporation, Japan

Junichi Sekiguchi
Department of Bioscience and Textile Technology, Interdisciplinary Graduate School of Science and Technology, Shinshu University, Japan

Witold Szaflarski, Michał Nowicki and Maciej Zabel
Department of Histology and Embryology, Poznań University of Medical Sciences, Poland

Moj Khaleghi
Department of Biology, Faculty of Sciences, Shahid Bahonar University, Kerman, Iran

Rouha Kasra Kermanshahi
Department of Biology, Faculty of Sciences, Alzahra University, Tehran, Iran

Bei-Wen Ying and Tetsuya Yomo
Osaka University, Japan

Maki Teramoto, Zilian Zhang, Takashi Kawasaki, Yutaka Kawarabayasi and Noriyuki Nakamura
National Institute of Advanced Industrial Science and Technology (AIST), Amagasaki, Hyogo

Motohiro Shizuma
Osaka Municipal Technical Research Institute (OMTRI), Joto-ku, Osaka, Japan

Elena S. Lobakova and Ivan A. Smirnov
Moscow State University, M.V. Lomonosov, Russia

Amanda Silva
Université Paris 7, France

Érica Silva-Freitas, Juliana Carvalho, Thales Pontes, Rafael Araújo-Neto, Kátia Silva, Artur Carriço and Eryvaldo Egito
Universidade Federal do Rio Grande do Norte, Brazil

Printed in the USA
CPSIA information can be obtained
at www.ICGtesting.com
JSHW011455221024
72173JS00005B/1078

9 781632 392169